Asymmetric Fluoroorganic Chemistry

Synthesis, Applications, and Future Directions

P. V. Ramachandran, EDITOR
Purdue University

American Chemical Society, Washington, DC

Library of Congress Cataloging-in-Publication Data

Asymmetric fluoroorganic chemistry : synthesis, applications, and future directions / P.V.
Ramachandran, editor.

    p.   cm.—(ACS symposium series ; 746)

    Includes bibliographical references and index.

    ISBN 0–8412–3639–9

    1. Organofluorine compounds—Congresses. 2. Asymmetric synthesis—
Congresses.

    I. Ramachandran, P.V. 1954– . II. Series

QD305.H15 A89 1999
547′.02—dc21

                      99–48380

Advisory Board

ACS Symposium Series

Foreword

The ACS Symposium Series was first published in 1974 to provide a mechanism for publishing symposia quickly in book form. The purpose of the series is to publish timely, comprehensive books developed from ACS sponsored symposia based on current scientific research. Occasionally, books are developed from symposia sponsored by other organizations when the topic is of keen interest to the chemistry audience.

Before agreeing to publish a book, the proposed table of contents is reviewed for appropriate and comprehensive coverage and for interest to the audience. Some papers may be excluded in order to better focus the book; others may be added to provide comprehensiveness. When appropriate, overview or introductory chapters are added. Drafts of chapters are peer-reviewed prior to final acceptance or rejection, and manuscripts are prepared in camera-ready format.

As a rule, only original research papers and original review papers are included in the volumes. Verbatim reproductions of previously published papers are not accepted.

ACS Books Department

Contents

Preface ix

1. **Fluorine-Containing Chiral Compounds of Biomedical Interest** 1
   Robert Filler

REAGENT-CONTROLLED ASYMMETRIC SYNTHESIS

2. **Asymmetric Synthesis of Fluoroorganic Compounds**
   via Chiral Organoboranes..22
   P. V. Ramachandran and Herbert C. Brown

3. **Catalytic Asymmetric Synthesis of Fluoroorganic Compounds:**
   Mukaiyama–Aldol and Henry Reactions..38
   Katsuhiko Iseki

4. **Asymmetric Fluorination of α-Amino Ketones: Synthesis of**
   Monofluoroketomethylene Peptide Isosteres and Related Compounds.......52
   Robert V. Hoffman and Junhua Tao

5. **Asymmetric Friedel–Crafts Reactions with Fluoral Catalyzed**
   by Chiral Binaphthol-Derived Titanium Catalysts 60
   Akihiro Ishii and Koichi Mikami

SUBSTRATE-CONTROLLED ASYMMETRIC SYNTHESIS

6. **Biomimetic, Reducing Agent-Free Reductive Amination**
   of Fluorocarbonyl Compounds: Practical Asymmetric Synthesis
   of Enantiopure Fluoroamines and Amino Acids....................................74
   Vadim A. Soloshonok

7. **Stereoselective and Enantioselective Synthesis of New Fluoroalkyl**
   Peptidomimetic Units: Amino Alcohols and Isoserines.........................84
   Ahmed Abouabdellah, Jean-Pierre Bégué, Danièle Bonnet-Delpon,
   Andrei Kornilov, Isabelle Rodrigues, and Truong Thi Thanh Nga

8. **Asymmetric Synthesis of Fluoroalkyl Amino Compounds via Chiral Sulfoxides**..98
Pierfrancesco Bravo, Luca Bruché, Marcello Crucianelli, Fiorenza Viani, and Matteo Zanda

9. **Studies Toward the Syntheses of Functionalized, Fluorinated Allyl Alcohols**..117
P. V. Ramachandran, M. Venkat Ram Reddy, and Michael T. Rudd

10. **Stereoselective Synthesis of β-Fluoroalkyl β-Amino Alcohols Units**.......127
Matteo Zanda, Pierfrancesco Bravo, and Alessandro Volonterio

11. **Diastereoselective Construction of Novel Sugars Containing Variously Fluorinated Methyl Groups as Intermediates for Fluorinated Aldols**..142
Takashi Yamazaki, Shuichi Hiraoka, and Tomoya Kitazume

SYNTHESIS OF FLUOROORGANIC TARGETS

12. **Synthesis of Enantiopure Fluorine-Containing Taxoids and Their Use as Anticancer Agents as well as Probes for Biomedical Problems**...158
Iwao Ojima, Tadashi Inoue, John C. Slater, Songnian Lin, Scott D. Kuduk, Subrata Chakravarty, John J. Walsh, Thierry Cresteil, Bernard Monsarrat, Paula Pera, and Ralph J. Bernacki

13. **The Synthesis of a Series of Fluorinated Tribactams**.............................182
Ferenc Gyenes, Andrei Kornilov, and John T. Welch

BIOORGANIC SYNTHESIS OF ASYMMETRIC FLUOROORGANIC COMPOUNDS

14. **Chemical and Biochemical Approaches to the Enantiomers of Chiral Fluorinated Catecholamines and Amino Acids**..........................194
K. L. Kirk, B. Herbert, S.-F. Lu, B. Jayachandran, W. L. Padgett, O. Olufunke, J. W. Daly, G. Haufe, and K. W. Laue

15. **Natural Products Containing Fluorine and Recent Progress in Elucidating the Pathway of Fluorometabolite Biosynthesis in _Streptomyces cattleya_**..210
D. O'Hagan and D. B. Harper

ASYMMETRIC FLUOROORGANIC CHEMISTRY IN MATERIALS CHEMISTRY

16. Synthetic Aspects of Fluorine-Containing Chiral Liquid Crystals............226
    Tamejiro Hiyama, Tetsuo Kusumoto, and Hiroshi Matsutani

17. Resolution of Racemic Perfluorocarbons through Self-Assembly
    Driven by Electron Donor–Acceptor Intermolecular Recognition..........239
    Maria Teresa Messina, Pierangelo Metrangolo,
    and Giuseppe Resnati

18. Catalytic Asymmetric Synthesis of Diastereomeric
    α- or β-CF₃ Liquid Crystalline Molecules: Conformational Probe
    for Anti-Ferroelectricity and Self-Assembly for Spontaneous Chiral
    Resolution of the Racemates 255
    Koichi Mikami

ASYMMETRIC FLUOROORGANIC CHEMISTRY IN AGROCHEMISTRY AND PHARMACY

19. Chiral 3-Aryl-4-halo-5-(trifluoromethyl)pyrazoles: Synthesis
    and Herbicidal Activity of Enantiomeric Lactate Derivatives
    of Aryl-Pyrazole Herbicides..272
    Bruce C. Hamper, Kindrick L. Leschinsky, Deborah A. Mischke,
    and S. Douglas Prosch

20. Chiral Fluorinated Anesthetics 282
    Keith Ramig

INDEXES

Author Index 294

Subject Index 295

vii

Preface

Fluoroorganic chemistry, a very important area of research, has attracted agricultural, materials, medicinal, and organic chemists alike. The recent flux of activities is well documented by several monographs and special issues of journals dedicated to this area. Due to the importance of enantiomerically pure pharmaceuticals, asymmetric organic synthesis has attracted the attention of organic and medicinal chemists alike. A combination of fluoroorganic and asymmetric synthesis has very great potential. This idea led to a symposium dedicated to the topic of *Asymmetric Synthesis of Fluoroorganic Compounds* that was held during the 216th National Meeting of the American Chemical Society (ACS) in Boston, Massachusettes, August 23 and 24, 1998. The symposium, sponsored by the ACS Divisions of Fluorine Chemistry and Organic Chemistry, was well received by organic chemists from both industry and academia and acted as a catalyst for this book.

This volume contains twenty chapters by leading fluoroorganic chemists worldwide who participated in the symposium. The book is organized into sections based on the symposium. One of the pioneers of asymmetric fluoroorganic chemistry, Robert Filler wrote the lead chapter discussing fluorine-containing chiral compounds of biomedical interest, which is followed by chapters discussing reagent- and substrate-controlled asymmetric synthesis, the synthesis of fluorine-containing target molecules, bioorganic synthesis, asymmetric fluoroorganic compounds in materials chemistry, agrochemistry, and pharmacy.

This book, which discusses the current state of the art in asymmetric fluoroorganic chemistry is intended for organic, materials, and medicinal chemists in academia and industry. This volume can be adapted for a graduate course in fluoroorganic chemistry.

Acknowledgments

I acknowledge all of the speakers who participated in the symposium, especially those who contributed to this volume. I thank Herbert C. Brown for his support and advice throughout the organization of the symposium and the production of this book. I thank the ACS Divisions of Fluorine Chemistry and Organic Chemistry for sponsoring the symposium. Financial support to the symposium by the ACS Corporate Associates and ACS Petroleum Research Fund are gratefully acknowledged. Financial support from the following companies also contributed to the success of the symposium: Dow Agrosciences, Monsanto Company, Aldrich Chemical Company, Inc., Eastman Kodak Company, Schering-Plough, Merck Research Laboratories, Central Glass International Inc. (Japan), Daikin Industries (Japan), Kanto Denka Kogyo Company (Japan), and Asahi Glass Company (Japan). I acknowledge the help provided by Venkat Ram Reddy during the editing process of the book. Last, but not least, I thank the ACS Books Department Staff, particularly Anne Wilson and Kelly Dennis, for their remarkable efforts in the production of this book.

P. V. RAMACHANDRAN
Department of Chemistry
Purdue University
West Lafayette, IN 47907–1393

Chapter 1

Fluorine-Containing Chiral Compounds of Biomedical Interest

Robert Filler

Department of Biological, Chemical, and Physical Sciences, Illinois Institute of Technology, Chicago, IL 60616–3793

In recent years, there has been a plethora of publications on asymmetric syntheses and enzymic methods leading to fluorine-containing chiral compounds. This review, based on a detailed survey of recent literature, focuses on compounds of interest in bio-and medicinal chemistry, e.g., amino acids, anticancer agents, sugars, nucleosides, central nervous system agents, and anesthetics. Methods involving asymmetric aldol reactions, asymmetric alkylation, enantioselective fluorinating agents, and enzymically controlled reactions are presented. Bioactive fluorine-containing compounds have been known since the 1930s and 1940s, but the remarkable developments of the 1950s, including the fluorosteroids, inhalation anesthetics, such as halothane, and the anticancer agent 5-fluorouracil, heralded the subsequent rapid advances we have witnessed during the ensuing forty years.

During the past decade, there has been an increased emphasis on new approaches to chiral compounds and asymmetric syntheses. This focus has been particularly pronounced in medicinal chemistry, where a specific enantiomer or diastereomer often exhibits enhanced therapeutic potency compared with the racemate. Organofluorine compounds have played a significant role in these advances. An earlier report emphasized a range of methods for the synthesis of chiral bioactive fluoroorganic compounds (1). Since the intent of this paper is to provide an overview which captures the scope and flavor of these recent developments, it seems quite appropriate to briefly cite the fascinating range of research studies of the other contributors to this book.

Bravo and co-workers report asymmetric syntheses of fluoroalkylamino compounds via chiral sulfoxides and the stereoselective synthesis of β-fluoroalkyl-β-amino alcohol units using chiral sulfoxides and the Evans aldol reaction. Bégúe and colleagues discuss the stereoselective and enantioselective synthesis of trifluoromethyl amino alcohols and fluoroalkyl isoserinates. Hoffman reports the aysmmetric fluorination of α-aminoketones, while

1

Soloshonok describes his continuing studies on a practical asymmetric methodology for the synthesis of enantiopure fluoro amines and amino acids. Chemical and biochemical approaches to non-racemic fluorinated catecholamines and amino acids are explored by Kirk and co-workers. Ramig discusses recent studies of the chiral fluorinated anesthetics, halothane, enflurane, isoflurane, and desflurane. Ojima and colleagues report exciting new results on the synthesis of enantiopure fluorine-containing analogs of paclitaxel and docetaxel and their use as anticancer agents and as important biochemical probes. In their studies of natural products containing fluorine, O'Hagan and Harper trace the biosynthetic origin of fluoroacetate and 4-fluorothreonine in a Streptomyces bacterium. Iseki reports efficient asymmetric syntheses of fluorinated aldols and nitro aldols of high enantiomeric purity, while Yamazaki discusses the construction of chiral aldols with fluorine-containing methyl groups starting from D-glucose. Hamper and co-workers at Monsanto describe the synthesis and herbicidal activity of enantiomeric lactate derivatives of trifluoromethyl-substituted pyrazole herbicides. Resnati and colleagues discuss the resolution of racemic perfluorocarbon halides via self-assembly involving intermolecular electron donor-acceptor interactions. A variety of reactions leading to fluorine-containing chiral molecules useful as liquid crystals are reported by Hiyama and co-workers. In the search for new antimicrobial agents bearing the β-lactam moiety, Welch and co-workers report the preparation of a chiral 3-fluoroazetidinone by cycloaddition of fluoroketene (from fluoroacetyl chloride and Et_3N) and an optically active arylimine. The azetidinone serves as a building block in a multistep synthesis of a new fluorine-containing tribactam (tricyclic β-lactam derivative). An important feature is the dominance of the electronic effect of fluorine over the steric influence of a methyl group on the same carbon. Ramachandran and Brown review their applications of chiral organoboranes derived from α-pinene in a three-pronged approach for the preparation of asymmetric organofluorine compounds in very high enantiomeric excess. Reactions include (1) asymmetric reduction of fluorinated ketones, (2) asymmetric allylboration of fluorinated aldehydes, and (3) asymmetric enolboration-aldolization of fluoro aldehydes and ketones. Using B-chlorodiisocampheylborane, the stereochemical outcomes in reactions (1) and (3) are apparently influenced by the presence of fluorines in the molecules. Ramachandran and colleagues review and compare the Morita-Baylis-Hillman (MBH) reaction and vinylmetalation (aluminum and copper) of fluorocarbonyls as routes to achiral and chiral fluorinated allyl alcohols. Use of terpenyl alcohols as chiral auxiliaries in asymmetric MBH and vinylmetalation reactions is explored. Mikami reports the asymmetric synthesis of diastereomeric α- or β- CF_3 liquid crystalline (LC) molecules by a carbonyl-ene reaction using fluoral in the presence of a chiral binaphthol-derived titanium catalyst (BINOL-Ti). These LCs function as conformational probes for ferroelectricity. The first example of spontaneous resolution of racemates in fluid LC phases has been observed with these diastereomers. Ishii and Mikami describe the first example of asymmetric catalysis (BINOL-Ti) in the Friedel-Crafts reaction of aryl compounds with fluoral to provide a practical route to chiral α-trifluoromethylbenzyl alcohols. Highly enantiopure β-trifluoroaldols have also been prepared from vinyl ethers via sequential diastereoselective reactions.

Fluoroamino Acids

Fluorine-substituted amino acids have often been used as probes to follow biochemical reactions. During the past few years, asymmetric syntheses of fluoro α-amino acids have been the subject of intense activity. Thus, Soloshonok and co-workers (2) reported highly diastereoselective asymmetric aldol reactions between prochiral trifluoromethyl ketones and the Ni (II)-complex of the monochiral Schiff base of glycine (Figure 1) and established reaction conditions for preparative syntheses of diastereo- and enantiomerically enriched trifluoromethyl-substituted serines. Sting and Seebach (3) converted enantiopure (S) - 4,4,4 - trifluoro - 3 - hydroxybutanoic acid in several steps to dioxanones, whose enolates were stereoselectively aminated with di- tert - butyl azodicarboxylate, DBAD. Acidic ring-opening and removal of the N-Boc groups provided the corresponding α-hydrazino -β - hydroxyesters which, on hydrogenolysis, gave the threonine analog (2R, 3S) - 2 - amino - 4,4,4 - trifluoro - 3- hydroxybutanoate methyl ester (1) (Figure 2). Previous synthetic and enzymatic approaches to trifluoromethyl analogs of threonine and allothreonine by other investigators are discussed in this paper (3). Enantiomerically enriched 3,3,3- trifluoroalanine (up to 63% ee) has been prepared by the asymmetric reduction of 2-(N-arylimino) - 3,3,3-trifluoropropanoic acid esters with a chiral oxazaborolidine catalyst (equation 1) (4).

In a novel approach, an enantiomerically pure derivative of 2-amino-4,4,4- trifluorobutanoic acid was synthesized via nucleophilic trifluoromethylation (Figure 3) (5). In the key step, Garner's aldehyde (2) (6), an oxazolidine derived from L-serine, reacted with the Ruppert-Prakash reagent, TMS-CF$_3$ (a trifluoromethide equivalent) (7) and tetrabutylammonium fluoride. (S) - 5,5,5,5′,5′,5′- Hexafluoroleucine (3) (88% ee) was prepared in 18% overall yield from hexafluoroacetone and ethyl bromopyruvate in seven steps (8). The highly enantioselective reduction of the keto carbonyl group of 4 to the hydroxyester 5 either by baker's yeast and sucrose (91% ee) or by catecholborane and an oxazaborolidine catalyst (99% ee) was the pivotal reaction of the sequence. These workers (9) also prepared (-)-(R)- 4,4,4,4′,4′,4′ -hexafluorovaline (6) (98% ee). The key step was the separation of the tosylate salts of the diastereomers formed by anti-Michael addition of (+)-(R)-

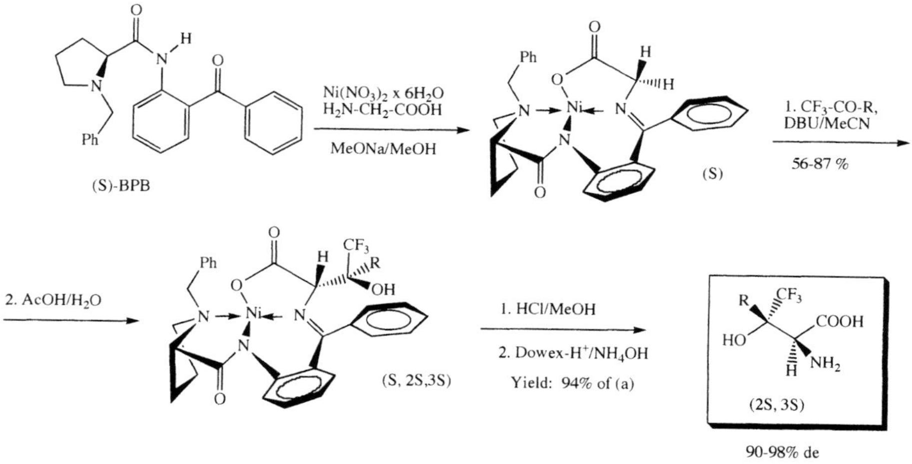

4

R= CH$_3$ (**a**), C$_4$H$_9$ (**b**), C$_7$H$_{15}$ (**c**), C$_8$H$_{17}$ (**d**), (CH$_2$)$_3$Ph (**e**), C≡C-Ph (**f**)

Figure 1. Highly diastereoselective asymmetric aldol reactions of chiral Ni(II)-complex of glycine with CF3COR (2).

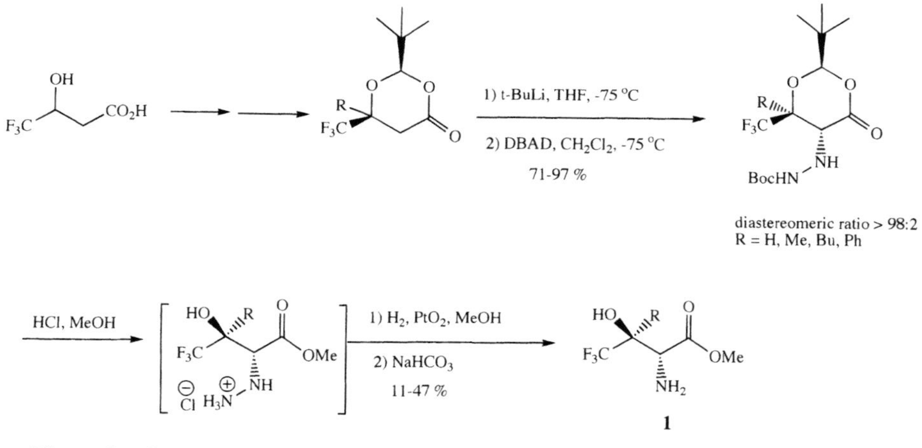

Figure 2. Synthesis of (2R, 3S)-2-amino-3-trifluoromethyl-3-hydroxy alkanoic acid derivatives (3).

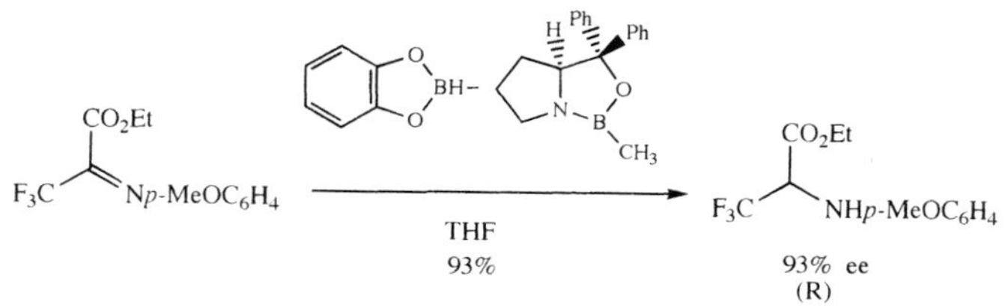

(1)

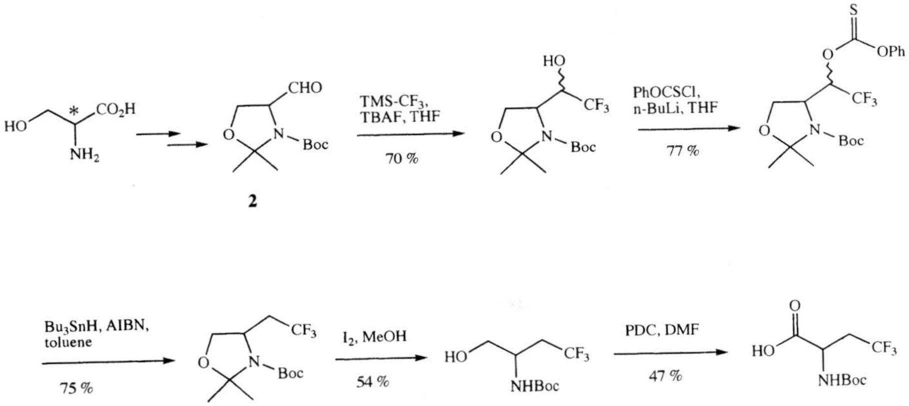

Figure 3. Synthesis of enantiomerically pure (2R)-N-Boc-2-Amino-4,4,4-Trifluorobutanoic acid via trifluoromethylation (5).

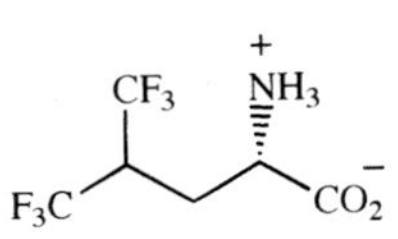

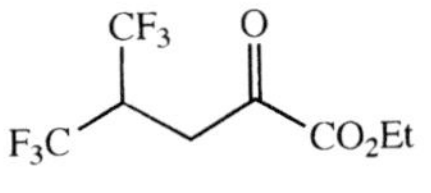

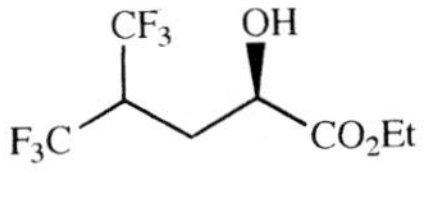

3 **4** **5**

1 -phenylethylamine to the benzyl ester of β,β- bis (trifluoromethyl) acrylic acid (Figure 4). The broad-spectrum antibacterial agent 3-fluoro-D-alanine (**7**), which causes irreversible inhibition of alanine racemase, was synthesized stereoselectively via a chiral sulfoxide (**10**). Two important reactions were the azidization of the α-fluoro-α'-sulfinyl alcohol under Mitsunobu conditions (equation 2) and the one-pot transformation of the N-Cbz α-sulfinylamine into the N-Cbz aminoalcohol through a non-oxidative Pummerer reaction (equation 3). Fluorine-containing β- amino acids of potential biomedical interest have also been investigated recently. Using a combined chemical-enzymatic method, the four stereoisomers of 3-amino-2-methyl-3-trifluoromethyl butanoic acid were isolated (Figure 5) (**11**). An enantioselective biomimetic transamination of β- keto carboxylic acid esters and amides via Schiff bases provide an efficient asymmetric synthesis of β- fluoroalkyl- β - amino acids, e.g., **8** (Figure 6) (**12**).

Fluoro Sugars

From a non-carbohydrate precursor, Davis and Qi achieved the asymmetric synthesis of 2-deoxy-2-fluoro-xylo-D-pyranose (**9**) (Figure 7) and 2-deoxy-2-fluoro-lyxo-L-pyranose (**13**). The key reaction was the highly diastereoselective fluorination of a chiral enolate using the electrophilic fluorinating agent, N-fluoro(benzenesulfonimide) (NFSi). In another study by Davis (**14**), the use of a chiral oxazolidinone adjuvant and fluorination with NFSi led to the chiral fluorohydrin **10** (>97% ee), which was oxidized by the Dess-Martin periodinane procedure to the non-racemic α- fluoroaldehyde **11** (94% ee). Conversion of **11** in four steps provided 1,2,3-tri-O-acetyl-4-deoxy-4-fluoro-D-arabinopyranose (**12**) as a 1:1 mixture of anomers which could not be separated by flash chromatography.

Fluoronucleosides

An unsaturated fluorinated lactone derived from D-mannose was catalytically reduced to the TBDMS-protected saturated lactone (Figure 8). The remarkably high diasteroselectivity (>99%) of this hydrogenation was the major factor in the success of the subsequent synthesis of 1-(2,3-dideoxy-2-fluoro-β-d-threo - pentafuranosylthymine) (β-2-F-ddT) **13**, a promising anti-HIV drug (**15**). The R- isomer of 5' - O - acetyl - 2' - deoxy - 3'- fluoro - 3'- S-(4-methoxyphenyl) - 3' - thiothymidine reacted with the electrophilic F- TEDA - BF$_4$ (SELECTFLUOR™) and triethylamine in acetonitrile to yield the C3'-fluoro compound as a mixture of diastereomers (R/S = 97/3) (equation 4) (**16**). A completely diastereoselective method for introducing fluorine into a chiral non-carbohydrate sugar ring precursor using NFSi, has been developed (**17**). The single diastereomer of the resulting fluorolactone was then converted into

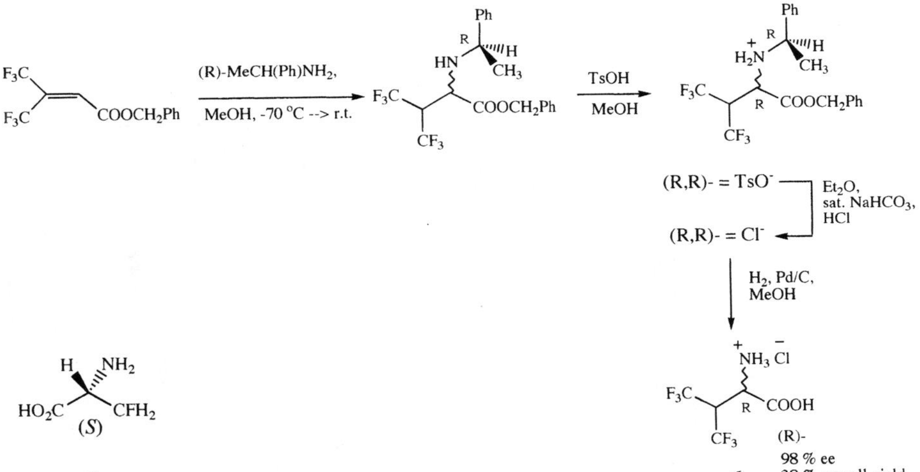

7 6

Figure 4. Synthesis of (R)-4,4,4,4′,4′,4′-hexafluorovaline (9).

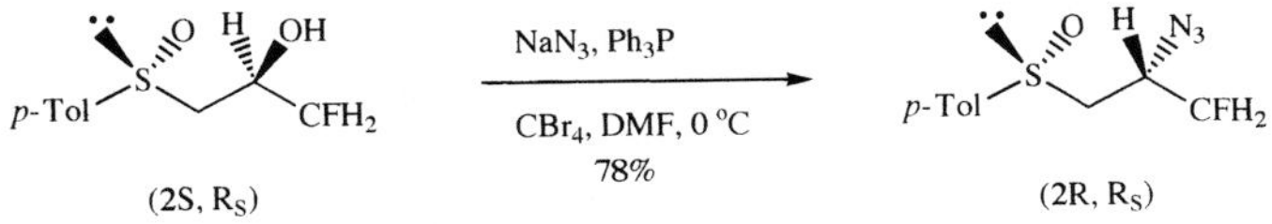

(2)

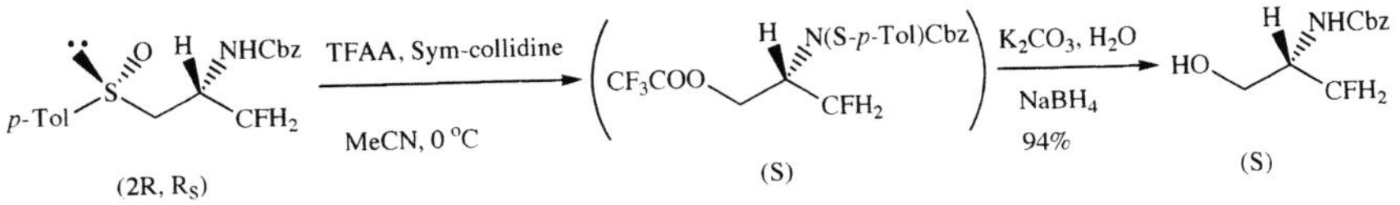

(3)

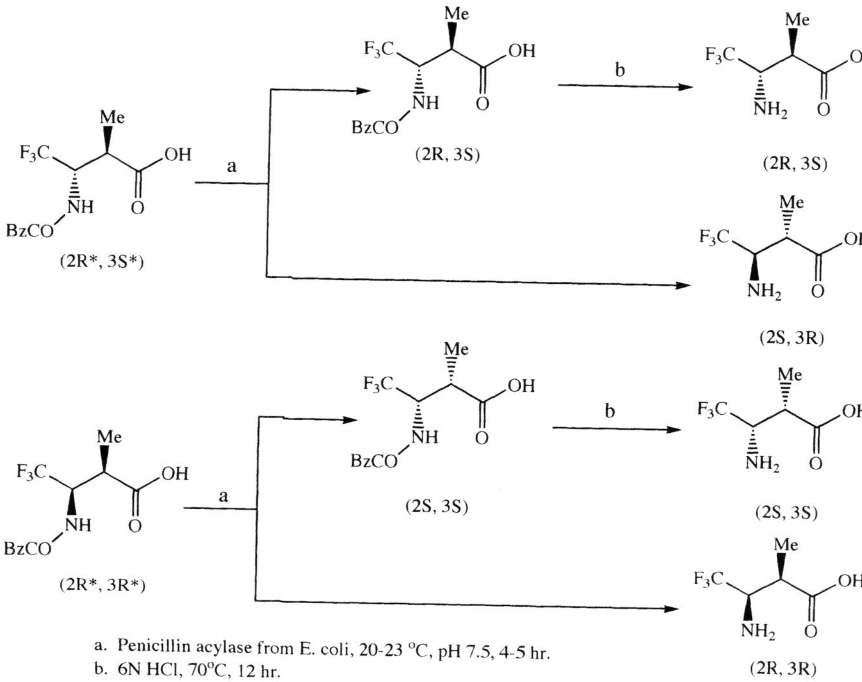

8

a. Penicillin acylase from E. coli, 20-23 °C, pH 7.5, 4-5 hr.
b. 6N HCl, 70°C, 12 hr.

Figure 5. Chemo-enzymatic approach to chiral β-trifluoromethyl-β-amino acids (11).

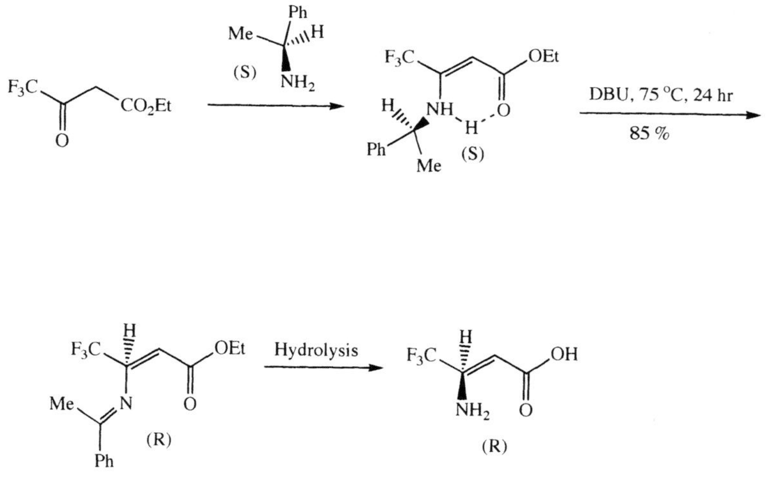

Figure 6. Asymmetric synthesis of β-(fluoroalkyl)-β-amino acids via transamination (12).

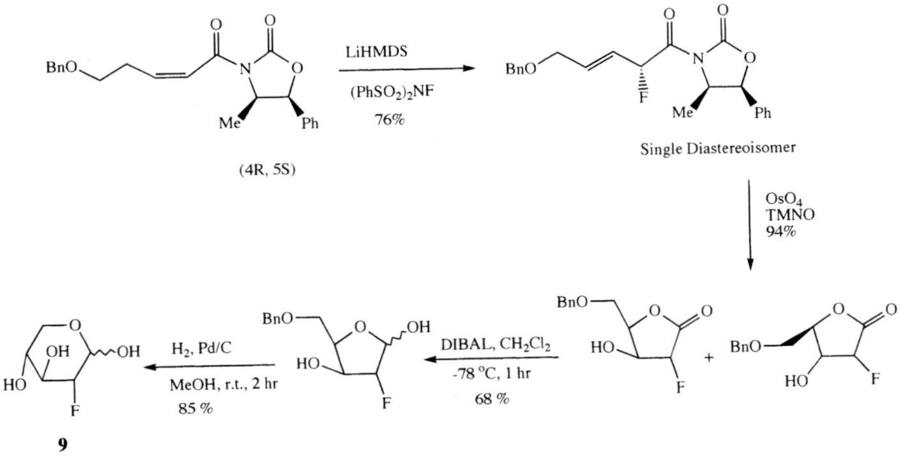

Figure 7. Non-carbohydrate synthesis of sugars (13).

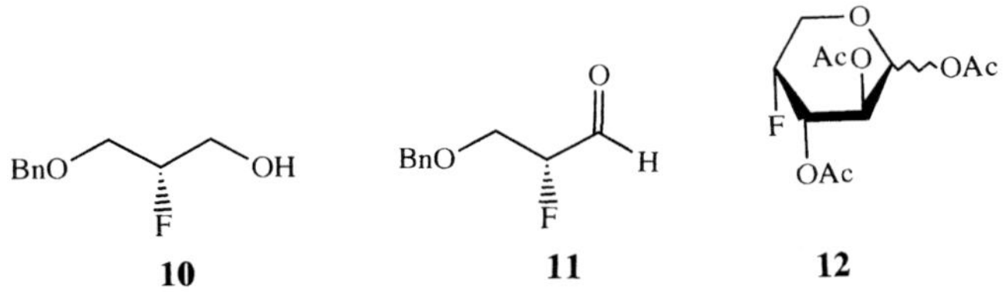

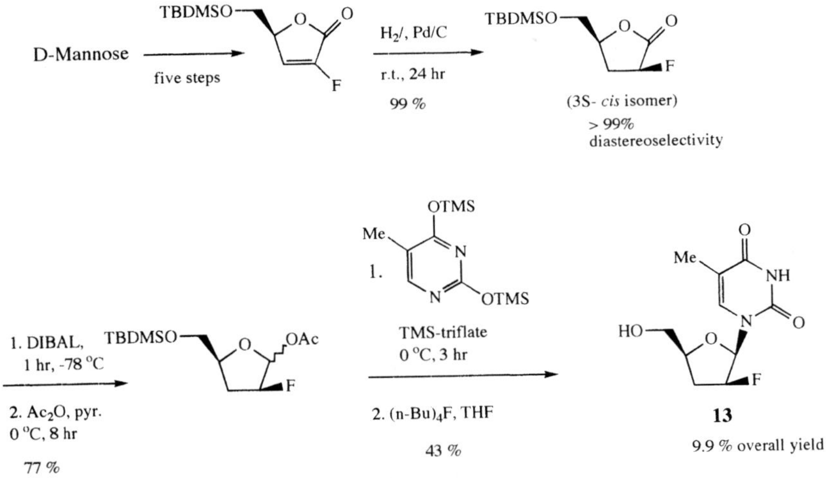

Figure 8. Synthesis of β-2-fluorodideoxy nucleosides (15).

$$\text{(4)}$$

R/S = 97/3

an anomeric acetate and subsequently, to novel α- 2'-fluoronucleosides (**14**) (Figure 9).

Anticancer Agents

The taxoid antitumor agents Taxol® (paclitaxel) and Taxotere® (docetaxel), a semisynthetic analog of paclitaxel, are among the leading anticancer drugs. With FDA approval, paclitaxel is currently in use for the treatment of advanced ovarian and breast cancer. Using the β- Lactam Synthon Method which they developed, Ojima and coworkers (18) have synthesized a series of new fluorine-containing analogs of paclitaxel (**15**) and docetaxel (**16**) (Figure 10). The key step in the synthesis involves the coupling of (3R, 4S)-1-acyl-β-lactams, e.g. **17**, of high enantiomeric purity, prepared via a chiral ester enolate-imine cyclocondensation, with protected baccatin III and its derivatives. The racemic form of Casodex® (biclutamide) (**18**), an androgen antagonist, is being used for the treatment of advanced prostate cancer. Very recently, the (R-) and (S-) enantiomers of structural analogs of biclutamide (**19**) have been synthesized (19). Several of the R-isomers have binding affinities to the androgen receptor (AR) higher than that of biclutamide. These analogs are the first reported examples of nonsteroidal AR agonists. They have potential use in male contraception or hormone replacement therapy. The chiral anthracyclinone, DA-125 (**20**), a fluoro analog of the anticancer drug doxorubicin (Adriamycin®) is in phase II clinical trials for treatment of breast, lung, and stomach tumors (20). Key structural features include the 2'-fluoro group which strengthens the glycosidic linkage, thereby reducing hydrolytic formation of a toxic aglycone and resulting cardiotoxicity. The incorporation of the β- alanine moiety on the A ring increases water solubility. The asymmetric synthesis of novel fluorine-containing anthracyclinones using chiral acetal auxiliaries has been described recently (21).

CNS Agents

(S)-(+) Dexfenfluramine (Redux®) **21**, a serotonin reuptake inhibitor, has been marketed in Europe and the United States as a very effective anti-obesity drug. This isomer, obtained by resolution of racemic fenfluramine using d-camphoric acid, exhibits a greater anorectic effect than the (R)-(-) and racemic forms, since it is more selective on serotonin as a 5-HT agonist (22). As a result of reports of undesirable side effects, e.g., valvular heart disease, Redux has been withdrawn from the market, while further studies continue. Racemic fluoxetine (Prozac®) (**22**) is widely used for treatment of major depression and is one of the most commonly prescribed medications. It is also approved for treatment of obsessive compulsive disorder and bulimia. Non-racemic fluoxetine and its intermediates have been prepared by chemical, enzymatic,

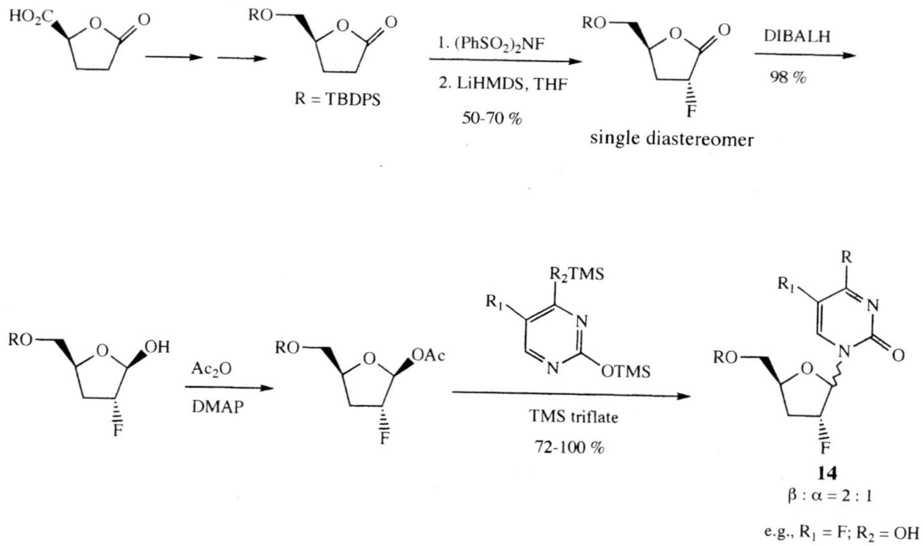

Figure 9. Complete diastereoselective fluorination in synthesis of 2-fluoronucleosides (17).

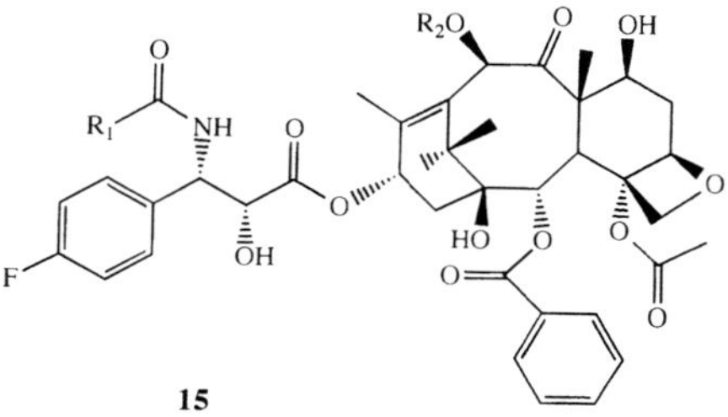

15

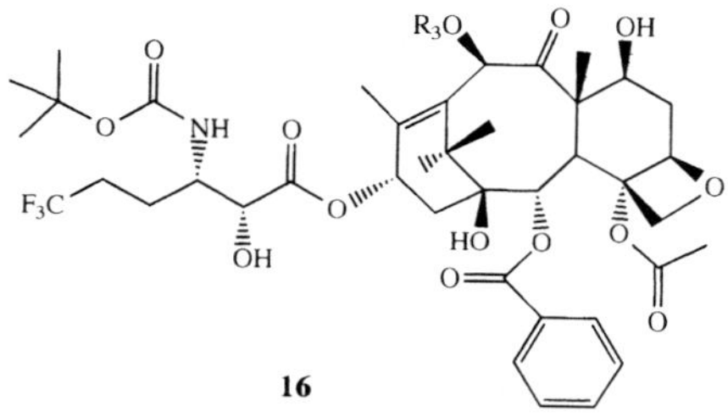

16

R_1 = Ph, R_2 = Ac
R_1 = 4-F-C_6H_4, R_2 = Ac
R_1 = *t*-BuOCO, R_2 = H
R_1 = *t*-BuOCO, R_2 = Ac

R_3 = H
R_3 = Ac

Figure 10. Fluorine-containing analogs of paclitaxel and docetaxel (18).

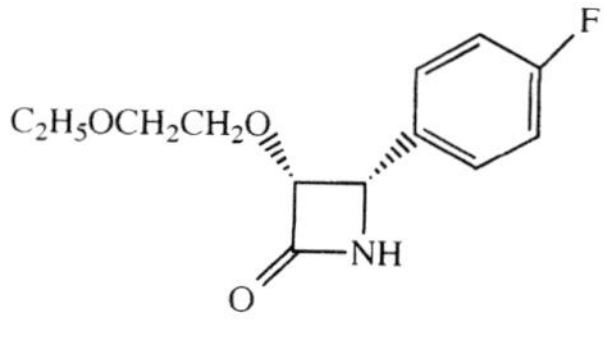

17

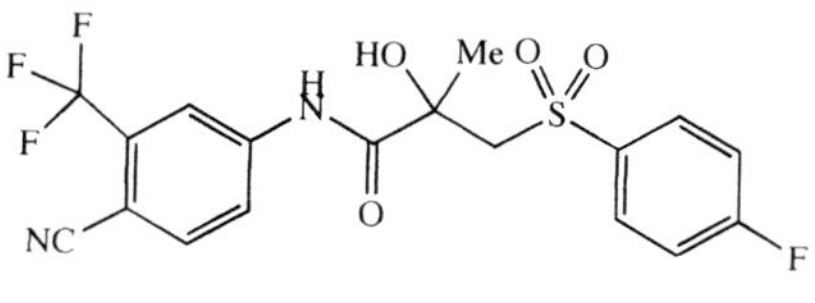

18

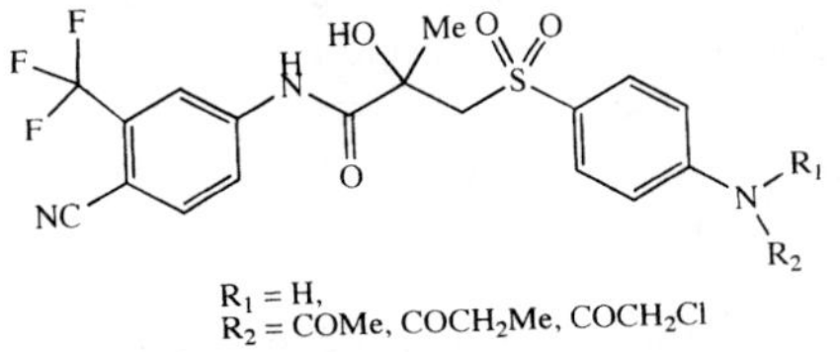

R₁ = H,
R₂ = COMe, COCH₂Me, COCH₂Cl

19

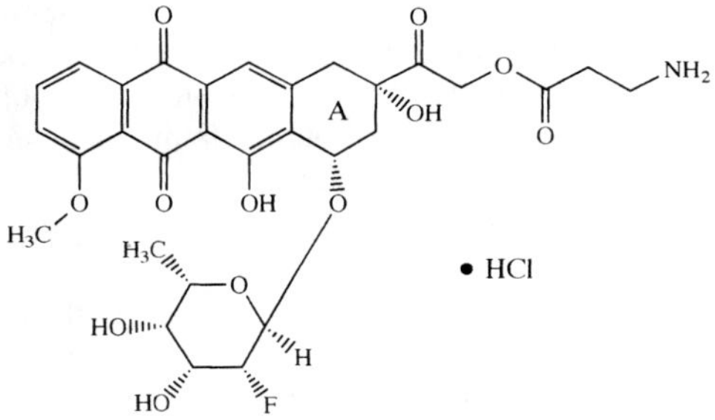

• HCl

20

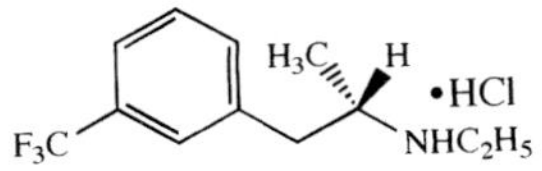

• HCl

21

• HCl

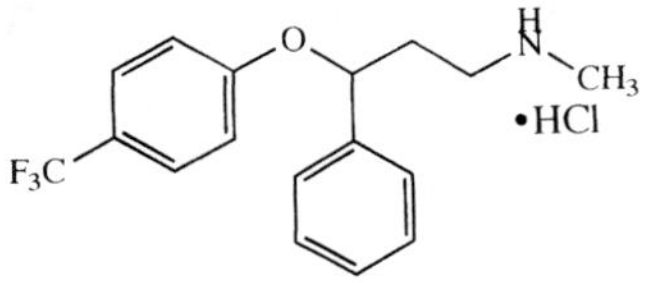

22

and microbiological methods. For example, asymmetry was introduced via an oxazaborolidine catalyzed ketone reduction (23) and by enantioselective hydrogenation of an amide using a chiral BINAP-Ru (II) catalyst (equation 5) (24). An asymmetric approach to (S)-fluoxetine involves a highly stereoselective coupling between racemic α-haloacids and lithium p-trifluoromethyl phenoxide, mediated by a pyrrolidine derived (S)- lactamide auxiliary (equation 6) (25). An efficient chemoenzymatic route to (R)-fluoxetine, which is about twice as effective biologically as the (S)- isomer, is depicted in Figure 11 (26).

Inhalation Anesthetics

An excellent review of the chemistry and pharmacology of fluorine-substituted volatile anesthetics has been presented (27). Among the newer compounds, the enantiomers of isoflurane have been isolated by alkaline decarboxylation of enantiomeric carboxylic acids, obtained by resolution of the racemic acid using (+) dehydroabietylamine and subsequent separation of the diastereomers (equation 7). The differences in the pharmacological properties of the enantiomers of isoflurane are significant. Thus, the (S)- (+) isomer binds to the acetylcholine receptor channel about 50% more tightly than the (R)- (-) enantiomer. The first enantioselective synthesis of desflurane involves the treatment of (R) - (-) isoflurane (23) with BrF_3 to give a 71% yield of (S) - (+) desflurane (24), with >96% inversion of configuration when the non-polar solvent bromine is used (equation 8) (28). It is likely that the bromine also assists in the removal of chloride in the S_N2 process. Preliminary testing on rodents suggest only slight differences between the (S) - (+) and (R)- (-) enantiomers of desflurane.

Cholesterol Absorption Inhibitors

Acyl-CoA: cholesterol O-acyltransferase [ACAT] catalyzes intracellular esterification of cholesterol and is thought to be involved in its accumulation in the arteries. ACAT inhibitors are being designed with the aim of preventing this accumulation. The synthesis and separation of potent orally active 4,4- bis (trifluoromethyl) imidazoline based ACAT inhibitors (Figure 12) has been described recently (29). The (R)- isomer (25) is 25 times more potent than the (S)- enantiomer (26).

Antifungal Agents

Fluconazole (Diflucan®) (27) is orally active and widely used for the treatment of systemic fungal infections. The emergence of fungi resistant to fluconazole has motivated a search for new antifungal agents. A number of fluorine-

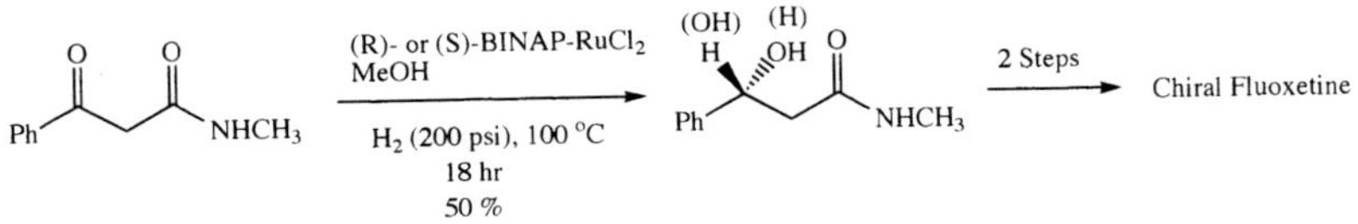

(5)

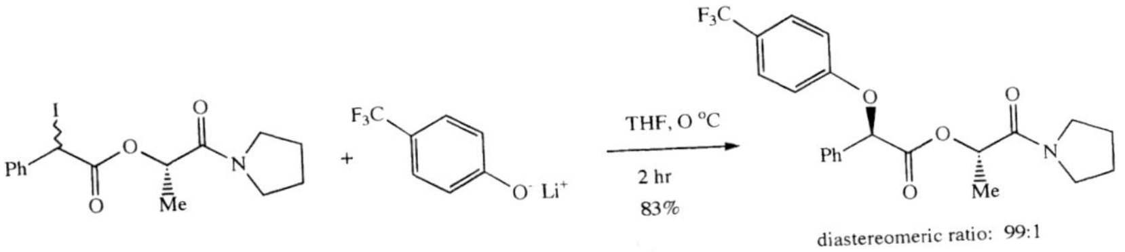

(6)

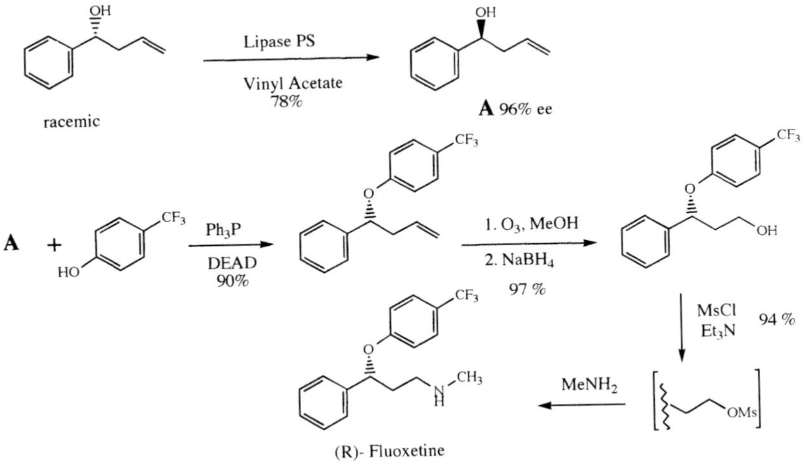

Figure 11. Chemoenzymatic route to (R)-fluoxetine (26).

(+)- $CF_3CHClOCF_2COOH$ + KOH $\longrightarrow$ (S)-(+)- $CF_3CHClOCHF_2$
98 % ee

(-)- $CF_3CHClOCF_2COOH$ + KOH $\longrightarrow$ (R)-(-)- $CF_3CHClOCHF_2$
98 % ee

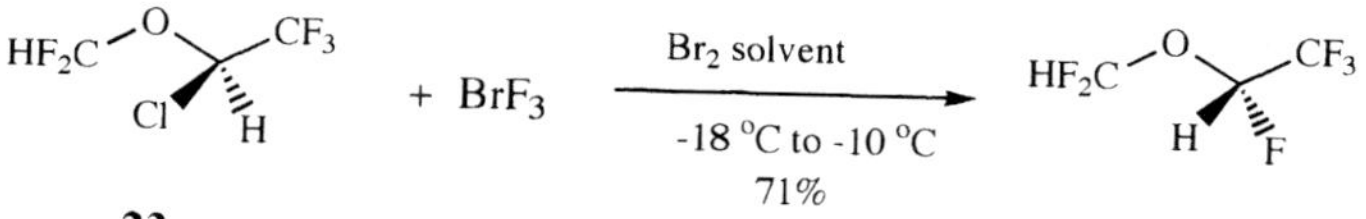

$$(7)$$

$$(8)$$

23
98.5% ee
(R)- (-)

24
91.7% ee
(S)- (+)

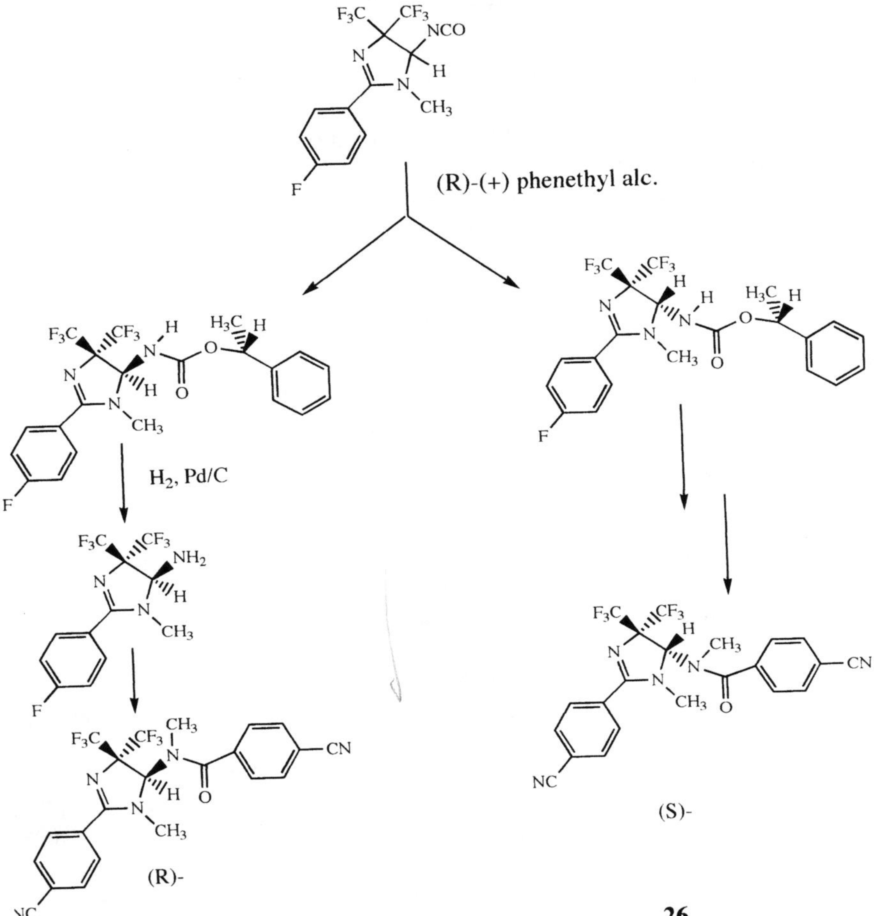

(R)-(+) phenethyl alc.

H$_2$, Pd/C

(S)-

(R)-

25

26

Figure 12. Acyl-CoA cholesterol O-transferase (ACAT) inhibitors (29).

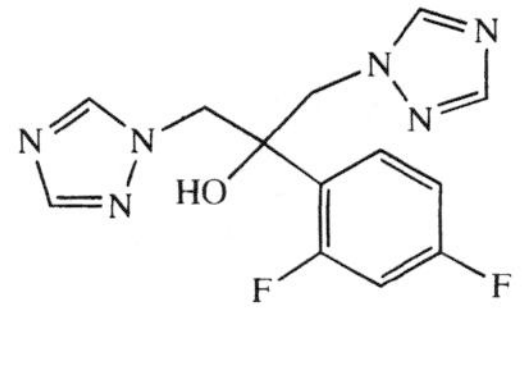

27

substituted optically active 1,2,4-triazolones have been investigated (30) and one of four stereoisomers, TAK 187 (**28**) exhibited potent antifungal activity in *in vitro* and *in vivo* assays (31).

HIV Inhibitor

The reaction of a cyclopropyl acetylide with a substituted trifluoromethyl aryl ketone in the presence of a chiral amino alcohol led to the chiral tertiary alcohol **29**, which reacted with phosgene to provide **30**, a highly potent HIV transcription inhibitor (Figure **13**) (32).

Enzymatically Controlled Reactions of Organofluorine Compounds

Kitazume and Yamazaki have published a stellar detailed review (33) which describes the transformations of organofluorine compounds into optically active functionalized fluorinated materials using microorganisms which are capable of discriminating between enantiomers. The review is strongly recommended to the reader.

Literature Cited

1. Resnati, G., Tetrahedron Report No. 341, Tetrahedron, **1993**, 49, 9385-9445.
2. Soloshonok, V.A., Avilov, D.V., Kukhar, V.P., Tetrahedron: Asymmetry, **1996**, 7, 1547-1550.
3. Sting, A.R., Seebach, D., Tetrahedron, **1996**, 52, 279-290.
4. Sakai, T., Yan, F., Kashino, S., Uneyama, K., Tetrahedron, **1996**, 52, 233-244.
5. Qing, F.-L., Peng, S., Hu, C.-M, J. Fluorine Chem., **1998**, 88, 79-81.
6. Garner, P., Park, J.M., J. Org. Chem., **1987**, 52, 2361-2364; Garner, P., Park, J.M., Organic Synthesis, in: Meyers, A.I. (Ed.), **1992**, 70, 18-22, Wiley.
7. Prakash, G.K.S., Kwshnamurti, R., Olah, G.A.; J. Am. Chem. Soc., **1989**, 111, 393.
8. Zhang, C., Ludin, C., Eberle, M.K., Stoeckli-Evans, H., Keese, R., Helv. Chim. Acta, **1998**, 81, 174-181.
9. Eberle, M.K., Keese, R., Stoeckli-Evans, H., Helv. Chim. Acta, **1998**, 81, 182-186.
10. Bravo, P., Cavicchio, G., Crucianelli, M., Poggiali, A., Zanda, M., Tetrahedron: Asymmetry, **1997**, 8, 2811-2815.
11. Soloshonok, V.A., Kirilenko, A.G., Fokina, N.A., Kukhar, V.P., Galushko, S.V., Svedas, V.K., Resnati, G., Tetrahedron: Asymmetry, **1994**, 5, 1225-1228.

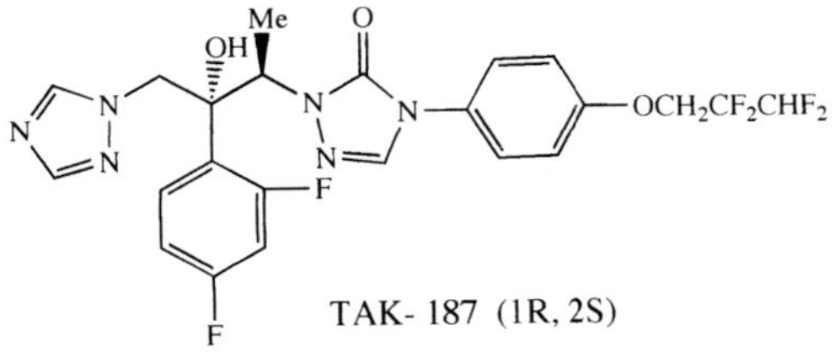

TAK- 187 (1R, 2S)

28

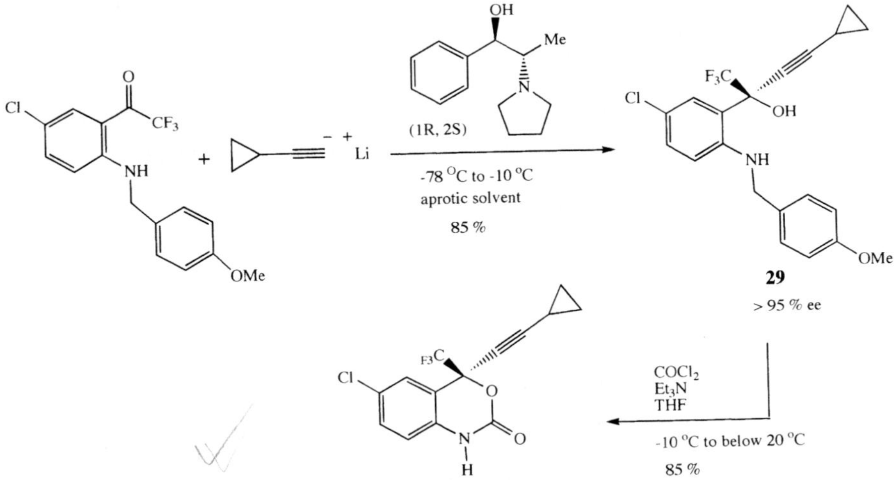

29

30

Figure 13. HIV reverse transcription inhibitor (32).

12. Soloshonok, V.A., Ono, T., Soloshonok, I.V., J.Org. Chem., **1997**, **62**, 7538-7539.
13. Davis, F.A., Qi, H., Tetrahedron Lett., **1996**, 37, 4345-4348.
14. Davis, F.A., Kasu, P.V.N., Sundarababu, G., Qi, H., J. Org. Chem., **1997**, 62, 7546-7547.
15. Patrick, T.B., Ye, W., J. Fluorine Chem., **1998**, 90, 53-55.
16. Lal, G.S., Synthetic Communications, **1995**, 25, 725-737.
17. McAtee, J.J., Schinazi, R.F., Liotta, D.C., J. Org. Chem., **1998**, 63, 2161-2167.
18. Ojima, I., Kuduk, S.D., Slater, J.C., Gimi, R.H., Sun, C.M., Tetrahedron, **1996**, 52, 209-224.
19. Kirkovsky, L., Yin, D., Zhu, Z., Dalton, J.T., Miller, D.D., Abstracts, 215th A.C.S. meeting, Dallas, Texas, Paper Medi. 055, March, 1998.
20. Drugs Fut., **1996**, 21, 782-786.
21. Lorimer, R.M. Rustenhoven, J.J. Rutledge, P.S., Chemistry in New Zealand, **1996**, 28-32.
22. Annual Reports in Medicinal Chemistry, Vol. 33, Bristol, J.A. ed., Academic Press, San Diego, CA, **1998**, p.332.
23. Corey, E.J. Reichard, G., Tetrahedron Lett., **1989**, 30, 5207-5210.
24. Huang, H.-L., Liu, L.T., Chen, S.-F, Ku, H., Tetrahedron: Asymmetry, **1998**, 9, 1637-1640.
25. Devine, P.N., Heid, R.M., Jr., Tschaen, D.M., Tetrahedron, **1997**, 53, 6739-6746.
26. Bracher, F., Litz, T., Bioorganic and Medicinal Chem., **1996**, 4, 877-880.
27. Halpern, D.F., in Organofluorine Compounds in Medicinal Chemistry and Biomedical Applications, Filler, R., Kobayashi, Y., and Yagupolskii, L.M., eds., Elsevier, Amsterdam, **1993**, pps. 101-133.
28. Rozov, L.A., Huang, C.G., Halpern, D.F., Vernice, G.G., Ramig, K., Tetrahedron: Asymmetry, **1997**, 8, 3023-3025.
29. Li, H.-Y., DeLucca, I., Drummond, S., Boswell, G.A., Tetrahedron, **1997**, 53, 5359-5372.
30. Kitazaki, T., Tamura, N., Tasaka, A., Matsushita, Y., Hayashi, R., Okonogi, K., Itoh, K., Chem. Pharm. Bull., **1996**, 44, 314-327.
31. Tasaka, A., Kitazaki, T., Tsuchimori, N., Matsushita, Y., Hayashi, R., Okonogi, K., Itoh, K., Chem Pharm. Bull., **1997**, 45, 321-326.
32. Thompson, A.S., Corley, E.G., Grabowski, E.J.J., Yasuda, N., U.S. Patent 5, 698, 741, December 16, 1997 (to Merck and Co.); Chem Abstr., **1997**, 126, 89382s.
33. Kitazume, T., Yamazaki, T., Topics in Curr. Chem., **1997**, 193, Chambers, R.D., ed., Springer, pps. 91-130.

REAGENT-CONTROLLED ASYMMETRIC SYNTHESIS

Chapter 2

Asymmetric Syntheses of Fluoroorganic Compounds via Chiral Organoboranes

P. V. Ramachandran and Herbert C. Brown

H. C. Brown and R. B. Wetherill Laboratories of Chemistry, Purdue University, West Lafayette, IN 47907–1393

A review of the current status of the applications of chiral organoboranes derived from α-pinene for the synthesis of fluoro-organic molecules has been made. Our three-pronged approach for the synthesis of asymmetric fluoro-organics involve: (i) asymmetric reduction of fluorinated ketones, (ii) asymmetric allylboration of fluorinated aldehydes, and (iii) asymmetric enolboration-aldolization of fluoro-ketones and aldehydes. It appears that the presence of fluorine atom(s) in the molecule influences the stereochemical outcome in asymmetric reduction and asymmetric enolboration-aldolization using *B*-chlorodiisopinocampheylborane.

Due to the benefits that it provides, the chemistry of fluorine containing molecules has been actively pursued by organic, agricultural, medicinal, and materials chemists (*1*). When asymmetric synthesis of organic molecules became a mature area during the last few decades (*2*), asymmetric fluoro-organic chemistry also began to develop (*3*). However, probably due to the unpredictable chemistry of fluorine-containing molecules (*4*) and the associated problems, the asymmetric synthesis of fluoro-organic compounds is yet to develop to its full potential in comparison to the progress made in other areas. The chemistry of non-fluorinated compounds is now being applied to fluorinated molecules in many research laboratories. We have been applying our chiral organoboranes (*5*) for the synthesis of enantiomerically pure fluoro-organic molecules using a three-prong approach: (i) asymmetric reduction, (ii) asymmetric allylboration, and (iii) asymmetric enolboration-aldolization. Some of our successes involving these methods are reviewed here.

Asymmetric Reduction

We became interested in the chemistry of fluoro-organic molecules during our program on asymmetric reduction of prochiral ketones, which provides a simple method for the synthesis of optically active *sec*-alcohols (*6*). Although non-enzymatic asymmetric reduction has been known for almost one-half century, it is only during the last two decades that efficient reagents and methods for the reduction of most classes of ketones have become available. Initially, most of the reagents synthesized were, at best, applicable only to the reduction of prochiral aralkyl ketones, such as acetophenone.

Pioneers in the asymmetric reduction area developed suitable reagents by modifying lithium aluminum hydride (LAH) (7) or by developing chiral versions of Meerwein-Ponndorf-Verley reductions with chiral aluminum alkoxides (8) or chiral Grignard reductions with chiral alkyl magnesium halides (9).

We developed B-chlorodiisopinocampheylborane (Aldrich: DIP-Chloride™, **1**) as an excellent reagent for the asymmetric reduction of aralkyl and α-hindered ketones (10). Examination of the reduction of 2,2,2-trifluoroacetophenone with (–)-**1** exhibited

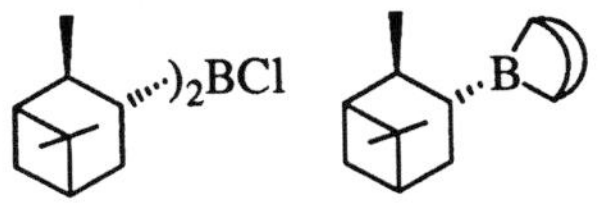

1   **2**

a slow rate, 24 h under neat condition at rt, when compared with the reduction of acetophenone itself, 5 h in THF or Et$_2$O (1M) at –25 °C. When compared with the product from acetophenone, the product from 2,2,2-trifluoroacetophenone was obtained with opposite stereochemistry (11) in 90% ee (S) (similar to the product of opposite stereochemistry from the reduction of pivalophenone) (Figure 1) (12). This result can be rationalized by considering the CF$_3$ group as the enantiocontrolling moiety providing the observed stereoselectivity (Figure 2). The inversion in stereochemistry is attributed to the probable electronic interaction between the fluorine atoms of the ketone and the chlorine atom of the reagent. Such interactions have been proposed previously as the influencing factor controlling the stereo-outcome of reductions with chiral Grignard (R*MgBr) (13) and R*OAlCl$_2$ (14) reagents.

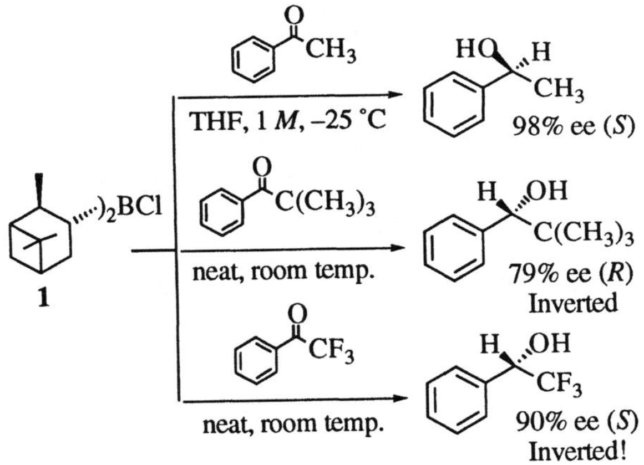

Figure 1. Comparison of asymmetric reduction of acetophenone, pivalophenone, and 2,2,2-trifluoroacetophenone with DIP-Chloride™

Mosher et al. had undertaken the asymmetric reductions of prochiral fluoro-ketones (15) on the basis of the early observations of McBee that such ketones undergo facile reductions with Grignard reagents (16). In addition to the preparation of optically pure fluorinated alcohols that was already proven to be useful as chiral solvating agents for spectroscopy and chiral stationery phases in chromatography (17), Mosher's group

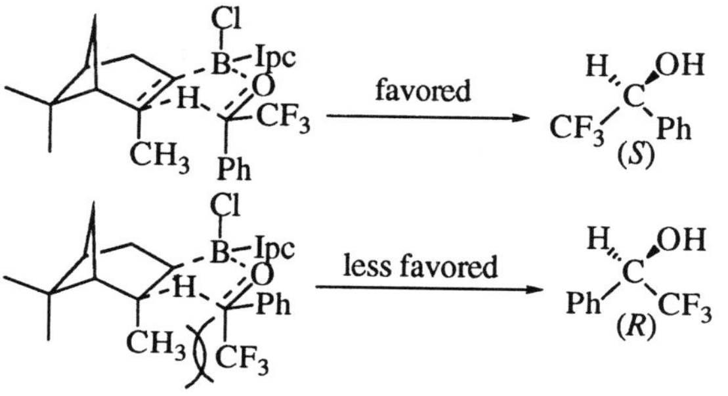

Figure 2. Mechanism of reduction of 2,2,2-trifluoroacetophenone with (–)-**1**

focused their attention to understanding the question of electronic versus steric effects of the CF_3 group. They reduced several trifluoromethyl ketones with chiral Grignard reagents of varying steric and electronic requirements and concluded that the steric environments of the reagents and the substrates play an important role in the stereochemical outcome (13). The appropriate interactions between the large group on the reagent and the large and small groups flanking the carbonyl moiety lead to the best results. However, in his comparison of the reduction of acetophenone and 2,2,2-trifluoroacetophenone with the same reagent, Mosher also observed a reversal in stereochemistry of the product alcohol from trifluoromethyl ketone.

Mosher compared the physical nature of the CH_3, CF_3 and $C(CH_3)_3$ groups and argued that the trifluoromethyl group should be very similar in overall sterics to a *tert*-butyl group (13). A recent report in the literature involving the kinetic resolution of fluoroalkyl vinyl carbinols also suggests that the enantiocontrolling effect of a CF_3 group tends to resemble that of a *tert*-alkyl group (18). However, a study of the dehydration of trifluoromethyl homoallylic alcohols suggest that the steric effect of a CF_3 as a substituent is as large as a cyclohexyl group and a little smaller than a *sec*-butyl group (19). Yet another investigation of the steric effect of a trifluoromethyl group *via* ene reaction of trifluoromethyl ketones revealed that a CF_3 group behaves as if it were a much larger than a phenyl or isobutyl group and as large as a *sec*-butyl group (20). In the literature, computational studies of the size of a CF_3 group have shown it to be only as bulky as an isopropyl group (21). However, an isopropyl group acts as a smaller moiety, less enantiocontrolling, compared to the effect of a phenyl group in chiral reductions with chiral Grignard reagents (12). The same conclusion can be reached in the reductions with **1** also.

Most of the chiral reductions of fluoro-ketones described in the literature discuss only the reduction of 2,2,2-trifluoroacetophenone. We have carried out a systematic investigation of the effect of fluorine in asymmetric reductions by reducing ketones with mono-, di-, and trifluoromethyl groups on one side of the carbonyl moiety and groups of differing electronic and steric requirements on the other side. Comparison of reductions with **1** and a similar reducing agent, *B*-isopinocampheyl-9-borabicyclo[3.3.1]nonane (Aldrich: Alpine-Borane®, **2**) were undertaken to understand the electronic effect of the chlorine atom in the reagent. Due to the limitation of space, we have restricted our discussions in this review to reductions with **1**. For a comprehensive summary of asymmetric reduction of fluorine containing ketones, we bring the reader's attention to an earlier review (22).

Reduction of Aryl Fluoroalkyl Ketones. As mentioned earlier, 2,2,2-trifluoroacetophenone is reduced to the product alcohol in 90% ee (*S*) with (–)-**1**.

Reduction of 1- and 2-trifluoroacetylnaphthalene with this reagent provided the product alcohols in 78% ee and 91% ee, respectively. 2,2,3,3,3-Pentafluoropropiophenone and 2,2,3,3,4,4,4-heptafluorobutyrophenone were reduced to the corresponding alcohols in 92% ee and 87% ee, respectively (Figure 3).

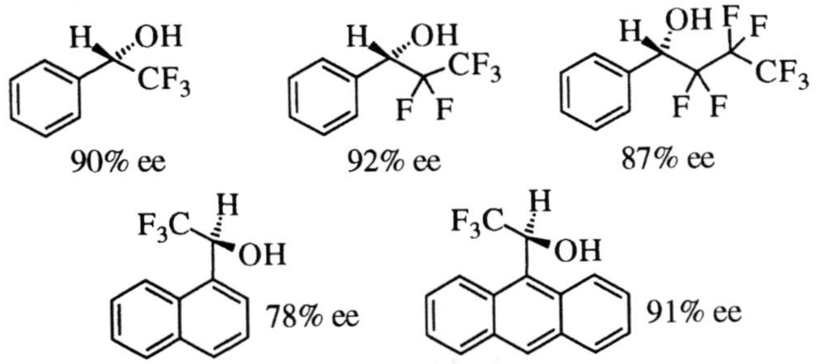

90% ee   92% ee   87% ee

78% ee   91% ee

Figure 3. Asymmetric reduction of aryl perfluoroalkyl ketones with **1**

In all of these cases the perfluoroalkyl group acts as the enantiocontrolling "larger" group as compared to the aryl groups. This produces alcohol products with consistent stereochemistry, opposite to those obtained from the reduction of the corresponding hydrogen analogs. It is interesting to note that the reduction of 2-fluoro- and 2,2-difluoroacetophenones with **1** provides the *R*-isomer of the product, similar stereochemistry as obtained from the reduction of propiophenone and isobutyrophenone (Figure 4).

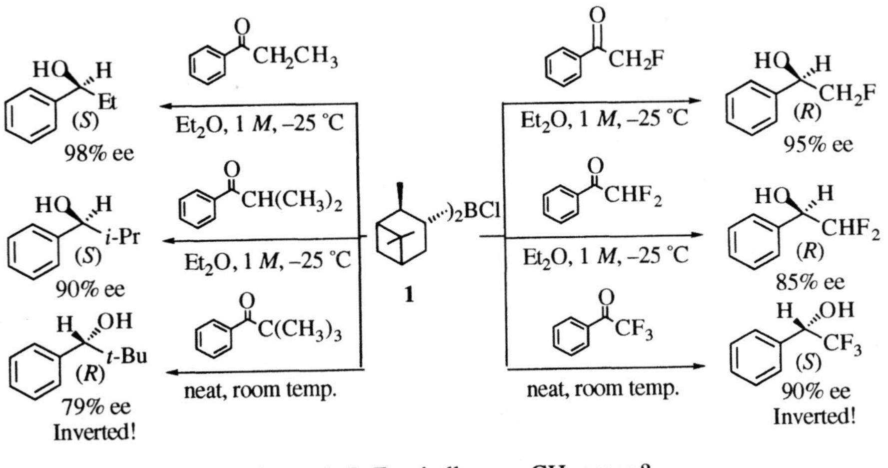

Figure 4. Is F as bulky as a CH_3 group?

Reduction of Alkyl Fluoroalkyl Ketones. Since **1** reduces non-fluorinated aralkyl ketones in high ee, a better understanding of the effects of fluorine atom in asymmetric induction with this reagent was sought from the reduction of a series of alkyl α-fluoroalkyl ketones. The fact that unhindered prochiral dialkyl ketones are typically reduced by **1** in relatively poor ee added value to this project. DIP-Chloride™ reduced alkyl trifluoromethyl ketones at a rate faster than that of the aryl derivatives

(*12*). However, the enantiomeric excesses of the product alcohols remained very high. Thus, 1,1,1-trifluoroacetone, 1,1,1-trifluorononan-2-one, 1,1,1-trifluorodecan-2-one, and cyclohexyl trifluoromethyl ketone were all reduced at rt within 4-12 h in 89% ee, 92% ee, 91% ee, and 87% ee, respectively (Figure 5) (*23*).

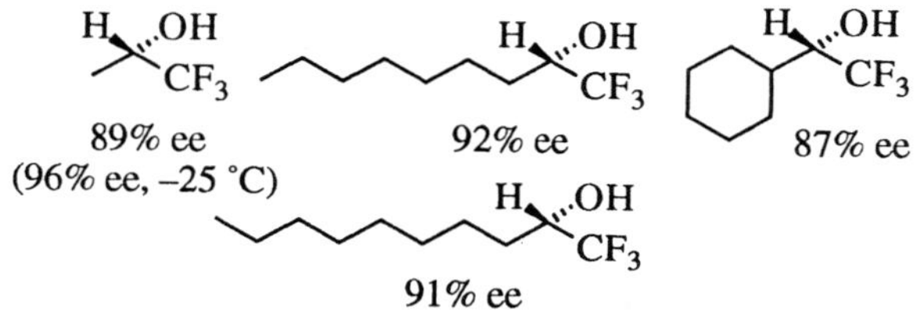

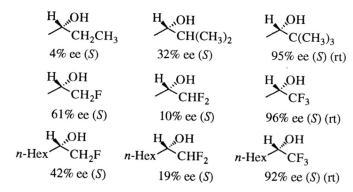

Figure 5. Asymmetric Reduction of alkyl trifluoromethyl ketones with **1**

The % ee achieved changed drastically depending on the number of α-fluorine atoms present in the ketone. The reduction of monofluoroacetone with **1** provided the product in 61% ee (Figure 6). (Gong, B.; Teodorovic´, A. V., unpublished results). Considering that **1** reduced 2-butanone to 2-butanol in only 4% ee, the reduction of monofluoroacetone to the product alcohol in such high ee probably highlights the electronic influence of the lone fluorine atom which has, theoretically, a steric size approximating that of a hydrogen atom. However, such an effect has not been observed for the reduction of 1,1-difluoroacetone. The high ee obtained for the product from 1,1,1-trifluoroacetone is normal for the reduction with **1**.

H, OH CH₂CH₃	H, OH CH(CH₃)₂	H, OH C(CH₃)₃
4% ee (*S*)	32% ee (*S*)	95% ee (*S*) (rt)
H, OH CH₂F	H, OH CHF₂	H, OH CF₃
61% ee (*S*)	10% ee (*S*)	96% ee (*S*) (rt)
n-Hex H, OH CH₂F	*n*-Hex H, OH CHF₂	*n*-Hex H, OH CF₃
42% ee (*S*)	19% ee (*S*)	92% ee (*S*) (rt)

Figure 6. Asymmetric reduction of alkyl fluoroalkyl ketones with **1**

The results were similar for the reduction of fluorooctanones. 1-Fluoro-2-octanone was reduced in 40% ee (*R*), whereas 1,1-difluoro- and 1,1,1-trifluoro-2-octanone were reduced in 32% (*S*), and 91% ee (*S*), respectively (Figure 6). These results also imply that the steric requirements of the trifluoromethyl group probably controls the enantioselectivity of these reductions.

α-Alkynyl α'-Fluoroalkyl Ketones. DIP-Chloride is a poor reagent for the chiral reduction of unhindered α-acetylenic ketones. (*10*). However, a systematic study of the asymmetric reduction of prochiral α-alkynyl α'-fluoroalkyl ketones with **1** revealed that perfluoroalkyl acetylenic ketones can be reduced in very high ee (92-≥ 99%) with this reagent (*22*). For example, 1,1,1-trifluoro-4-phenyl-3-butyn-2-one,

1,1,1,2,2-pentafluoro-5-phenyl-4-pentyn-3-one, and 4,4,5,5,6,6,6-heptafluoro-1-phenyl-1-hexyn-3-one are all reduced with **1** in Et₂O at −25 °C within 0.25-2 h in 98%, 96%, and 94% ee, respectively. Similarly, 1,1,1-trifluoro-3-octyn-2-one, 1,1,1,2,2-pentafluoro-4-nonyn-3-one, and 1,1,1,2,2,3,3-heptafluoro-5-decyn-4-one are reduced with **1** in ≥ 92% ee (Figure 7).

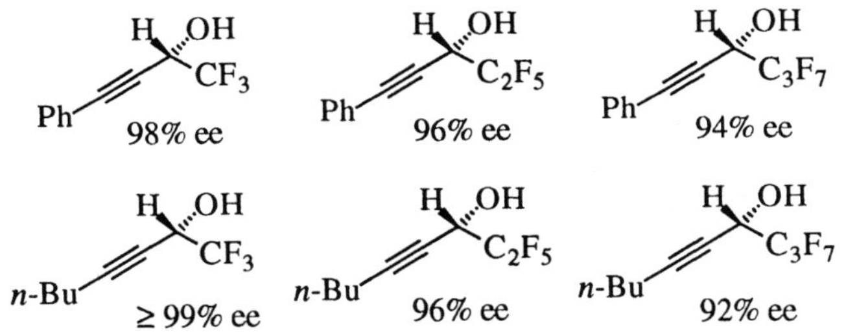

Figure 7. Asymmetric reduction of α-acetylenic α'-perfluoroalkyl ketones

The successful reduction of perfluoroalkyl alkynyl ketones with DIP-Chloride, similar to the reduction of hindered acetylenic ketones, again suggests that the steric requirements of a perfluoroalkyl group might be comparable to those of a *tert*-alkyl group.

It requires the "bulk" of three fluorine atoms to realize the high ee for the acetylenic ketone reduction with **1** which is ineffective for the reduction of di- and monofluoromethyl acetylenic ketones. In all of the above reductions, the fluoroalkyl group acts as the enantiocontrolling group with one exception. A remarkable inversion in selectivity is observed in the reduction of monofluoromethyl acetylenic ketones with **1** (Figure 8), indicating that in the transition state the acetylenic moiety acts as the enantiocontrolling group instead of the anticipated monofluoromethyl group. These results highlight the combined effects of both the electronic and steric influences of the fluorine in controlling both the rate and the enantioselectivity in the asymmetric reduction of prochiral fluorinated acetylenic ketones (*24*).

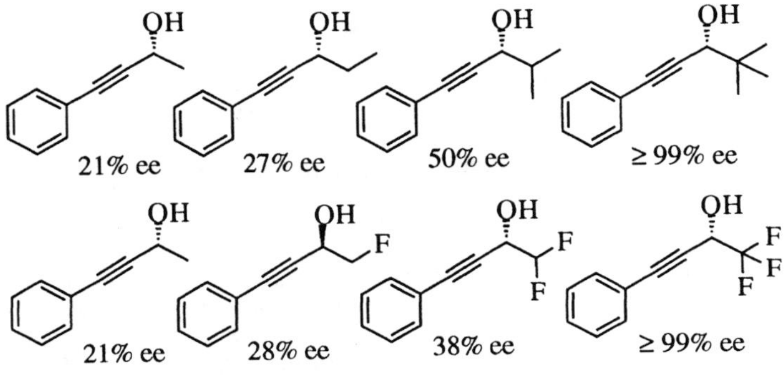

Figure 8. Asymmetric reduction of α-acetylenic α'-fluoroalkyl ketones with **1**

Bisperfluoroalkyl Aryl Diketones. The importance of C_2-symmetric chiral auxiliaries (*25*), including diols as the chiral director in asymmetric synthesis persuaded us to undertake a systematic study of the convenient synthesis of such compounds by the DIP-Chloride™ reduction of representative diacetyl and bistrifluoroacetyl aromatic compounds (26). The reductions of the fluorinated diketones were complete in 3-7 d and the products were obtained in high yield in essentially enantiomerically pure form (Figure 9). The rates of reductions were slower and the product stereochemistries are opposite compared to those of the corresponding methyl analogs. The synthesis of these fluorinated diols provides an opportunity to study the effect of fluorine atoms in these chiral directors (27).

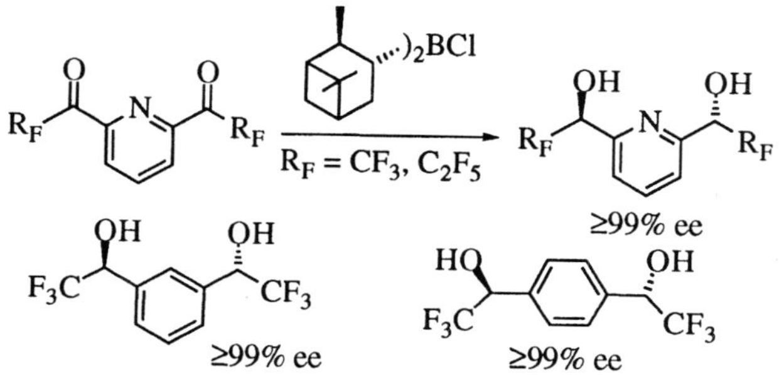

Figure 9. Asymmetric reduction of bistrifluoroacetyl aryl ketones with 1

Synthesis and Ring-Cleavage Reactions of Trifluoromethyloxirane. Optically active 3,3,3-trifluoropropeneoxide is an excellent chiral building block for asymmetric fluoro-organic synthesis (28). We exploited the capability of 1 to reduce perfluoroalkyl ketones in high ee by preparing the enantiomers of 1-bromo-3,3,3-trifluoro-2-propanol in 96% ee via the asymmetric reduction of the corresponding ketone with either (+)- or (–)-isomer of the reagent (Figure 10) (29).

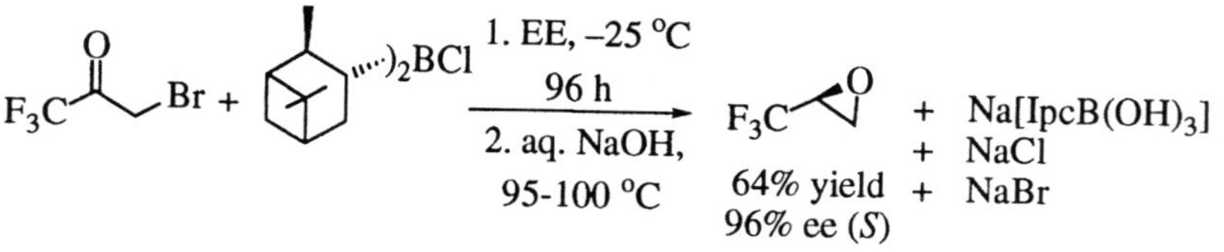

Figure 10. Preparation of trifluoromethyloxirane

The ring cleavage reactions of the trifluoromethyloxirane with the appropriate nucleophiles provided a general synthesis of chiral trifluoromethyl carbinols, such as 1-amino-3,3,3-trifluoro-2-propanol, 1-azido-3,3,3-trifluoro-2-propanol, 1-diethylamino-3,3,3-trifluoro-2-propanol, 1-cyano-3,3,3-trifluoro-2-propanol, 1,1,1-trifluoro-2-propanol, 1,1,1-trifluoro-2-octanol, 1-phenyl-3,3,3-trifluoro-2-propanol, 1-ethoxy-3,3,3-trifluoro-2-propanol, and 1,2-dihydroxy-3,3,3-trifluoropropane, in 61-88% yields without loss of optical activity (Figure 11) (29).

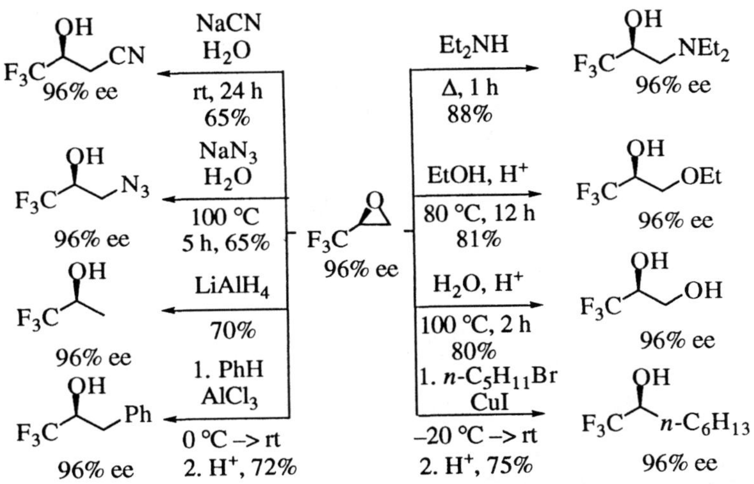

Figure 11. Ring-opening reactions of trifluoromethyloxirane

The application of this fluoro-epoxide for the synthesis of dimeric liquid crystals which exhibited stable anti ferroelectric chiral Smetic C Phase and were good dopants for anti-ferroelectric liquid crystals was shown by Hiyama and co-workers (Figure 12) (30).

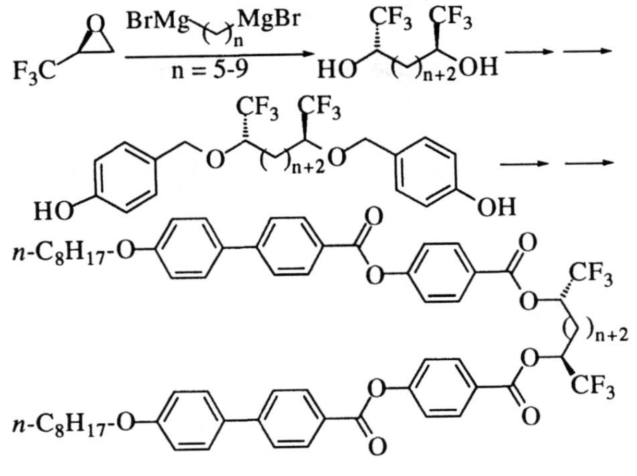

Figure 12. Application of trifluoromethyloxirane:
synthesis of ferroelectric liquid crystals

Ring-Substituted Acetophenones. Kirk and coworkers have shown the importance of fluorine substitution in the aromatic ring of several ring-fluorinated neuroactive amines (31). Due to their apparent importance, we undertook the asymmetric reduction of ring-fluorinated acetophenones with 1. The reduction of 2',3',4',5',6'-pentafluoroacetophenone with 1 provided the corresponding alcohol in unusually low ee (44%, *S*) for the reduction of an aralkyl ketone. Probably, the interaction of the pentafluorophenyl group with the chlorine atom of the reagent lowers the ee. 2,6-difluoroacetophenone was reduced to the corresponding alcohol in 74% ee.

All of the other mono-, di-, tri-, and tetrafluoro substituted acetophenones that were reduced yielded ≥94% ee for the corresponding α-phenethanols (Figure 13) (Ramachandran, P. V. unpublished results).

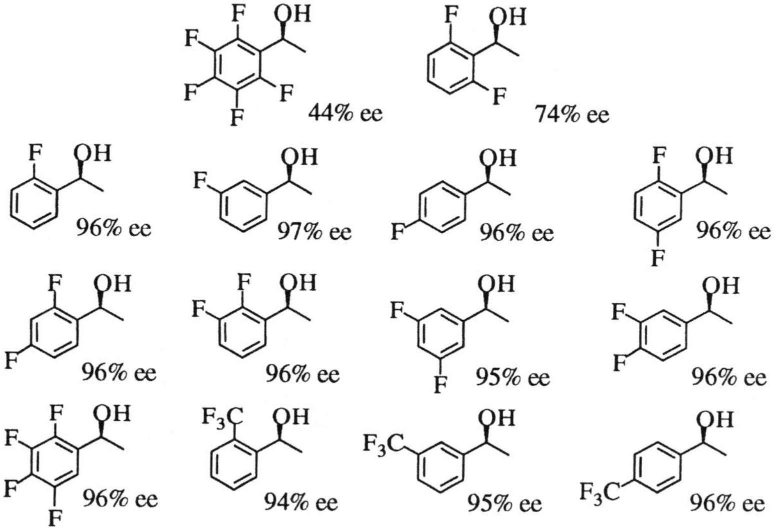

Figure 13. Preparation of ring-fluorinated α-phenethanols via reduction with 1

Allylboration of Fluorinated Aldehydes with Chiral Borane Reagents

Allylic organometallic reactions producing homoallylic alcohols has attained considerable importance in the art of asymmetric synthesis of highly sophisticated conformationally non-rigid systems (*32*). The product contains an alkene moiety which could be further transformed into other functional groups as exemplified in the synthesis of macrolide and ionophore antibiotics with a plethora of stereodefined *vic*-diols or β-methyl alcohols (*33*). Here, not only the enantioselectivity, but the diastereoselectivity of the reaction is also highly important. Accordingly, numerous searches for the most efficient reagent that can achieve both these selectivities in a single step have been made (*34*). Chiral organoboranes have revealed their uniqueness and advantages for achieving these desired transformations.

Two decades ago Mikhailov reported that allylboron derivatives, in marked contrast to the saturated trialkylboranes, undergo a fast addition to carbonyl groups with allylic rearrangement (*35*). α-Pinene was put to test for allylboration and it provided another major application of a diisopinocampheylborane reagent in organic synthesis (Fig 14). The syntheses of *B*-allyldiisopinocampheylborane (Ipc₂BAll, **3**) and *B*- allyldi-2-isocaranylborane (2-Icr₂BAll, **4**) are simple and the reaction with aldehydes at –78 or –100 °C, followed by either alkaline hydrogen peroxide or ethanolamine work-up, provides the homoallylic alcohols in very good yields in very high enantiomeric excesses (*36*, *37*).

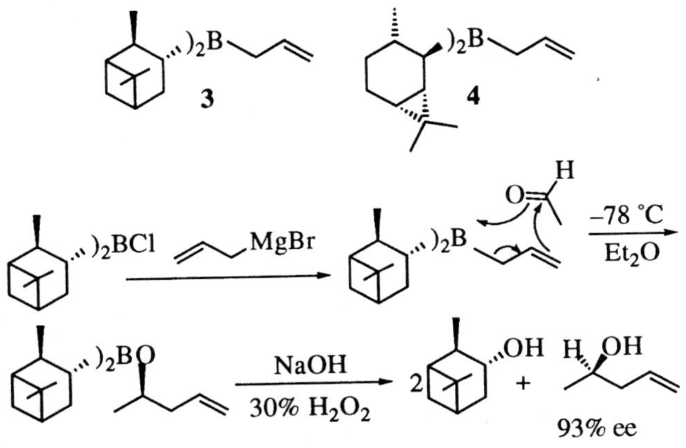

Figure 14. Synthesis and reaction of *B*-allyldiisopinocampheylborane

We applied **3** and **4** for the allylboration of ring-fluorinated benzaldehydes and obtained 95-≥99% ee for the product homoallyl alcohols (Figure 15) (Sreedharan, M.; Kumar, J. S. D., unpublished results).

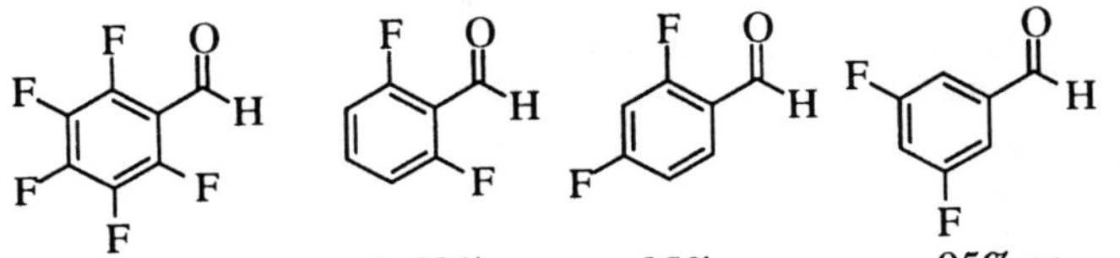

Ipc_2BAll:	≥ 99% ee	≥ 99% ee	95% ee	95% ee
2-Icr_2BAll:	96% ee	97% ee	96% ee	98% ee

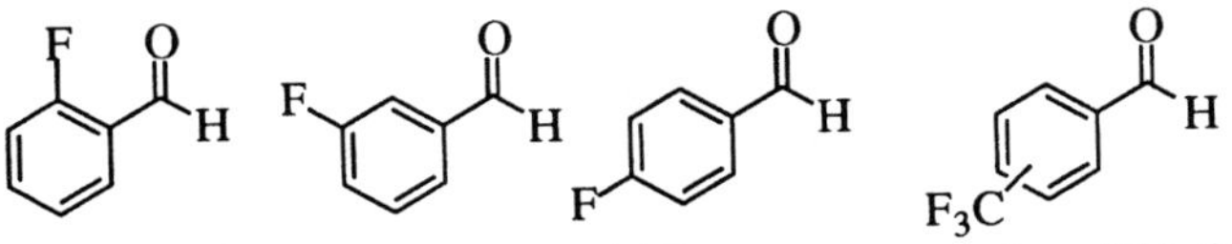

Ipc_2BAll:	92% ee	92% ee	94% ee	*o*-, *m*-, and *p*-: 97-≥ 99% ee
2-Icr_2BAll:	98% ee	97% ee	96% ee	*o*-, *m*-, and *p*-: 95-97% ee

Figure 15. Asymmetric allylboration of ring-fluorinated benzaldehydes with Ipc_2BAll and 2-Icr_2BAll

We then turned our attention to the allylboration of perfluoroalkyl aldehydes. Both **3** and **4** provided the perfluoroalkyl homoallyl alcohols in ≥ 99% ee. We converted these alcohols to the corresponding 4-perfluoroalkyl-γ-butyrolactones via a protection, hydroboration-oxidation, deprotection, cyclization procedure (Figure 16) (Kumar, J. S. D., unpublished results). One of these, 4-trifluoromethyl-γ-butyrolactone, which is a building block for the synthesis of fluoro-analogs of natural products, such as brefeldin, sulcatol, zearalenone, etc. (*38*), has been prepared in optically active form by Seebach (*38*) and Kitazume (*39*).

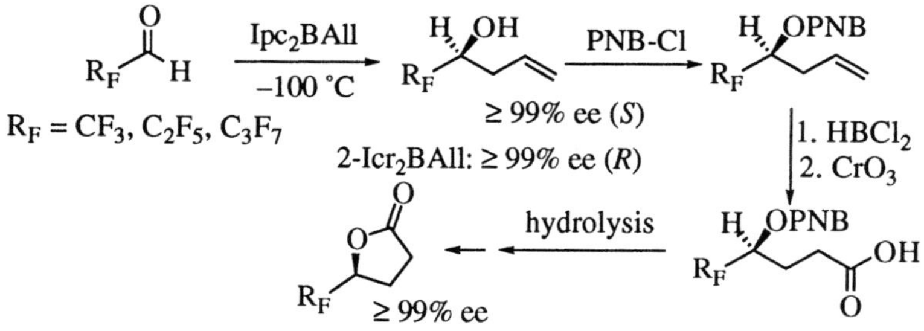

Figure 16. Synthesis of 4-perfluoroalkyl-γ-butyrolactones

Enolboration-Aldolization of Fluoro-ketones and Aldehydes

Stereoselective aldol reactions have been known for quite some time (40) Of all the elements studied for application in enolization reactions, boron offers special advantages and has been used extensively (Figure 17) (41). Asymmetric enolboration received considerable attention this past decade and several researchers have developed different chiral auxiliaries to induce diastereoselectivities and enantioselectivities in the reaction [42]. For example, Paterson developed Meyers' diisopinocampheylboron triflate (Ipc$_2$BOTf, **5**) (43) as an excellent asymmetric enolborating agent (44). We have carried out a systematic study of the influence of all of the parameters, such as the effect of the ketone, amines, and the leaving group in the reagents (45). We had also shown the effect of steric requirements of ketones and aldehydes in asymmetric enolborations of methyl ketones with DIP-Chloride™ (46).

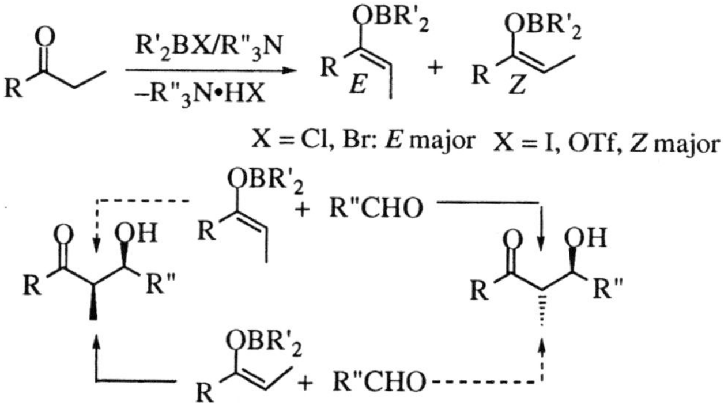

Figure 17. Stereochemistry of enolboration-aldolization

Recently, Iseki, Kobayashi and co-workers had shown the influence of fluorine in aldolizations with boron enolates (47). CF$_3$CHO provides products involving a si-face attack compared to a re-face attack in the case of CH$_3$CHO in the aldolization reaction shown in Figure 18.

We were interested in applying DIP-Chloride™ for the enolboration-aldolization of ring-fluorinated acetophenones and benzaldehydes. During this research, we encountered a very interesting phenomenon. When we carried out the aldolization of diisopinocampheylboron enolate from 2',3',4',5',6'-pentafluoroacetophenone with pentafluorobenzaldehyde at –78 °C, we obtained the product aldol, within 5 d, in 79%

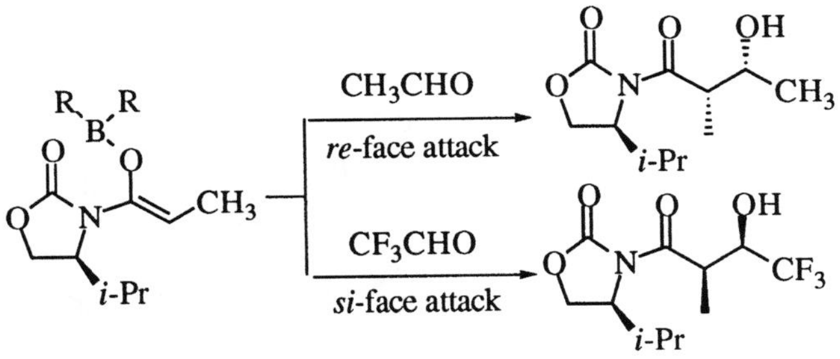

Figure 18. Effect of fluorine-substitution in enolboration:
Reversal of π-face selectivity

yield and 82% ee. When we warmed the reaction to –25 °C to accelerate the rate, the reaction was complete in 24 h and the product was obtained in 79% yield and 94% ee (Figure 19) (Wang, B., unpublished results). We obtained higher ee by performing the aldolizations at higher temperature! This is unusual in asymmetric synthesis.

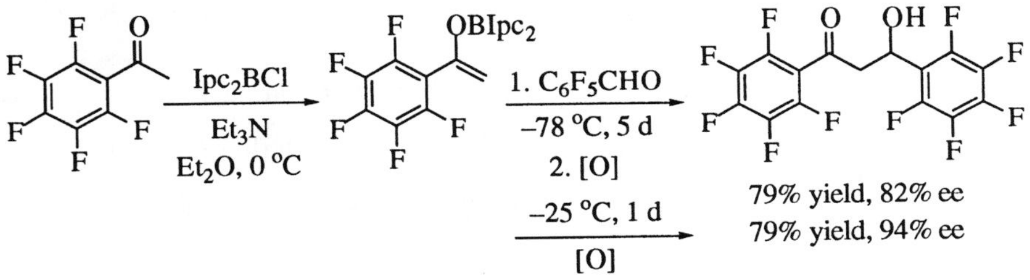

Figure 19. Increase in enantioselectivity at higher temperature in enolboration-aldolization of pentafluoroacetophenone-pentafluorobenzaldehyde

The results can be rationalized by considering a tandem enolboration-aldolization-reduction reaction that we have observed in non-fluorinated systems with DIP-Chloride™ (Figure 20) (Lu, Z. H.; Xu, W. C., unpublished results).

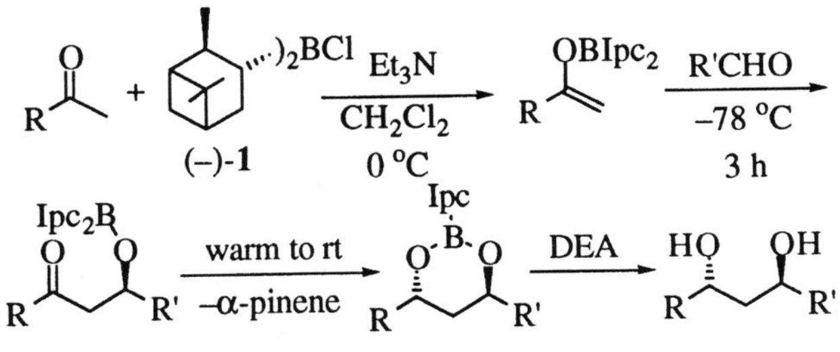

Figure 20. Tandem enolboration-aldolization-intramolecular reduction

Invoking this tandem reaction, it can be seen that a kinetic resolution takes place at higher temperature. The minor isomer of the aldolate undergoes an intramolecular reduction leaving the major aldol in improved enantiomeric excess (Figure 21).

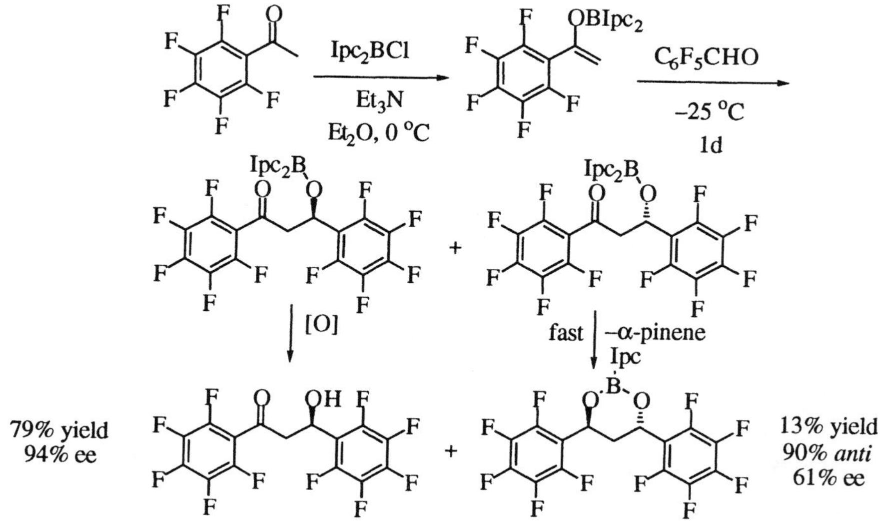

Figure 21. Kinetic resolution in Enolboration-aldolization

This effect is observed only with 2',3',4',5',6'-pentafluoro- and 2',6'-difluoroacetophenone as can be seen from the summary of results from the tandem enolboration-reduction shown in Figure 22. In all of the cases the *anti*-diol is obtained as the major isomer.

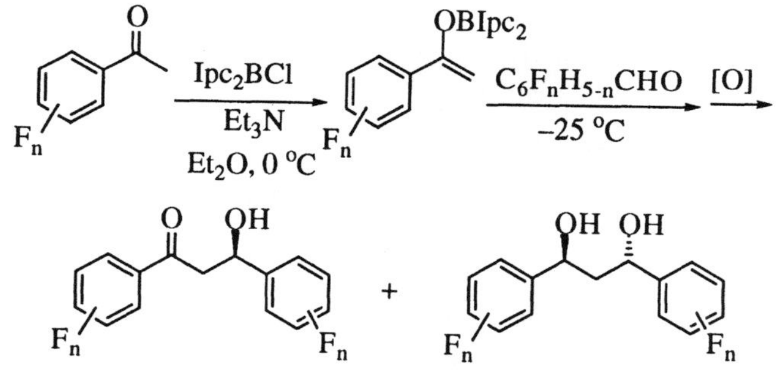

n	% yield	% ee	% yield	anti/syn	% ee (anti)
5	79	94	13	90/10	61
2 (2,6-)	60	86	14	86/14	70
1 (4-)	60	20	13	95/5	74
1 (3-)	50	62	30	>99/<1	70
1 (2-)	27	92	60	>99/<1	9*
0			63	92/8	60

Figure 22. Effect of fluorine on tandem enolboration-aldolization-reduction

The tandem enolboration-aldolization-intermolecular reduction normally provides the product *syn*-diols in high diastereoselectivity (*48*). However, in the case of the reduction of the enolboration-aldolization product from trifluoroacetone-acetaldehyde combination with $LiBH_4$, we observed that a 1:1 mixture of syn/anti mixture is obtained. The intramolecular reduction process still gave 94% *anti*-product (Figure 23) (Wang, B., unpublished results).

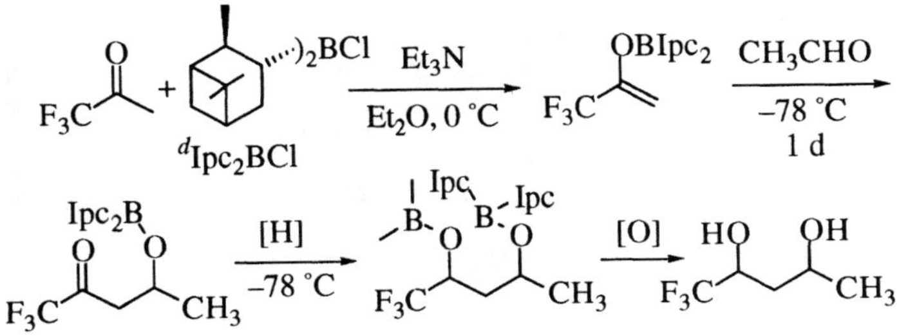

[H]: $LiBH_4$: *Syn/anti* : 50/50, ee: *syn* 72%, *anti* : 29%
[H]: BMS : *Syn/anti* : 63/37, ee: *syn* 79%, *anti* : 74%
[H]: intramolecular: *Syn/anti* : 6/94, ee: *syn*: 99%, *anti*: 53%

Figure 23. Effect of fluorine on enolboration-aldolization-intermolecular reduction

Conclusion

In conclusion, we have demonstrated that our chiral organoborane reagents can be effectively applied for the preparation of asymmetric fluoro-organic compounds in very high enantiomeric excess. The unpredictable nature of fluorine containing substrates (Flustrates) (*49*) is manifested in all of the reactions other than allylborations. Apart from providing the products in high ee, this research has led to several interesting questions of theoretical interest. Fluoro-organic chemistry is really fascinating!

Acknowledgment

The financial support from the United States Army Research Office is gratefully acknowledged.

Literature Cited

1.   For several recent reviews see: *Biomedical Frontiers of Fluorine Chemistry*, Ojima, I.; McCarthy, J. R.; Welch, J. T. Eds. ACS Symposium Series 639, American Chemical Society, Washington DC, **1996**.

2.   Advances in Asymmetric Synthesis, Hassner, A. Ed. JAI Press, Greewnwich, CT, **1995**.

3.   Resnati, G.*Tetrahedron* **1993**, *49*, 9385.

4.   Schlosser, M. *Angew. Chem. Int. Ed. Engl.* **1998**, *39*, 1496.

5.   (a) Brown, H. C.; Ramachandran, P. V. In ref. 2 (a), Chapter 5, pp 147-210. (b) Brown, H. C. Ramachandran, P. V. *J. Organometal. Chem.* **1995**, *500*, 1.

6.   *Reductions in Organic Chemistry*. American Chemical Society Symposium Series # 641. Abdel-Magid, A. F. Ed., American Chemical Society, Washington, DC, **1996**.

36

7. (a) Bothner-By, A. A. *J. Am. Chem. Soc.* **1951**, *73*, 846. (b) Landor, S. R.; Miller, B. J.; Tatchell, A. R. *Proc. Chem. Soc.* **1964**, 227.

8. (a) Doering, W. von E.; Young, R. W. *J. Am. Chem. Soc.* **1950**, *72*, 631. (b) Jackman, L. M.; Mills, J. A.; Shannon, J. S. *J. Am. Chem. Soc.* **1950**, *72*, 4814.

9. Vavon, G.; Riviere, C.; Angelo, B. *Compt. rend.* **1946**, *222*, 959.

10. (a) Brown, H. C.; Chandrasekharan, J.; Ramachandran, P. V. *J. Am. Chem. Soc.* **1988**, *110*, 1539. DIP-Chloride™ is the Trade Mark of the Aldrich Chemical Company. (b) Brown, H. C.; Ramachandran, P. V. *Acc. Chem. Res.* **1992**, *25*, 16. (c) Ramachandran, P. V.; Brown, H. C. in ref. 6. Chapter pp 84-97.

11. The *S*-configuration of the fluoro alcohol has opposite stereochemistry compared to the *S*-isomer of α-phenethanol. This is an artifact of Cahn-Ingold-Prelog priority rules. Cahn, R. S.; Ingold, C.; Prelog, V. *Angew. Chem. Int. Ed. Engl.* **1966**, *5*, 385.

12. Ramachandran, P. V.; Teodorovic´, A. V.; Brown, H. C. *Tetrahedron* **1993**, *49*, 1725.

13. For an early extensive discussion of this topic, see: Morrison, J. D.; Mosher, H. S. *Asymmetric Organic Reactions*, Prentice Hall: Englewood Cliffs, N.J., 1971, pp 160-218.

14. Nasipuri, D.; Bhattacharya, P. K. *J. Chem. Perkin Tans. I* **1977**, 576.

15. Mosher, H. S.; Stevenot, J. E.; Kimble, D. O. *J. Am. Chem. Soc.* **1956**, *78*, 4374.

16. (a) McBee, E. T.; Pierce, O. R.; Higgins, J. F. *J. Am. Chem. Soc.* **1952**, *74*, 1736. (b) McBee, E. T.; Pierce, O. R.; Meyer, D. D. *J. Am. Chem. Soc.* **1955**, *77*, 83.

17. Pirkle, W. H.; Finn, J. in *Asymmetric Synthesis* Morrison, J. D. Ed. Academic Press: New York,. 1983, Vol. 1. Chapter 6.

18. Hanzawa, Y.; Kawagoe, K.; Kobayashi, Y. *Chem. Pharm. Bull.* **1987**, *35*, 2609.

19. Nagai, T.; Nishioka, G.; Koyama, M.; Ando, A.; Miki, T.; Kumadaki, I. *J. F. Chem.* **1992**, *57*, 229.

20. Nagai, T.; Nishioka, G.; Koyama, M.; Ando, A.; Miki, T.; Kumadaki, I. *Chem. Pharm. Bull.* **1992**, *40*, 593.

21. Bott, G.; Field, L. D.; Sternhell, S. *J. Am. Chem. Soc.* **1980**, *102*, 5618.

22. Ramachandran, P. V.; Brown, H. C. in *EPC-Synthesis of Fluoro-Organic Compounds*, Soloshonok, V. A. Ed. Wiley: Chichester, U. K., **1999** (in press).

23. Ramachandran, P. V.; Teodorovic´, A. V.; Gong, B.; Brown, H. C. *Tetrahedron: Asym.* **1994**, *5*, 1075.

24. Ramachandran, P. V.; Gong, B.; Teodorovic´, A. V.; Brown, H. C. *Tetrahedron: Asym.* **1994**, *5*, 1061.

25. Whitesell, J. K. *Chem. Rev.* **1989**, *89*, 1581.

26. Ramachandran, P. V.; Chen, G. M.; Lu, Z. H.; Brown, H. C. *Tetrahedron Lett.* **1996**, *37*, 3795.

27. Brown, H. C.; Chen, G. M.; Ramachandran, P. V. *Chirality*, **1997**, *9*, 506.

28. Katagiri, T.; Obara, F.; Toda, S.; Furuhashi, K. *Synlett* **1994**, 507.

29. Ramachandran, P. V.; Gong, B.; Brown, H. C. *J. Org. Chem.* **1995**, *60*, 41.

30. Suzuki, Y.; Isozaki, T.; Kusumoto, T.; Hiyama, T. *Chem. Lett.* **1995**, 719.

31. Kirk, K. L. In ref. 1 Chapter 9.

32. (a) Ley, S. V.; Merritt, A. T. *Nat. Prod. Rep.* **1992**, *9*, 243. (b) Danishefsky, S. J. *Aldrichim. Acta* **1986**, *19*, 59. (c) Masamune, S.; Choy, W. *Aldrichim. Acta.* **1982**, *15*, 47.

33. (a) Hoffmann, R. W. *Angew. Chem. Int. Ed. Engl.* **1992**, *31*, 1124. (b) Martin, S. F.; Guinn, D. E. *Synthesis* **1991**, 245. (c) Hoffmann, R. W. *ibid.* **1987**, *26*, 489. (d) Masamune, S. *Angew. Chem. Int. Ed. Engl.* **1985**, *24*, 1.

34. Yamamoto, Y.; Asao, N. *Chem. Rev.* **1993**, *93*, 2207.

35. Mikhailov, B. M. *Organometal. Chem. Rev. A.* **1972**, *8*, 1.

36. Brown, H. C.; Jadhav, P. K. *J. Am. Chem. Soc.* **1983**, *105*, 2092.

37. Brown, H. C.; Randad, R. S.; Bhat, K. S.; Zaidlewicz, M.; Racherla, U. S. *J. Am. Chem. Soc.* **1990**, *112*, 2389.

38. Seebach, D.; Renaud, P. *Helv. Chim. Acta* **1985**, *68*, 2342.

39. Kitazume, T.; Ishikawa, N. *Chem. Lett.* **1984**, 1815.

40. For reviews on the aldol reactions, see: (a) Heathcock, C. H. in ref. 2, Vol.3, Chapter 2. (b) Evans, D. A.; Nelson, J. V.; Taber, T. R.; *Top. Stereochem.* **1982**, *13*, 1.

41. For reactions of achiral boron enolates, see: (a) Fenzl, W.; Koster, R. *Liebigs Ann. Chem.* **1975**, 1322. (b) Brown, H. C.; Dhar, R. K.; Bakshi, R. K.; Pandiarajan, P. K.; Singaram, B. *J. Am. Chem. Soc.* **1989**, *111*, 3441. (c) Mukaiyama, T. *Org. React.* **1982**, *28*, 203. (d) Evans, D. A.; Nelson, J. V.; Vogel, E.; Taber, T. R. *ibid.* **1981**, *103*, 3099. (e) Masamune, S.; Mori, S.; Van Horn, D.; Brooks, D. W. *Tetrahedron Lett.* **1989**, 1665.

42. For the asymmetric aldol reactions of enolborinates using different chiral auxiliaries, see: (a) Corey, E. J.; Choi, S. *Tetrahedron Lett.* **1991**, *32*, 2857. (b) Enders, D.; Lohray, B. B. *Angew. Chem. Int. Ed. Engl.* **1988**, *27*, 581. (c) Masamune, S.; Choy, W.; Kerdesky, F. A.; Imperiali, B. *J. Am. Chem. Soc.* **1981**, *103*, 1566. (d) Evans, D. A.; Bartroli, J.; Shih, T. L. *ibid.* **1981**, *103*, 2127.

43. Meyers, A. I.; Yamamoto, Y. *J. Am. Chem. Soc.* **1981**, *103*, 4278. *ibid. Tetrahedron* **1984**, 40, 2309.

44. Paterson, I. *Org. React.* **1997**, *51*, 1.

45. Brown, H. C.; Ganesan, K. ; Dhar, R. K. *J. Org. Chem.* **1993**, *58*, 147.

46. Ramachandran, P. V.; Xu, W. C.; Brown, H. C. *Tetrahedron Lett.* **1996**, *37*, 4911.

47. Iseki, K.; Oishi, S.; Kobayashi, Y. *Tetrahedron* **1996**, *52*, 71.

48. (a) Paterson, I.; Channon, J. A. *Tetrahedron Lett.* **1992**, *33*, 797. (b) Bonini, C.; Racioppi, R.; Righi, G.; Rossi, L. *Tetrahedron: Asym.* **1994**, *5*, 173.

49. Seebach, D. *Angew. Chem. Int. Ed. Engl.* **1990**, *29*, 1320.

Chapter 3

Catalytic Asymmetric Synthesis of Fluoroorganic Compounds: Mukaiyama–Aldol and Henry Reactions

Katsuhiko Iseki

MEC Laboratory, Daikin Industries, Ltd., Miyukigaoka 3, Tsukuba, Ibaraki 305–0841, Japan

The efficient catalytic asymmetric syntheses of fluorinated aldols and nitroaldols of high enantiomeric purity are described. Difluoroketene and bromofluoroketene trimethylsilyl ethyl acetals react with various aldehydes in the presence of chiral Lewis acids at -78°C to afford the corresponding desired aldols with up to 99% ee. It is noteworthy that the aldol reactions of the fluorine-substituted acetals at -78°C and at higher temperatures (-45 or -20°C) provide the (+) and (-) aldols, respectively, with excellent-to-good enantioselectivity. The Henry reaction of 2,2-difluoroaldehydes with nitromethane is carried out using lanthanoid-lithium-BINOL catalysts to give the nitroaldols with up to 95% ee. The enantiotopic face selection for the fluorine-containing aldehydes is the reverse of that for nonfluorinated aldehydes.

The synthesis of optically active chiral fluoroorganic compounds, which play an important role in the research on biological chemistry and in the development of medicines, is now one of the most fascinating aspects of modern organofluorine chemistry in view of fluorine's unique influence on biological activity (1-12). Such compounds have also been applied to the preparation of materials such as liquid crystals (13-15). Although catalytic asymmetric synthesis has attracted the most attention as methods available for preparing chiral molecules, fluorine-containing molecules with unexpected and generally unusual reactivity frequently incapacitate asymmetric catalyses developed for nonfluorinated compounds from working (16). The strongly electronegative nature of fluorine often alters the reaction courses established for hydrocarbon patterns to invert the stereochemical results. Thus, fluoroorganic compounds remain as extremely difficult but challenging problems to be solved for catalytic enantiocontrolled synthesis.

Fluorine, due to its extremely high electronegativity, has a considerable electronic effect on its neighboring groups in a molecule. For example, the introduction of a difluoromethylene moiety into a bioactive peptide has led to the discovery of potent competitive and reversible protease inhibitors mimicking the transition state for amide bond cleavage such as HIV protease and renin inhibitors (17-21).

HIV protease inhibitor (Sham *et al.*, 1991)   renin inhibitor (Rosenberg, 1989)

This review describes the first catalytic asymmetric Mukaiyama-aldol reaction of fluorine-substituted ketene silyl acetals with aldehydes and the catalytic asymmetric Henry (nitroaldol) reaction of 2,2-difluoroaldehydes with nitromethane to provide the optically active aldols and nitroaldols, respectively, which must be versatile synthetic intermediates for the fluorinated protease inhibitors.

Catalytic Asymmetric Aldol Reaction of Fluorine-Substituted Ketene Silyl Acetals

Only a few enantioselective approaches to optically active 2,2-difluoro-3-hydroxy esters, key synthetic intermediates for a variety of important chiral fluorinated molecules, have been reported. Braun *et al.* and Andrés *et al.* have independently reported that the Reformatsky reagents generated from bromodifluoroacetates react with aromatic aldehydes in the presence of stoichiometric amounts of chiral amino alcohols such as *N*-methylephedrine to afford the corresponding desired aldols in good optical yields (*22, 23*). However, these methods are not catalytic, and the decrease in quantity of the chiral ligands dramatically suppresses the enantioselectivity. Thus, we became very interested in developing an unprecedented catalytic asymmetric aldol reaction of difluoroketene silyl acetals promoted by chiral Lewis acids.

Preparation of Difluoroketene and Bromofluoroketene Trimethylsilyl Ethyl Acetals. Although some difluoroketene silyl acetals have been prepared *in situ* by treatment of 2-halo-2,2-difluoroacetate with zinc metal and trialkylchlorosilane and then applied to the synthesis of several useful racemates (*24-27*), the acetals have never been isolated in pure form. These impure acetal solutions containing zinc salts are not thought to be practical for the asymmetric aldol reaction using chiral Lewis acids because the zinc salts also act as Lewis acids. Thus, we began our studies with the preparation of a salt-free, pure difluoroketene silyl acetal (*28, 29*). The Reformatsky reagent, prepared in THF from ethyl bromodifluoroacetate and activated zinc powder according to the reported procedure (*22*), was treated with chlorotrimethylsilane to give the difluoroketene acetal solution which was diluted with *n*-pentane and filtered to remove the zinc salt. The filtrate was then concentrated *in vacuo*. After the dilution-filtration-concentration process was repeated two more times, the oily residue was distilled under reduced pressure to afford the pure difluoroketene trimethylsilyl ethyl acetal **1** as a colorless oil. In a similar manner, bromofluoroketene trimethylsilyl ethyl acetal **2** without zinc salts was obtained from ethyl dibromofluoroacetate as a 62:38 mixture of the *E*- and *Z*-isomers (*30*). The minor isomer was determined to be the *Z*-form by NOE observation between the fluorine and the methylene proton of the ethoxyl group (*31*).

$E/Z = 62/38$

1 (b.p. 51-53°C/45 mmHg)   **2** (b.p. 37-39°C/1.2 mmHg)

Uncatalyzed Aldol Reaction of Fluorine-Substituted Ketene Silyl Acetals. We next examined the reactivity of the fluorine-substituted ketene silyl acetals in the absence of a Lewis acid. As shown in Scheme 1, the difluoroketene acetal **1** was found to react with benzaldehyde in dichloromethane even at -78°C, while the dimethylketene trimethylsilyl methyl acetal **3** gave only a trace amount of the corresponding aldol under the same conditions (29). The bromofluoroketene acetal **2** is also more reactive than the fluorine-free acetal **3** and provided a mixture of *syn*- and *anti*-aldols in 79% yield at 40°C (Iseki, K.; Kuroki, Y.; Kobayashi, Y., MEC Laboratory, Daikin Industries, Ltd., unpublished data.). The *syn/anti* ratio (60/40) at 40°C may be ascribed to the *E/Z* ratio (62/38) of the acetal **3** (Scheme 1).

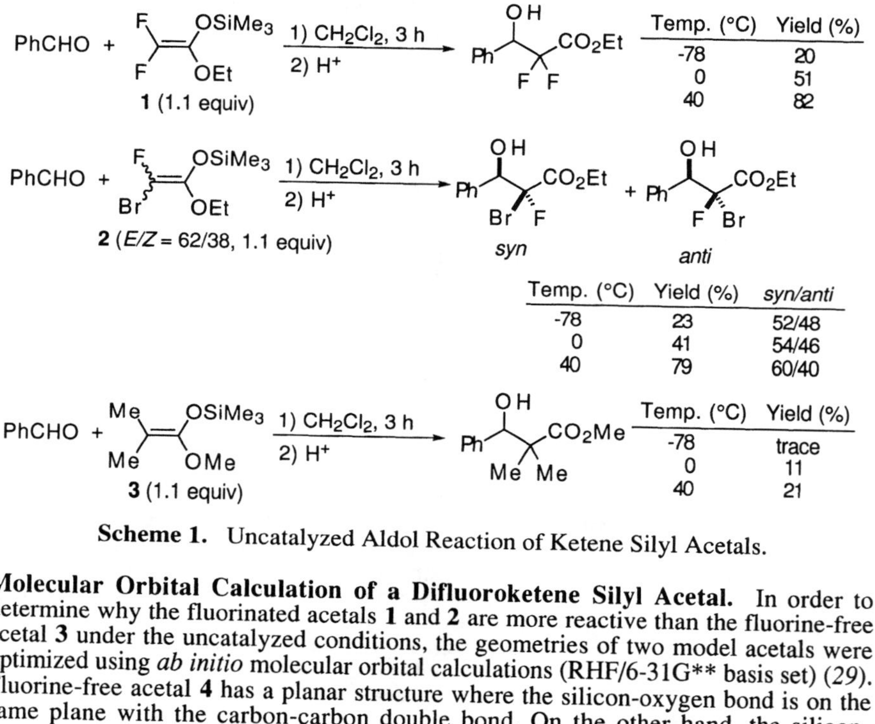

PhCHO +

Temp. (°C)	Yield (%)
-78	20
0	51
40	82

PhCHO +

Temp. (°C)	Yield (%)	syn/anti
-78	23	52/48
0	41	54/46
40	79	60/40

PhCHO +

Temp. (°C)	Yield (%)
-78	trace
0	11
40	21

Scheme 1. Uncatalyzed Aldol Reaction of Ketene Silyl Acetals.

Molecular Orbital Calculation of a Difluoroketene Silyl Acetal. In order to determine why the fluorinated acetals **1** and **2** are more reactive than the fluorine-free acetal **3** under the uncatalyzed conditions, the geometries of two model acetals were optimized using *ab initio* molecular orbital calculations (RHF/6-31G** basis set) (29). Fluorine-free acetal **4** has a planar structure where the silicon-oxygen bond is on the same plane with the carbon-carbon double bond. On the other hand, the silicon-oxygen bond in difluoroketene acetal **5** is out of the carbon-carbon double bond plane. This torsion of the silicon-oxygen bond in **5** seems to be ascribed to the +Iπ effect of the fluorine atoms.

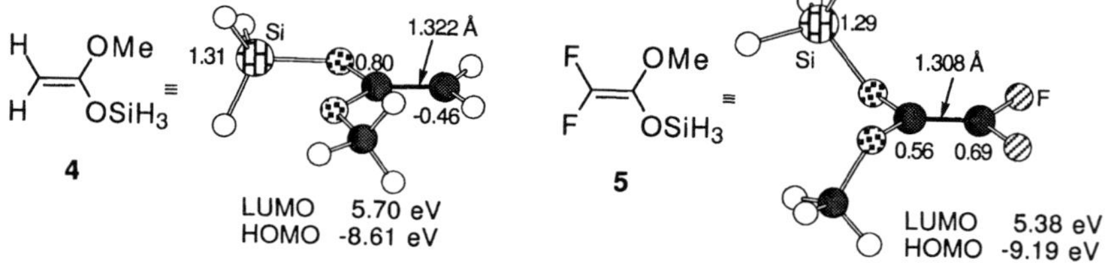

4
LUMO 5.70 eV
HOMO -8.61 eV

5
LUMO 5.38 eV
HOMO -9.19 eV

The uncatalyzed aldol reaction of highly reactive enoxysilanes such as enoxysilacyclobutanes has previously been reported to proceed through six-membered cyclic transition states (*32, 33*), and the torsional structure of the difluoroketene acetal **5** is considered to be more suitable for the uncatalyzed aldol reaction than the planar geometry of **4** because the acetal **5** has a geometrical similarity to the ketene acetal in the cyclic transition states. As mentioned above (Scheme 1), a 60:40 mixture of the *syn-* and *anti*-aldols was obtained at 40°C from the bromofluoroketene silyl acetal **2** (*E/Z* =62/38), suggesting that the uncatalyzed aldol reaction of the fluorine-substituted ketene acetals **1** and **2** proceeds preferentially through boat-like cyclic transition states. Denmark *et al.* proposed that the boat-like transition states are extremely predominant in the uncatalyzed aldol reaction of enoxysilacyclobutanes and trichlorosilyl enolates (*33, 34*).

Asymmetric Aldol Reaction of Difluoroketene Silyl Acetal 1 Catalyzed by Chiral Lewis Acids.

We turned our attention to evaluating several chiral Lewis acid catalysts, which are known to be capable of serving as asymmetric catalysts in the aldol reaction of nonfluorinated ketene silyl acetals, for their usefulness in the reaction of the difluoroketene acetal **1** with aldehydes. A couple of boron complexes, Masamune's catalyst **6** (*35, 36*) and Kiyooka's catalyst **7** (*37, 38*), were found to be effective for our study. For Masamune's catalyst **6**, the reaction was carried out by adding an aldehyde in nitroethane to a solution of the acetal **1** and the catalyst **6** in the same solvent over 3 h at -78°C with stirring at that temperature for an additional hour prior to quenching. With Kiyooka's catalyst **7**, an aldehyde in nitroethane was added to a solution of the acetal **1** and the catalyst **7** in nitroethane at -45°C for 5 min, followed by stirring at -45°C for 2 h (*28, 29*). Nitroethane is the best medium for the enantioselectivity.

As shown in Table I, the use of 20 mol% of the catalysts smoothly promotes the aldol reaction to provide excellent-to-good chemical and optical yields. For example, the reaction of benzaldehyde using a catalyst **6** afforded (*R*)-ethyl 2,2-difluoro-3-hydroxy-3-phenylpropanoate in 99% chemical yield and with 97% enantiomeric excess (entry 1). The catalyst **6** seems to be superior to the catalyst **7** for the enantioselectivity of aldehydes bearing an aromatic ring (entries 1-3). The highest enantioselectivity, 98% ee, was obtained with benzyloxyacetaldehyde using the catalyst **6** (entry 3). The catalyst **7** is more suitable for the enantioselection of secondary aliphatic aldehydes, than the catalyst **6**. The aldol reactions of cyclohexanecarboxaldehyde and 2-ethylbutanal provided greater than 90% optical yields (entries 7 and 8).

Table I. Catalytic Asymmetric Aldol Reaction of Difluoroketene Acetal 1

$$RCHO + \underset{\substack{F \\ F}}{\overset{\substack{F \quad OSiMe_3 \\ OEt}}{\diagdown}} \xrightarrow[EtNO_2]{\text{catalyst 6 or 7 (20 mol\%)}} R\underset{F\ F}{\overset{O\ H}{\diagdown}}CO_2Et$$

1 (1.2 equiv)

Entry	RCHO	Catalyst	Temp. (°C)	ee (%)	Yield (%)
1	C_6H_5CHO	6	-78	97	>99
2	(E)-$C_6H_5CH=CHCHO$	6	-78	96	99
3	$C_6H_5CH_2OCH_2CHO$	6	-78	98	94
4	n-$C_9H_{19}CHO$	6	-78	92	93
5	n-C_3H_7CHO	7	-45	94	90
6	$(CH_3)_2CHCH_2CHO$	7	-45	96	85
7	$(C_2H_5)_2CHCHO$	7	-45	95	90
8	c-$C_6H_{11}CHO$	7	-45	94	97

Asymmetric Aldol Reaction of Bromofluoroketene Silyl Acetal 2 Catalyzed by Lewis Acids 6. We next examined the aldol reaction of the bromofluoroketene silyl acetal **2** mediated by the catalyst **6** (*30*). The reaction was carried out by the addition of an aldehyde in nitroethane to a solution of 1.2 equivalents of the acetal **2** and 20 mol% of the catalyst **6** in the same solvent over 3 h at -78°C and stirring at that temperature for an additional hour prior to quenching. As shown in Table II, the reaction of benzaldehyde afforded a 69:31 mixture of (2*S*,3*R*)- and (2*R*,3*R*)-2-bromo-2-fluoro-3-hydroxy-3-phenylpropanoates. The enantiomeric excess of the *syn*-isomer is 98% ee and that of the *anti*-isomer is 90% ee (entry 1). Although the reactions are not diastereoselective in all cases (*syn/anti* = 69/31 to 46/54), all *syn*- and *anti*-aldol products were obtained with excellent-to good chemical and optical yields.

Table II. Catalytic Asymmetric Aldol Reaction of Bromofluoroketene Acetal 2

$$RCHO + \underset{\substack{Br}}{\overset{\substack{F \quad OSiMe_3 \\ OEt}}{\diagdown}} \xrightarrow[EtNO_2,\ -78°C]{\text{catalyst 6 (20 mol\%)}} R\underset{\substack{Br\ F \\ syn}}{\overset{O\ H}{\diagdown}}CO_2Et + R\underset{\substack{F\ Br \\ anti}}{\overset{O\ H}{\diagdown}}CO_2Et$$

$E/Z = 62/38$
2 (1.2 equiv)

Entry	RCHO	syn/anti	ee (syn, %)	ee (anti, %)	Yield (%)
1	C_6H_5CHO	69/31	98	90	90
2	(E)-$C_6H_5CH=CHCHO$	57/43	83	83	96
3	$C_6H_5CH_2CH_2CHO$	46/54	98	98	89
4	$C_6H_5CH_2OCH_2CHO$	57/43	97	97	81
5	n-C_3H_7CHO	46/54	97	98	90
6	$(CH_3)_2CHCH_2CHO$	48/52	98	98	96
7	$(C_2H_5)_2CHCHO$	54/46	99	98	70
8	c-$C_6H_{11}CHO$	52/48	94	89	74

Effects of Reaction Temperature on the Stereoselection during the Aldol Reaction with Difluoroketene Silyl Acetal 1. As shown in Table III, the enantioface selection for aldehydes depends on the reaction temperature during the aldol reaction of difluoroketene silyl acetal **1** mediated by the catalyst **6** (*29*). The reaction was carried out by adding an aldehyde to a solution of the acetal **1** and the catalyst **6** over 3 h at the indicated temperature and stirring at that temperature for 1 h prior to quenching. Benzyloxyacetaldehyde provided the corresponding (+)-aldol as the major enantiomer with 98% ee and 32% ee, respectively, at -78 and -60°C (entries, 1 and 2). On the other hand, the reactions at -45, -30 and 0°C preferentially gave the (-)-aldol with significant degrees of enantioselectivity ranging from 81% ee to 85% ee (entries 3-5). In the case of cyclohexanecarboxaldehyde, the reaction at -78°C gave the corresponding (+)-aldol with 76% ee, while that at -45°C provided the (-)-enantiomer with 92% ee (entries 6 and 7). Butanal, 3-methylbutanal and 2-ethylbutanal also gave (+)- and (-)-aldols, respectively, at 78 and -45°C with excellent-to-good enantiomeric excesses (entries 8-13). However, the dimethylketene silyl acetal **3** did not cause such a reversal in enantioface selection (entries 16-18).

Table III. Effects of Reaction Temperature on the Stereoselection (1)

$$RCHO + \underset{\substack{\text{F} \quad \text{OEt} \\ \textbf{1} (1.2 \text{ equiv})}}{\overset{\text{F} \quad \text{OSiMe}_3}{\diagup\diagdown}} \xrightarrow[\text{EtNO}_2]{\text{catalyst } \textbf{6} (20 \text{ mol\%})} R\underset{\text{F F}}{\overset{\text{O H}}{\diagdown\diagup}} CO_2Et$$

Entry	RCHO	Temp. (°C)	ee (%)	Yield (%)
1	$C_6H_5CH_2OCH_2CHO$	-78	98 (+)	94
2		-60	32 (+)	88
3		-45	81 (-)	92
4		-30	85 (-)	88
5		0	81 (-)	88
6	$c\text{-}C_6H_{11}CHO$	-78	76 (+)	87
7		-45	92 (-)	90
8	$n\text{-}C_3H_7CHO$	-78	97 (+)	91
9		-45	79 (-)	92
10	$(CH_3)_2CHCH_2CHO$	-78	82 (+)	85
11		-45	75 (-)	93
12	$(C_2H_5)_2CHCHO$	-78	64 (+)	81
13		-45	88 (-)	92
14	C_6H_5CHO	-78	97 (R)	99
15		-45	33 (S)	94

$$RCHO + \underset{\substack{\text{Me} \quad \text{OMe} \\ \textbf{3} (1.2 \text{ equiv})}}{\overset{\text{Me} \quad \text{OSiMe}_3}{\diagup\diagdown}} \xrightarrow[\text{EtNO}_2]{\text{catalyst } \textbf{6} (20 \text{ mol\%})} R\underset{\text{Me Me}}{\overset{\text{O H}}{\diagdown\diagup}} CO_2Me$$

Entry	RCHO	Temp. (°C)	ee (%)	Yield (%)
16	$C_6H_5CH_2OCH_2CHO$	-78	83 (+)	97
17		-45	52 (+)	88
18		0	35 (+)	70

Effects of Reaction Temperature on the Stereoselection during the Aldol Reaction with Bromofluoroketene Silyl Acetal 2. The enantioface selection for aldehydes during the aldol reaction of bromofluoroketene silyl acetal **2** using the catalyst **6** also depends on the reaction temperature (*31*). The aldol additions of 3-methylbutanal at -78°C and higher temperatures (-45, -20 and -10°C) afforded both the *syn*- and *anti*-aldols showing opposite signs of optical rotation (Table IV, entries 1-4). In all cases shown in Table IV, the *syn*- and *anti*-aldols obtained at -78°C show dextrorotation, and those at -20°C are levorotatory. While the reaction carried out at -78°C is not diastereoselective and gives both *syn*- and *anti*-2-bromo-2-fluoro-3-hydroxy esters with excellent enantioselectivity, it is noteworthy that the reaction at -20°C shows *anti* selectivity and provides the *anti*-aldols in high optical yields. For example, butanal afforded ethyl (+)-*anti*-2-bromo-2-fluoro-3-hydroxyhexanoate of 98% ee at -78°C, while the reaction at -20°C gave the (-)-*anti*-aldol with 93% ee (entries 7 and 8). The ratio of the *syn*- and *anti*-aldols at -20°C was 11:89 (entry 8).

Table IV. Effects of Reaction Temperature on the Stereoselection (2)

$$RCHO + \underset{\substack{\text{Br} \quad \text{OEt} \\ \textbf{2}\ (E/Z=62/38,\ 1.2\ \text{equiv})}}{\overset{\substack{\text{F} \quad \text{OSiMe}_3}}{\diagdown\diagup}} \xrightarrow[\text{EtNO}_2]{\text{catalyst }\textbf{6}\ (20\ \text{mol\%})} \underset{syn}{R\diagup\diagdown CO_2Et} + \underset{anti}{R\diagup\diagdown CO_2Et}$$

Entry	RCHO	Temp. (°C)	syn/anti	ee (syn) (%)	ee (anti) (%)	Yield (%)
1	(CH₃)₂CHCH₂CHO	-78	48/52	98 (+)	98 (+)	96
2		-45	51/49	14 (-)	24 (-)	94
3		-20	11/89	31 (-)	91 (-)	87
4		-10	16/84	31 (-)	86 (-)	80
5	c-C₆H₁₁CHO	-78	52/48	94 (+)	89 (+)	74
6		-20	20/80	18 (-)	81 (-)	90
7	n-C₃H₇CHO	-78	46/54	97 (+)	98 (+)	90
8		-20	11/89	49 (-)	93 (-)	87
9	(C₂H₅)₂CHCHO	-78	54/46	99 (+)	98 (+)	70
10		-20	23/77	21 (-)	74 (-)	85
11	C₆H₅CH₂CH₂CHO	-78	46/54	98 (+)	98 (+)	89
12		-20	13/87	48 (-)	92 (-)	85

Reaction Mechanism. We were very interested in clarifying the reason why the fluorine-substituted ketene acetals **1** and **2** enantioface selection is dependent on reaction temperature. After the addition of an aldehyde over 3 h at -78°C and stirring at -78°C for 1 h in the aldol reaction with the bromofluoroketene acetal **2** and the catalyst **6**, the reaction mixture stirred at -20°C for 2 h was found not to affect both the *syn/anti* ratio and the enantioselectivity (*31*). With the difluoroketene acetal **1**, elevating the reaction temperature to -45°C and holding it at the same temperature for 2 h after the aldehyde addition over 3 h and an additional hour at 78°C gave almost the same enantiomeric excess compared to the reaction without the elevation of the temperature and did not reverse the enantioface selection (Iseki, K.; Kuroki, Y.; Kobayashi, Y., MEC Laboratory, Daikin Industries, Ltd., unpublished data.). These results suggest that the reversal of the enantioselectivity and the *anti* stereoselection at -20°C are not caused by isomerization of the aldol products *via* equilibrium. Although an aldehyde was added to the acetal (**1** or **2**) and the catalyst **6** in the reaction mentioned above, the addition of the bromofluoroketene acetal **2** to an aldehyde and the catalyst **6** over 3 h at -20°C did not cause any reversal in the enantiofacial selection (*31*). Such a reversal of the addition order was also examined in the reaction with the difluoroketene acetal **1** at -45°C but was found not to influence the enantioface selection (Iseki, K.; Kuroki, Y.; Kobayashi, Y., MEC Laboratory, Daikin Industries, Ltd., unpublished data.).

Figure 1 shows the ^{19}F NMR of a 1:1 mixture of the difluoroketene acetal **1** and the catalyst **6** in nitroethane-d_5 at -78 and -20°C (Iseki, K.; Kuroki, Y.; Kobayashi, Y., MEC Laboratory, Daikin Industries, Ltd., unpublished data.). The appearance of new peaks at the higher temperature may suggest the formation of a boron enolate. A similar change was observed in the ^{19}F NMR of a 1:1 mixture of the bromofluoroketene acetal **2** and **6** (*31*). On the contrary, the ^{13}C NMR of a 1:1 mixture of ^{13}C-labeled dimethylketene silyl acetal **3** in nitroethane-d_5 at -20°C provided almost the same spectrum as that at -78°C.

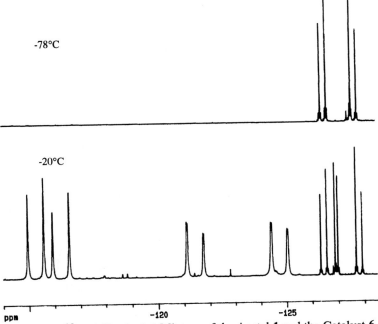

-78°C

-20°C

ppm   -120   -125

Figure 1. ^{19}F NMR of a 1:1 Mixture of the Acetal **1** and the Catalyst **6**.

We propose two different reaction courses for the aldol reactions of the fluorine-substituted ketene silyl acetals **1** and **2** at -78°C and at higher temperatures (-45°C and -20°C). As shown in Scheme 2, the reaction at -78°C preferentially proceeds through the extended open transition states in which the attack of the acetals occurs on the *Si* face of the aldehyde (*29,31*). From transition state **8**, the *anti*-aldol is obtained, while transition state **9** provides the *syn*-isomer.

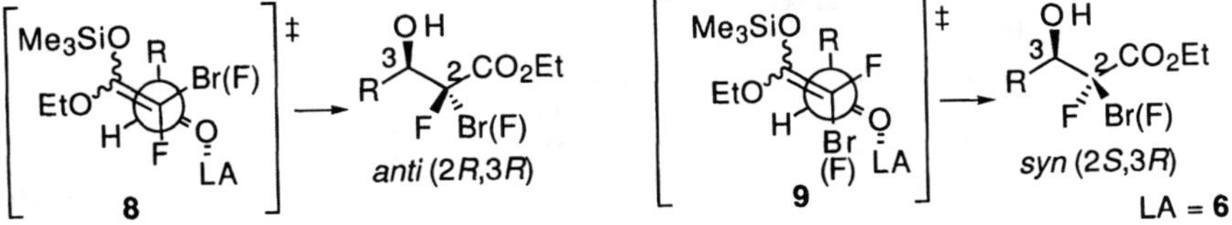

$$\left[\begin{array}{c} Me_3SiO \quad R \\ EtO \quad\quad Br(F) \\ H \quad F \quad O \\ \quad LA \end{array}\right]^{\ddagger} \longrightarrow \begin{array}{c} O \; H \\ R \underset{F \; Br(F)}{\overset{3 \quad 2}{\bigwedge}} CO_2Et \\ anti\,(2R,3R) \end{array}$$

8

$$\left[\begin{array}{c} Me_3SiO \quad R \\ EtO \quad\quad F \\ H \quad Br \quad O \\ (F) \quad LA \end{array}\right]^{\ddagger} \longrightarrow \begin{array}{c} O \; H \\ R \underset{F \; Br(F)}{\overset{3 \quad 2}{\bigwedge}} CO_2Et \\ syn\,(2S,3R) \end{array}$$

9
                                                 LA = **6**

Scheme 2. Extended Open Transition States.

We propose six-membered chair-like closed transition states for the reaction at the higher temperatures (-45°C and -20°C) as shown in Scheme 3 (*31*). The reaction proceeds exclusively with *Re* facial enantioselection to the aldehyde. The *anti* selectivity may be caused by the predominant formation of the (*E*)-boron enolate **11** and/or by its higher reactivity than the (*Z*)-isomer **10**. From the (*E*)-isomer, the corresponding *anti*-aldol should be obtained via the cyclic transition state **13**.

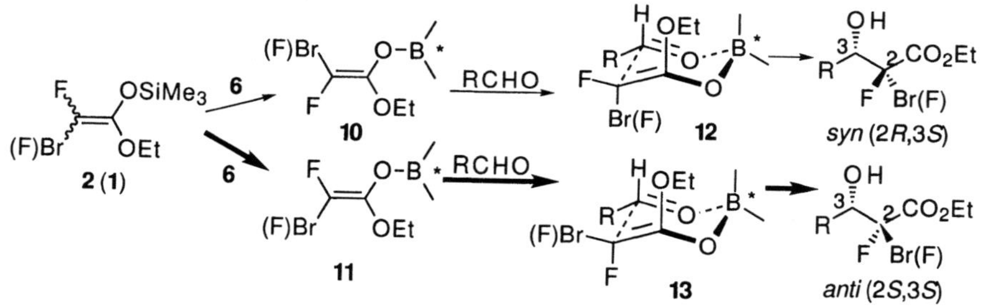

Scheme 3. Closed Chair-Like Transition States.

Catalytic Asymmetric Nitroaldol (Henry) Reaction of 2,2-Difluoroaldehydes

Shibasaki and coworkers have developed lanthanoid-lithium-BINOL complexes (LLB catalysts) as efficient catalysts for the asymmetric nitroaldol (Henry) reaction (*39-46*). The heterobimetallic asymmetric catalysts effectively mediate the reaction of a variety of aldehydes with nitroalkanes to afford the corresponding desired nitroaldols with high enantioselectivity (Scheme 4). We examined the capability of the LLB complexes as asymmetric catalysts for the nitroaldol reaction of 2,2-difluoroaldehydes with nitromethane (*47*).

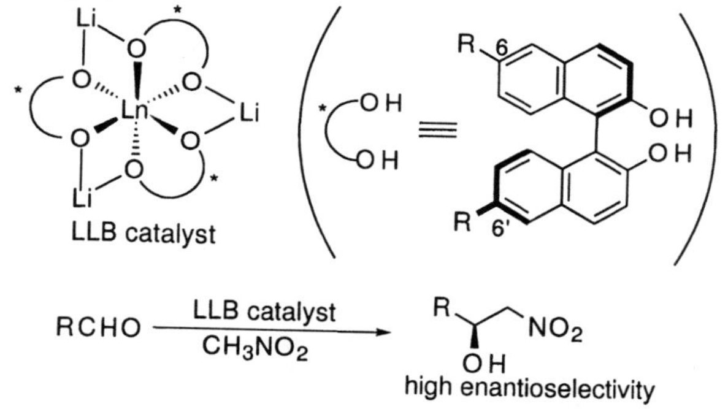

Scheme 4. LLB Catalysts for Asymmetric Nitroaldol (Henry) Reaction.

Assessment of Rare-Earth Elements for Henry Reaction of 2,2-Difluoroaldehydes.

As shown in Table V, we began by surveying various rare earth metals to optimize their chemical and optical yields. Several heterobimetallic catalysts were prepared *in situ* from Ln(O*i*-Pr)$_3$, (*R*)-BINOL (3 molar equiv), BuLi (3 molar equiv) and water (1 molar equiv) (*46*) and then evaluated using 5-phenyl-2,2-difluoropentanal and nitromethane. The reaction was carried out in the presence of 20 mol% catalyst in THF at -40°C for 96 h, and all LLB complexes gave the corresponding (*S*)-nitroaldol in good chemical and optical yields. It is notable that the highest enantioselectivity (87% ee) was obtained with the samarium complex (entry 2) because the lanthanum complex generally provides the best enantioselectivity result for nonfluorinated aliphatic aldehydes (*39-46*). The relationship between the optical yields given by the 2,2-difluoroaldehyde and the ionic radii of the rare earth metals rather resembles that of an aromatic aldehyde, benzaldehyde in the reaction when the europium complex gives the highest enantioselection (*42*).

Table V. Asymmetric Nitroaldol Reaction of 5-Phenyl-2,2-difluoropentanal

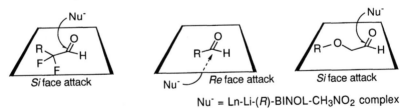

Entry	Catalyst (mol%)	ee (%)	Yield (%)
1	La-Li-(*R*)-BINOL (20)	55	74
2	Sm-Li-(*R*)-BINOL (20)	87	77
3	Eu-Li-(*R*)-BINOL (20)	86	75
4	Gd-Li-(*R*)-BINOL (20)	83	69
5	Yb-Li-(*R*)-BINOL (20)	69	82

In the lanthanoid-lithium-(*R*)-BINOL-catalyzed nitroaldol reaction of 2,2-difluoroaldehydes, the nitronates were found to preferentially react on the *Si* face of the aldehydes. On the contrary, (*R*)-LLB generally causes attack with the *Re* facial selection as ahown in Figure 2 (*39,40,42*). Therefore, the enantiotopic face selection for 2,2-difluoroaldehydes is opposite to that for nonfluorinated aldehydes. This stereoselectivity is identical with that of β-oxaaldehydes, suggesting that the fluorine atoms at the α position exert a significant influence on the enantioface selection. The fluorine atoms may coordinate with the rare earth or the lithium of LLB complexes.

Nu⁻ = Ln-Li-(*R*)-BINOL-CH_3NO_2 complex

Figure 2. Enantioface Selection in Asymmetric Nitroaldol Reaction.

Asymmetric Nitroaldol Reaction of Various 2,2-Difluoroaldehydes. The catalytic asymmetric nitroaldol reaction of several 2,2-difluoroaldehydes with nitromethane was carried out using 5 or 8 mol% of Sm-Li-(*R*)-6,6'-bis((triethylsilyl)ethynyl)-BINOL complex in THF at -40°C for 168 h to afford the corresponding nitroaldols in excellent-to-good optical yields as shown in Table VI (*47*). The highest enantioselectivity (95% ee) was obtained with 2-cyclohexyl-2,2-difluoroacetaldehyde (entry 5).

Table VI. Asymmetric Nitroaldol Reaction of Various 2,2-Difluoroaldehydes

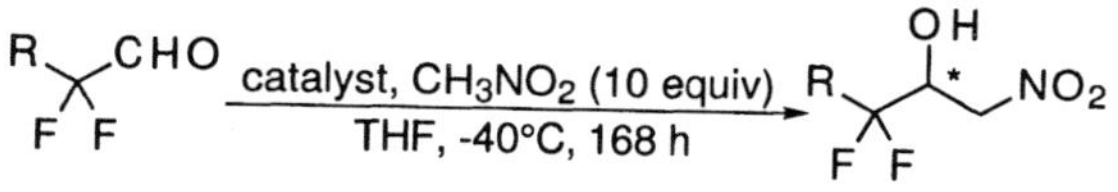

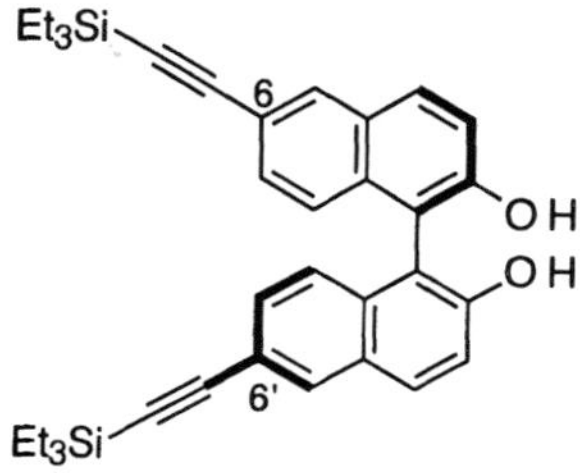

catalyst: Sm-Li-(R)-6,6'-bis((triethylsilyl)ethynyl)-BINOL

(R)-6,6'-bis((triethylsilyl)ethynyl)-BINOL

Entry	R	Catalyst (%)	ee (%)	Yield (%)
1	$CH_3(CH_2)_6$	5	74	55
2	$C_6H_5CH_2OCH_2CH_2$	5	80	52
3	$(CH_3)_2CHSCH_2$	5	85	55
4	$4\text{-}(CH_3OC_2H_4)C_6H_4O$	8	70	73
5	$c\text{-}C_6H_{11}$	8	95	58

Preparation of a Fluorinated Analog of the β-Adrenergic Blocking Agent, Metoprolol. The catalytic asymmetric nitroaldol reaction of 2,2-difluoroaldehydes was applied as a key step to the synthesis of a biologically significant molecule in enantiomerically pure form. The introduction of fluorine atoms into the β-position to a hydroxyl group enhances its acidity, making it a better proton donor to the active sites of enzymes and receptors (48). β,β-Difluoro peptidyl alcohols also act as reversible protease inhibitors. We designed and synthesized a difluorinated analog **14** of a β_1-selective β-adrenergic blocker, metoprolol, which is an effective drug for hypertension (49).

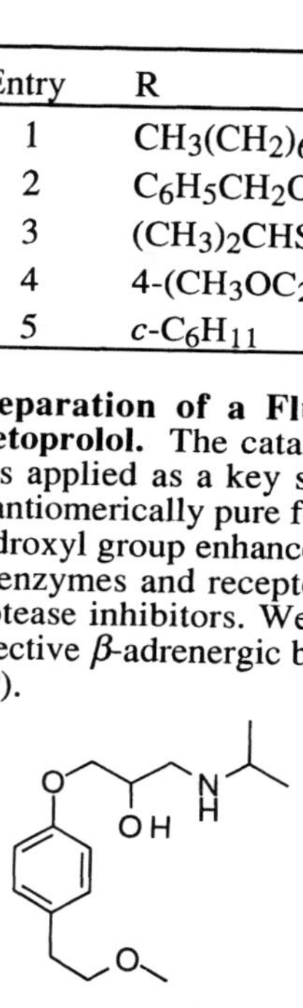

metoprolol

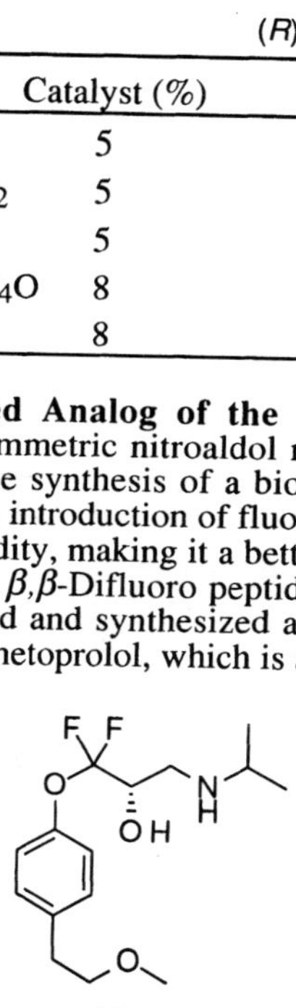

(S)-**14**

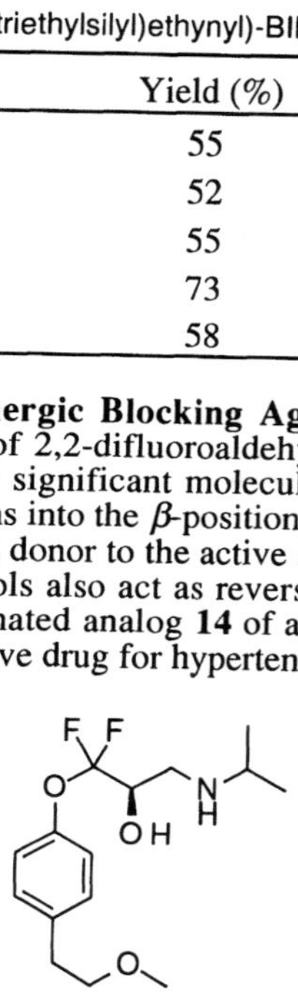

(R)-**14**

As shown in Scheme 5, the enantiocontrolled synthesis of the (S)-**14** was started from 4-(methoxyethyl)phenol **15** which was converted to the 2,2-difluoroester **16** in 73% yield by treatment with chlorodifluoroacetic acid and sodium in refluxing dioxane (50) followed by esterification with iodoethane. The ester **16** was reduced with diisobutylaluminum hydride to afford the 2,2-difluoro aldehyde **17** in 80% yield. The aldehyde **17** was allowed to react with nitromethane at -40°C in the presence of 8 mol% of the Sm-Li-(R)-BINOL complex to provide the difluoro nitroaldol (S)-**18** with 75% ee in 52% yield. After a single recrystallization, the enantiomerically pure nitroaldol (S)-**18** (>99% ee) was recovered from the mother liquor in 65% yield. Reductive alkylation of the homochiral nitroaldol (S)-**18** was accomplished by PtO$_2$-catalyzed hydrogenation in the presence of acetone in methanol to give (S)-**14** in 89% yield (51). In the same manner, the nitroaldol (R)-**18** obtained using Sm-Li-(S)-BINOL was converted to the enantiomerically pure (R)-**14**.

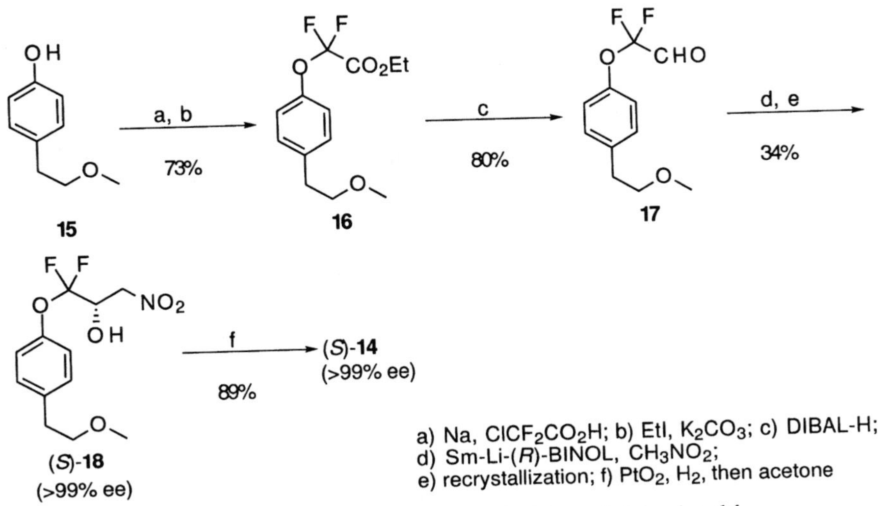

a) Na, ClCF$_2$CO$_2$H; b) EtI, K$_2$CO$_3$; c) DIBAL-H;
d) Sm-Li-(R)-BINOL, CH$_3$NO$_2$;
e) recrystallization; f) PtO$_2$, H$_2$, then acetone

Scheme 5. Preparation of (S)-Difluorinated Metoprolol Analog **14**.

Contrary to our expectations, the fluoro analog (S)-**14** showed slighly lower β$_1$- and β$_2$-adrenergic activities compared to metoprolol. The activities of (R)-**14** were significantly reduced.

Conclusion

In conclusion, we have succeeded in the preparation of difluoroketene and bromofluoroketene silyl acetals in pure form. These acetals were found to enantioselectively react with aldehydes in the presence of Lewis acids to provide biologically interesting 2,2-difluoro-3-hydroxy and 2-bromo-2-fluoro-3-hydroxy esters, respectively, with high enantiomeric excesses. Interestingly, the enantioface selection to aldehydes depends on the reaction temperature in the aldol reactions, being specific to the fluorine-substituted ketene silyl acetals. We have also developed the catalytic asymmetric nitroaldol (Henry) reaction of 2,2-difluoroaldehydes with nitromethane mediated by lanthanoid-lithium-BINOL complexes, and the corresponding, biologically significant nitroaldols can now be obtained in high optical yields. The enantiotopic face selection in the nitroaldol reaction of 2,2-difluoroaldehydes is the reverse of that for nonfluorinated aldehydes, suggesting that the fluorine atoms at the β-position play a decisive role in controlling the stereochemistry. We are currently studying to improve the catalyst turnover and chemical and optical yields of these aldol and nitroaldol reactions.

Significant progress has been lately made in the development of catalytic asymmetric synthesis of chiral fluorine-containing molecules (52-55). Although chiral fluoroorganic compounds still have a lot of problems to be overcome for enantiocontrolled synthesis, I am confident that such compounds will be produced using chiral catalysts on an industrial scale in the near future.

Acknowledgements

It is a great pleasure to acknowledge Professor Masakatsu Shibasaki of the University of Tokyo for collaboration on the catalytic asymmetric nitroaldol reaction. The author is grateful to Professor Yoshiro Kobayashi for the encouraging discussions and thanks

50

also go to Y. Kuroki, D. Asada, M. Kuroki and S. Oishi for their excellent technical assistance and S. Kishimoto for obtaining the molecular orbital calculations.

References

1. *Biological Aspects of Fluorine Chemistry;* Filler, R.; Kobayashi, Y., Eds.; Kodansha Ltd: Tokyo and Elsevier Biomedical Press: Amsterdam, 1982.
2. Welch, J. T.; Eswarakrishnan, S. *Fluorine in Bioorganic Chemistry;* John Wiley & Sons: New York, 1991.
3. *Selective Fluorination in Organic and Bioorganic Chemistry;* Welch, J. T., Ed.; ACS Symp. Ser. 456; American Chemical Society: Washington, D.C., 1991.
4. *Organofluorine Compounds in Medicinal Chemistry and Biomedical Applications;* Filler, R.; Kobayashi, Y.; Yagupolskii, L. M., Eds; Elsevier: Amsterdam, 1993.
5. *Biomedical Frontiers of Fluorine Chemistry;* Ojima, I.; McCarthy, J. R.; Welch, J. T., Eds.; ACS Symp. Ser. 639; American Chemical Society: Washington, D.C., 1996.
6. Welch, J. T. *Tetrahedron* **1987**, *43*, 3123-3197.
7. Bravo, P.; Resnati, G. *Tetrahedron: Asymmetry* **1990**, *1*, 611-692.
8. Resnati, G. *Tetrahedron* **1993**, *49*, 9385-9445.
9. Iseki, K.; Kobayashi, Y. *J. Synth. Org. Chem. Jpn.* **1994**, *52*, 40-48.
10. *Enantiocontrolled Synthesis of Fluoro-Organic Compounds;* Hayashi, T.; Soloshonok, V. A., Eds.; Tetrahedron: Asymmetry, Special Issue; *Tetrahedron: Asymmetry* **1994**, *5*, issue N 6.
11. Iseki, K.; Kobayashi, Y. *Rev. Heteroatom Chem.* **1995**, *12*, 211-237.
12. *Fluoroorganic Chemistry: Synthetic Challenges and Biomedical Rewards;* Resnati, G.; Soloshonok, V. A., Eds.; Tetrahedron Symposium in Print, 58; *Tetrahedron* **1996**, *52*, issue N 1.
13. Bömelburg, J.; Heppke, G.; Ranft, A. *Z. Naturforsch.* **1989**, 1127-1131.
14. Buchecker, R.; Kelley, S. M.; Fünfschilling, J. *Liquid Cryst.* **1990**, *8*, 217-227.
15. Arakawa, S.; Nito, K.; Seto, J. *Mol. Cryst. Liq. Cryst.* **1991**, *204*, 15-25.
16. Seebach, D. *Angew. Chem. Int. Ed. Engl.* **1990**, *29*, 1320-1367.
17. Gelb, M. H.; Svaren, J. P.; Abeles, R. H. *Biochemistry* **1985**, *24*, 1813-1817.
18. Gelb, M. H. *J. Am. Chem. Soc.* **1986**, *108*, 3146-3147.
19. Kirk, K. L. In *Fluorine-Containing Amino Acids;* Kukhar', V. P.; Soloshonok, V. A., Eds.; John Wiley & Sons: New York, 1995, pp 343-401.
20. Thaisrivongs, S.; Pals, D. T.; Kati, W. M.; Turner, S. R.; Thomasco, L. M.; Watt, W. *J. Med. Chem.* **1986**, *29*, 2080-2087.
21. Sham, H. L. In *Fluorine-Containing Amino Acids;* Kukhar', V. P.; Soloshonok, V. A., Eds.; John Wiley & Sons: New York, 1995, pp 333-342.
22. Braun, M.; Vonderhagen, A.; Waldmüller, D. *Liebigs Ann. Chem.* **1995**, 1447-1450.
23. Andrés, J. M.; Martínez, M. A.; Pedrosa, R.; Pérez-Encabo, A. *Synthesis* **1996**, 1070-1072.
24. Kitagawa, O.; Taguchi, T.; Kobayashi, Y. *Tetrahedron Lett.* **1988**, *29*, 1803-1806.
25. Taguchi, T.; Kitagawa, O.; Suda, Y.; Ohkawa, S.; Hashimoto, A.; Iitaka, Y.; Kobayashi Y. *Tetrahedron Lett.* **1988**, *29*, 5291-5294.
26. Burton, D. J.; Easdon, J. C. *J. Fluorine Chem.* **1988**, *38*, 125-129.
27. Kitagawa, O.; Hashimoto, A.; Kobayashi, Y.; Taguchi, T. *Chem. Lett.* **1990**, 1307-1310.
28. Iseki, K.; Kuroki, Y.; Asada, D.; Kobayashi, Y. *Tetrahedron Lett.* **1997**, *38*, 1447-1448.

29. Iseki, K.; Kuroki, Y.; Asada, D.; Takahashi, M.; Kishimoto, S.; Kobayashi, Y. *Tetrahedron* **1997**, *53*, 10271-10280.
30. Iseki, K.; Kuroki, Y.; Kobayashi, Y. *Tetrahedron Lett.* **1997**, *38*, 7209-7210.
31. Iseki, K.; Kuroki, Y.; Kobayashi, Y. *Synlett* **1998**, 437-439.
32. Myers, A. G.; Kephart, S. E.; Chen, H. *J. Am. Chem. Soc.* **1992**, *114*, 7922-7923.
33. Denmark, S. E.; Griedel, B. D.; Coe, D. M.; Schnute, M. E. *J. Am. Chem. Soc.* **1994**, *116*, 7026-7043.
34. Denmark, S. E.; Wong, K.-T.; Stavenger, R. A. *J. Am. Chem. Soc.* **1997**, *119*, 2333-2334.
35. Parmee, E. R.; Tempkin, O.; Masamune, S. *J. Am. Chem. Soc.* **1991**, *113*, 9365-9366.
36. Parmee, E. R.; Hong, Y.; Tempkin, O.; Masamune, S. *Tetrahedron Lett.* **1992**, *33*, 1729-1732.
37. Kiyooka, S.; Kaneko, Y.; Komura, M.; Matsuo, H.; Nakano, M. *J. Org. Chem.* **1991**, *56*, 2276-2278.
38. Kiyooka, S.; Kaneko, Y.; Kume, K. *Tetrahedron Lett.* **1992**, *33*, 4927-4930.
39. Sasai, H.; Suzuki, T.; Arai, S.; Arai, T.; Shibasaki, M. *J. Am. Chem. Soc.* **1992**, *114*, 4418-4420.
40. Sasai, H.; Suzuki, T.; Itoh, N.; Shibasaki, M. *Tetrahedron Lett.* **1993**, *34*, 851-854.
41. Sasai, H.; Itoh, N.; Suzuki, T.; Shibasaki, M. *Tetrahedron Lett.* **1993**, *34*, 855-858.
42. Sasai, H.; Suzuki, T.; Itoh, N.; Arai, S.; Shibasaki, M. *Tetrahedron Lett.* **1993**, *34*, 2657-2660.
43. Sasai, H.; Kim, W.-S.; Suzuki, T.; Shibasaki, M.; Mitsuda, M.; Hasegawa, J.; Ohashi, T. *Tetrahedron Lett.* **1994**, *35*, 6123-6126.
44. Sasai, H.; Yamada, Y. M. A.; Shibasaki, M. *Tetrahedron* **1994**, *50*, 12313-12318.
45. Sasai, H.; Suzuki, T.; Itoh, N.; Tanaka, K.; Date, T.; Okamura, K.; Shibasaki, M. *J. Am. Chem. Soc.* **1993**, *115*, 10372-10373.
46. Sasai, H.; Tokunaga, T.; Watanabe, S.; Suzuki, T.; Itoh, N.; Shibasaki, M. *J. Org. Chem.* **1995**, *60*, 7388-7389.
47. Iseki, K.; Oishi, S.; Sasai, H.; Shibasaki, M. *Tetrahedron Lett.* **1996**, *37*, 9081-9084.
48. Patel, D. V.; Rielly-Gauvin, K.; Ryono, D. E.; Free, C. A.; Smith, S. A.; Petrillo Jr, E. W. *J. Med. Chem.* **1993**, *36*, 2431-2447.
49. Iseki, K.; Oishi, S.; Sasai, H.; Shibasaki, M. *Bioorg. Med. Chem. Lett.* **1997**, *7*, 1273-1274.
50. Yagupolskii, L. M.; Korinko, V. A. *Zh. Obshch. Khim.* **1969**, *39*, 1747-1751.
51. Sasai, H.; Suzuki, T.; Itoh, N.; Shibasaki, M. *Appl. Organomet. Chem.* **1995**, *9*, 421-426.
52. For a recent review see: Iseki, K. *Tetrahedron* **1998**, *54*, 13887-13914.
53. Mikami, K.; Yajima, T.; Takasaki, T.; Matsukawa, S.; Terada M.; Uchimaru, T.; Maruta, M. *Tetrahedron* **1996**, *52*, 85-98, and references therein.
54. Soloshonok, V. A.; Hayashi, T. *Tetrahedron: Asymmetry* **1994**, *5*, 1091-1094, and references therein.
55. Mikami, K.; Takasaki, T.; Matsukawa, S.; Maruta, M. *Synlett* **1995**, 1057-1058, and references therein.

Chapter 4

Asymmetric Fluorination of α-Amino Ketones: Synthesis of Monofluoroketomethylene Peptide Isosteres and Related Compounds

Robert V. Hoffman and Junhua Tao[1]

Department of Chemistry and Biochemistry, New Mexico State University, North Horseshoe Drive, Las Cruces, NM 88003–8001

A simple, stereocontrolled synthesis of monofluoro ketomethylene dipeptide isosteres has been developed. N-Tritylated ketomethylene dipeptide isosteres, prepared from N-tritylated amino acids, are converted to their Z-TMS enol ethers and fluorinated with Selectfluor. There is cooperative stereocontrol between the allylic N-tritylamine group and the alkyl group at C-2. The method is short (6 steps) and diastereoselective (85->95%) and enantioselective (>95%).

The introduction of fluorine into organic molecules has been an important component of modern medicinal and agricultural chemistry because fluorine is known to impart a variety of unusual properties when introduced into biologically active compounds.(*1,2*) A very interesting group of such molecules can be classed as densely functionalized compounds. These are generally saturated molecules which contain two or more contiguous functional groups and one or more stereogenic centers (Figure 1). The synthesis of such compounds is an interesting enterprise because it demands that chemoselectivity, regioselectivity, and stereoselectivity be controlled throughout the course of the synthesis.

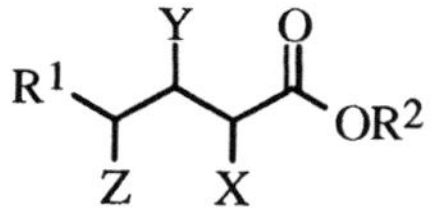

$$X, Y, Z = H, R, O, N, S, F$$

Figure 1. Densely functionalized compound types

Fluorinated targets can be prepared either by manipulating fluorinated precursors(*3*) or by introducing fluorine(s) into an unfluorinated intermediate.(*4*) The latter strategy is inherently more versatile and convergent because the same reaction process can potentially be used to introduce fluorine into a large number of related compounds.

Methods that are used to introduce fluorine into a densely functionalized molecule

[1]Current address: Agouron Pharmaceuticals, 3565 General Atomics Court, San Diego, CA 92121–1121.

must address the same selectivity issues common to the preparation of any densely functionalized compound. A common approach is to use electrophilic fluorination for this purpose. There are quite a few electrophilic fluorinating agents that have been developed.(5) Most all require that the substrate contain an electron rich carbon atom (enolate, enol, enol derivative, electron rich π-system, etc.) to serve as a nucleophilic site for the attachment of the electrophilic fluorine. Thus α-fluoro carbonyl compounds produced from enol derivatives have been very common products obtained from electrophilic fluorinations.

If the final fluorine-containing targets are densely functionalized, an issue to be constantly reckoned with is stereoselectivity in the fluorination step. Because the electrophilic fluorination of an enol or enolate is inherently non-stereoselective, given the planar geometry of the carbon nucleophile, stereoselectivity can be achieved only by using a scalemic fluorinating agent to induce stereoselection, or by using a chiral precursor or chiral auxiliary to control stereoselection. The use of scalemic fluorinating agents, examined most extensively by Davis' group,(6) have proven to be only moderately successful with ee's ranging up to 75%. For most purposes, this is not an acceptable level of stereocontrol.

An alternate strategy is to use asymmetric induction by a preexisting chiral center in the substrate to direct the fluorination step. This approach has not been widely employed and the best results are those of Enders who reported that scalemic α-silyl ketones can be stereoselectively fluorinated at the α'-position.(7) After desilylation, α-fluoro ketones can be obtained in high ee's for cyclic systems. Unfortunately, open chain systems give poorer results. In this symposium Davis has reported recent results using a chiral auxiliary as a stereocontrol element for the synthesis of α-fluoro esters in high enantiomeric purity.(8)

In the last several years we have developed a new general strategy for the synthesis of peptide isosteres that utilizes an acetoacetic ester synthesis to assemble the γ-ketoester core unit.(9-11) By using Cbz-protected amino acids 1 and scalemic α–triflyloxy esters 2 as alkylating agents, the densely functionalized, fully elaborated ketomethylene peptide isosteres 3 can be assembled quickly, efficiently and with good stereocontrol at both C-2 and C-5 (Scheme 1).(12)

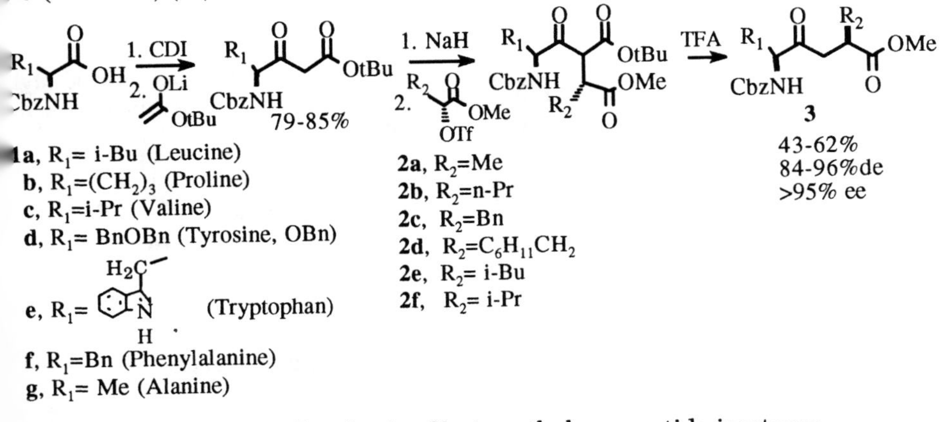

1a, R_1= i-Bu (Leucine)
b, R_1=$(CH_2)_3$ (Proline)
c, R_1=i-Pr (Valine)
d, R_1= BnOBn (Tyrosine, OBn)

e, R_1= [structure] (Tryptophan)

f, R_1=Bn (Phenylalanine)
g, R_1= Me (Alanine)

2a, R_2=Me
2b, R_2=n-Pr
2c, R_2=Bn
2d, R_2=$C_6H_{11}CH_2$
2e, R_2= i-Bu
2f, R_2= i-Pr

79-85%

43-62%
84-96%de
>95% ee

Scheme 1. Synthesis of ketomethylene peptide isosteres

In order to extend this methodology to include hydroxyethylene peptide isosteres, a nitrogen protecting group was needed which would provide good stereocontrol at C-4 in the reduction of the aminoketone product 3. The trityl protecting group for nitrogen was chosen because it is easily attached to amino acids. Moreover it is quite bulky and insures an open Felkin-Ahn transition state in the reduction step. The value of bulky protecting groups as stereocontrol elements in aminoketone reactions was first demonstrated by Reetz for N,Ndibenzylamino protecting groups(*13*) and used effectively in our first stereocontrolled hydroxyethylene peptide synthesis.(*14*) In fact the N-trityl group gives even better stereocontrol in the reduction of N-tritylated α-amino-β-keto esters to statine analogs and ultimately sphingosines.(*15*) In the event, reduction of the adjacent ketone group in the N-trityl aminoketone intermediate **4** provides the related hydroxyethylene peptide isostere **5** with good stereocontrol as well (Scheme 2).(*16*)

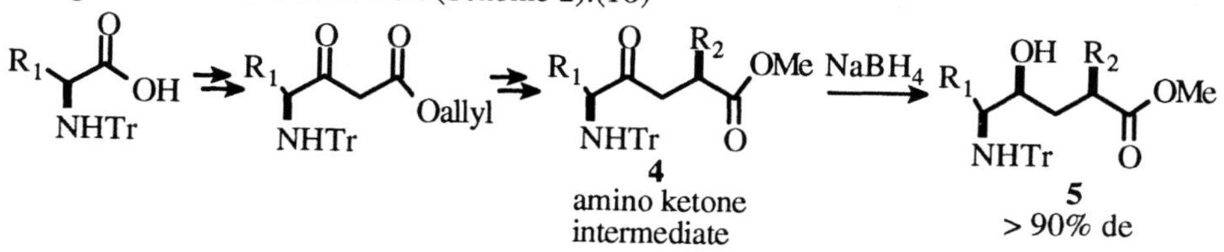

Scheme 2. Synthesis of hydroxyethylene peptide isosteres.

In the course of adapting this chemistry to the synthesis of other densely functionalized compounds, several observations pertinent to the fluorination of these compounds were made. First enolate formation in N-tritylated aminoketones **6** occurs distal from the amino group by removal of the α'-protons without loss of stereochemistry at the α-position.(*16-18*) The resulting enolate can be silylated to give the silyl enol ether **7** (Scheme 3). Moreover, recent work by Xie suggests that the Z,E-regiochemistry of the enol ether might be controlled effectively by the choice of the base.(*19*)

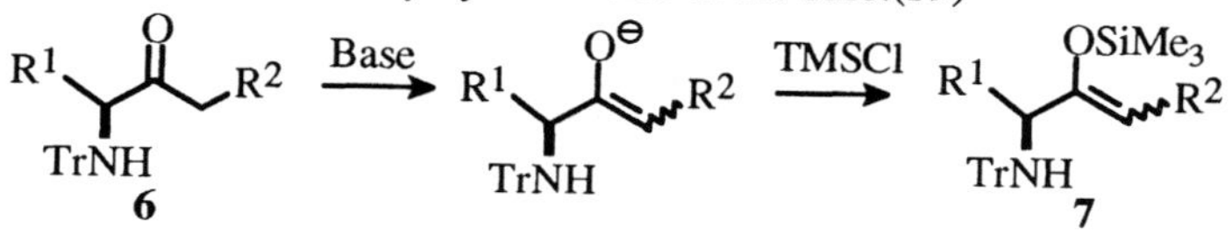

Scheme 3. Enolate formation in N-tritylated aminoketones

Secondly, it possible to use the reagent Selectfluor to electrophilically fluorinate silyl enol ethers under very mild conditions.(*4*) We have used this route to access β-fluoro-α-keto esters **8** as intermediates to fluorinated dehydroamino esters **9** (Scheme 4).(*20*)

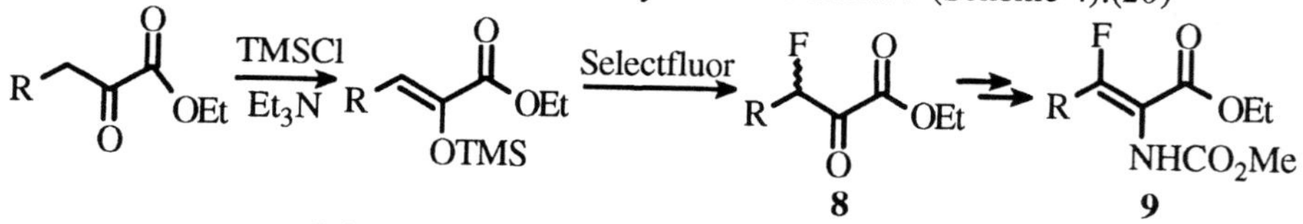

Scheme 4. Synthesis of β-fluoro-α-ketoesters **8**.

These observations led us to test the idea that an N-tritylamino group of an aminoketone could be used as a 1,3-allylic stereocontrol element for the electrophilic fluorination of silyl enol ethers. This was deemed feasible since the approach trajectory of nucleophiles to the carbonyl group of N-tritylated aminoketones along the Dunitz angle is no all that different from the perpendicular approach trajectory of electrophiles to the π-bond of the enol ether and therefore might be subject to the same type of stereochemical control (Figure 2).

These assumptions were checked using a series of amino ketones **10** prepared from N-tritylated amino acids by the normal route.(*16*) These were converted predominantly to the Z-TMS enol ethers **11** with NaHMDS/ TMSCl.(*19*) A 10:1 ratio of Z,E isomers was evident

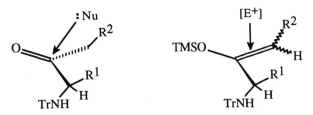

Figure 2. Comparison of nucleophilic carbonyl addition with electrophilic enol addition.

from the signal for the vinyl proton in the ^{1}H nmr of **11** It is assumed that the major isomer produced under these conditions is the Z-isomer.(*19*) Fluorination of this mixture with Selectfluor gave monofluoroaminoketone **12** in a 10 : 1 ratio of diastereomers. The fact that the diastereomeric ratio of products is identical to the diastereomeric ratio of starting materials suggests that the fluorination of the enol ether is stereospecific.

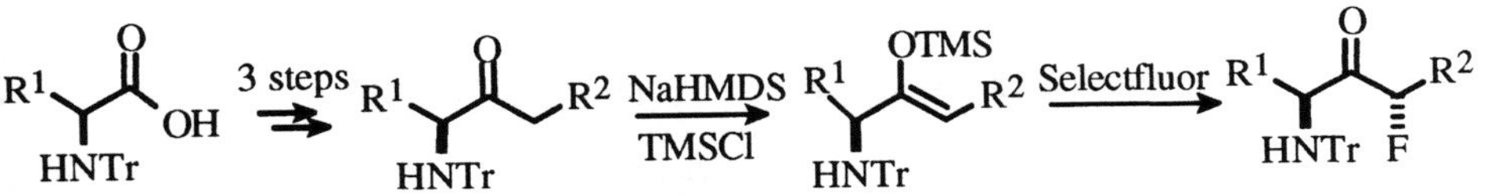

R^1 $\overset{O}{\underset{HNTr}{\bigwedge}}$ OH $\overset{3 \text{ steps}}{\Longrightarrow}$ R^1 $\overset{O}{\underset{HNTr}{\bigwedge}}$ R^2 $\overset{NaHMDS}{\underset{TMSCl}{\longrightarrow}}$ R^1 $\overset{OTMS}{\underset{HNTr}{\bigwedge}}$ R^2 $\overset{Selectfluor}{\longrightarrow}$ R^1 $\overset{O}{\underset{HNTr \ F}{\bigwedge}}$ R^2

10 **11** **12**
Z/E 10:1 73%
R$_1$= CH$_3$, R$_2$=CH$_2$CH$_2$Ph 82% de
R$_1$= i-Bu, R$_2$=CH$_2$CH$_2$Ph (10:1)
R$_1$=OBn, R$_2$=C$_{13}$H$_{27}$

Scheme 5. Electrophilic fluorination of Z-TMS enol ethers of N-trityl amino ketones

N-trityl aminoketones **10** were next converted to the E-TMS enol ethers **13** with lithium *tert*-butyl trimethylsilyl amide (LTBTMS/TMSCl) .(*19*) The E-isomer predominated by a ratio of 8:1 over the Z-isomer as indicated by the vinyl signals in the ^{1}H nmr. Fluorination of this mixture with Selectfluor gave fluoroaminoketone **14** in an 8:1 diastereomeric ratio (Scheme 6). In this case the major product **14** was identical with the minor diastereomer in the fluorination of **11**. Moreover the minor diastereomer in the fluorination of **13** is found to be **12**, the major diastereomer from the fluorination of **11**. Thus again the electrophilic fluorination of the E-enol ether is also stereospecific and gives fluorinated products with the opposite stereochemistry as those obtained from the Z-enol ether. These results show clearly that 1,3-allylic stereocontrol by an N-tritylamino group is

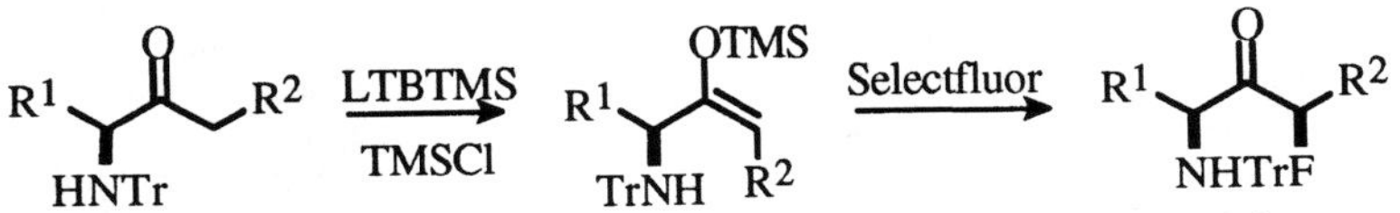

R^1 $\overset{O}{\underset{HNTr}{\bigwedge}}$ R^2 $\overset{LTBTMS}{\underset{TMSCl}{\longrightarrow}}$ R^1 $\overset{OTMS}{\underset{TrNH \ R^2}{\bigwedge}}$ $\overset{Selectfluor}{\longrightarrow}$ R^1 $\overset{O}{\underset{NHTrF}{\bigwedge}}$ R^2

10 **13** **14**
Z/E 1:8 70%
R$_1$= CH$_3$, R$_2$=CH$_2$CH$_2$Ph 78% de
R$_1$= i-Bu, R$_2$=CH$_2$CH$_2$Ph (8:1)

Scheme 6. Electrophilic fluorination of E-TMS enol ethers of N-trityl amino ketones quite effective in controlling electrophilic fluorination of the π-bond for both Z- and E-TMS enol derivatives N-trityl aminoketones.

These model studies provided the impetus to extend this methodology to the preparation of monofluoro ketomethylene peptide isosteres. Monofluoro ketomethylene peptide isosteres **15** (Figure 3) could be very interesting peptide mimetics since they are true peptide isosteres whose binding region can be extended in both the P and P' directions by standard methods. Moreover the ketone carbonyl group is rendered more electrophilic by the fluoro substituent but yet is only partially hydrated (≈ 20-50%) upon treatment with water so that interaction with several different types of proteases is possible. The first synthesis of a monofluoro ketomethylene peptide isostere utilized a homoenolate equivalent to assemble the isostere skeleton and fluorination of a silyl enol ether to introduce the C-3 fluorine substituent.(20) Overall this route was long (>10 steps) and the introduction of fluorine was completely non-stereoselective.

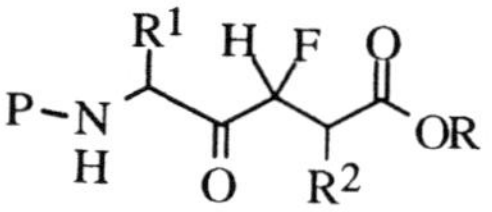

15

Figure 3. Monofluoroketomethylene peptide isostere

A series of (2R, 5S)-N-tritylated ketomethylene peptide isosteres **4a-e** were converted to their corresponding TMS-enol ethers **16a-e** with NaHMDS/ TMSCl (Scheme 7). Only a single enol geometry could be detected in the nmr and it was assigned as the Z-isomer because of the base employed.(19) These were fluorinated with Selectfluor to give monofluoro ketomethylene peptide isosteres **17a-e** in good yields.

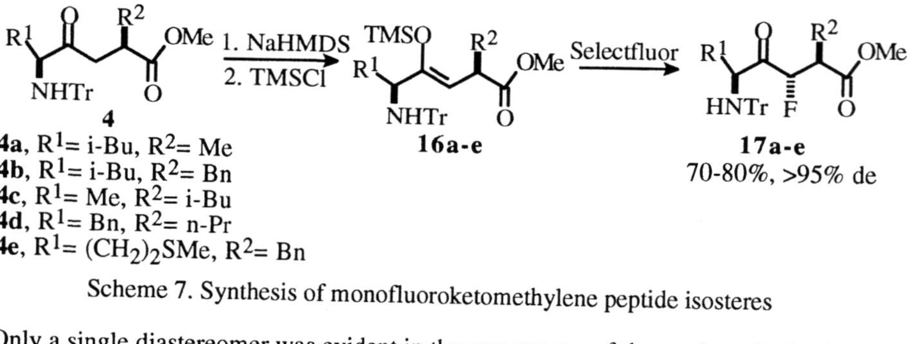

4a, R^1= i-Bu, R^2= Me
4b, R^1= i-Bu, R^2= Bn
4c, R^1= Me, R^2= i-Bu
4d, R^1= Bn, R^2= n-Pr
4e, R^1= (CH$_2$)$_2$SMe, R^2= Bn

Scheme 7. Synthesis of monofluoroketomethylene peptide isosteres

Only a single diastereomer was evident in the nmr spectra of the products indicating that the fluorination of **16** was highly stereoselective. The configuration was assigned 3S as shown in Scheme 7 based on nmr coupling constants of the H-2 and H-3 protons.

The single diastereomer of **17a-e** produced from the fluorination of **16a-e** had J$_{2,3}$= 7-8 Hz. Since the configurations at C-2 and C-5 are known, then the products can only be epimeric at C-3. Thus the two possible diastereomers of **17a-e** are the (2S, 3S, 5S) or the (2S, 3R, 5S) diastereomers (Figure 4). Using R^1 and R^2= methyl for calculational simplicity, a conformational analysis around the C2-C3, C3-C4, and C4-C5 bonds was undertaken at the AM1 level of theory.(21) For **17** 2S,3S,5S there were three low energy conformations which were 1.7-3.0 kcal /mole more stable than the next lowest energy conformer and they all had a dihedral angle of 179.5° between the C-2 and C-3 protons. For **17** 2S,3R,5S the conformation of lowest energy was also 2.5 kcal more stable than the next lowest energy conformation, but it had a dihedral angle of only 83.2° between the C-2 and C-3 protons. The substantial coupling constant of J_{H-H} = 7-8 Hz found in the product is consistent the 179.5° dihedral angle found for the major conformers of the **17** 2S,3S,5S diastereomer thus establishing its configuration.

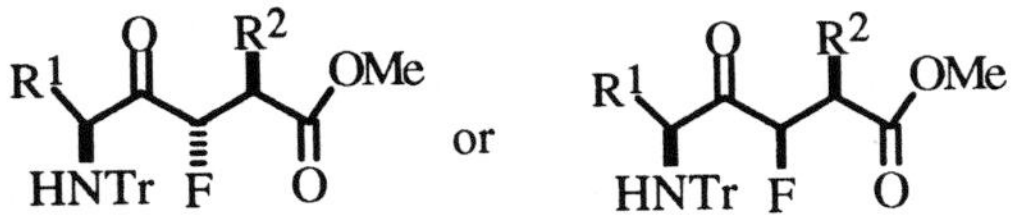

17a-e (2S, 3S, 5S)   **17a-e** (2S, 3R, 5S)

Figure 4. C-3 epimers of 2S, 5S- **17**

When ketomethylene isosteres **4f,g** (2S, 5S) which have the opposite (and unnatural) 2S stereochemistry at C-2, were used in the same sequence, comparable yields of monofluoro ketomethylene peptide isosteres were obtained, however the diastereoselectivity was much lower (Scheme 8). Since only a single TMS-enol ether **16** was produced on enolization, the reduced diastereoselectivity must occur in the fluorination step. The major diastereomer was assigned the 3S configuration on the basis of nmr.

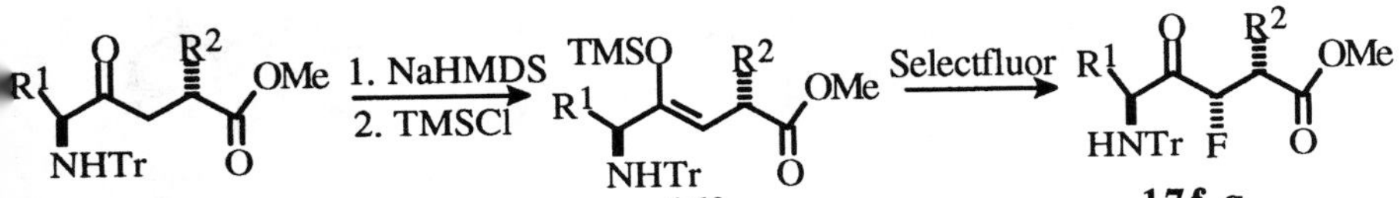

4f, R^1= i-Pr, R^2= Bn
4g, R^1= CH2OBn, R^2= i-Bu

16f, g

17f,g
70-80%
≈ 2:1

Scheme 8. Electrophilic fluorination of 2S epimers of **4**.

Using a similar analysis, the two possible diastereomers of **17f,g** have either the (2R, 3S, 5S) or the (2R, 3R, 5S) configurations (Figure 5). AM1 calculations on these diastereomers of **17** (R^1, R^2= Me) were carried out. Again each diastereomer was found to have one conformation preferred by about 2.5 kcal/mol over the next lowest ones. The most stable conformer of **17** (2R, 3R, 5S) was found to have a dihedral angle between the C-2 and C-3 protons of 170° and the most stable conformer of **17** (2R, 3S, 5S) was found to have a dihedral angle of 77° between these protons. Thus the major isomer which has J_{H-H} =7.8-8.7 Hz can be assigned as the **17** (2R, 3R, 5S) isomer and the minor diastereomer which has J_{H-H} 4.6-4.8 Hz can be assigned as the **17** (2R, 3S, 5S) isomer.

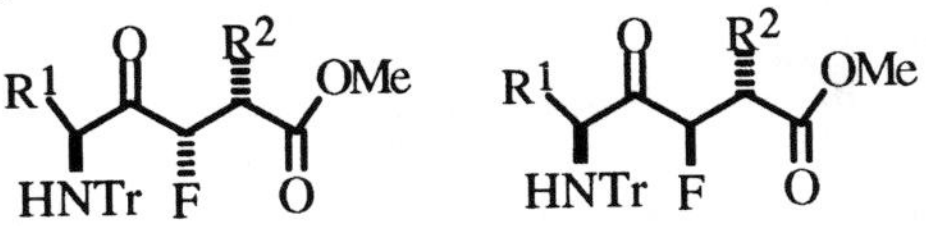

17f,g (2R, 3S, 5S)   **17f,g** (2R, 3R, 5S)

Figure 5. C-3 epimers of 2R, 5S - **17**

These stereochemical results show that in the syn-2R, 5S isomers of **16a-e**, 1,3-allylic stereocontrol by the N-trityl amine group and 1,2 stereocontrol by the chiral center at C-2 are complementary in dictating the S-configuration at C-3, and thus high diastereoselectivity is observed in the fluorination step (Scheme 7). In the anti-2S, 5S diastereomers of **16f,g**, 1,3-allylic stereocontrol afforded by the N-trityl amine group is opposed by the 1,2 stereocontrol of the chiral center at C-2 and diastereomeric mixtures are obtained.

A structural rationale for the stereocontrol afforded by the N-tritylamino group in the fluorinations of TMS-enol ethers **11** and **13** derived from simple amino ketones **10** is available from AM1 modeling.*(21)* For Z-enol ether **11**, conformation **18** which minimizes $A_{1,3}$ interactions is favored by some 3 kcal/mol (Figure 6). This favored conformer has the bulky N-trityl group blocking one face of the π-bond so that the electrophile must deliver the fluoro group to the opposite face, thus producing a single product. For the corresponding E-enol ether **13**, conformation **19**, which also minimizes $A_{1,3}$ interactions, is favored by nearly 3.0 kcal/mole. Again the bulky trityl group shields one face of the π-bond thus producing the opposite configuration upon fluorination. The experimental data show that stereocontrol by the trityl group is very efficient and consistent with this model.

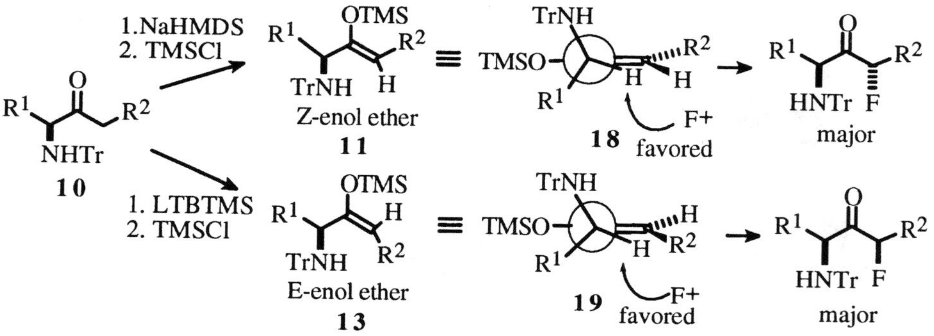

Figure 6. 1,3-Allylic stereocontrol by an N-tritylamine group.

A similar conformational analysis was carried out on the *syn*-Z-TMS enol ethers *syn-***16** derived from ketomethylene peptide isosteres **4a-e** (Figure 6). Using R^1, $R^2 =$ Me it was found that one conformation **20** was preferred by more than 3 kcal/mol over the next lowest-energy conformer. This conformation minimizes $A_{1,2}$ interactions and has both the N-trityl group at C-5 and the C-2 methyl group above one face of the enol double bond. Approach of the fluorinating agent from the face opposite these groups leads to the observed stereochemistry.

In *anti*-Z-TMS enol ether *anti-***16**, the preferred conformation **21**, which also lies 3 kcal/mol below the next lowest conformer, has the N-trityl group at C-5 and the methyl group at C-2 on opposite faces of the double bond (Figure 7). The 1,2 stereocontrol of the methyl group must be comparable to 1,3-allylic stereocontrol exerted by the N-trityl group and a mixture of diastereomers is produced. This rationale is consistent with the observation that replacement of the C-2 methyl group of **4f** with a slightly larger C-2 isobutyl group in **4g** leads to a small increase in the diastereoselectivity as well.

More work is needed to further pin down the stereochemistry of the fluorination and confirm the validity of these models, but they do provide a paradigm for understanding the stereocontrol that is observed. In addition, more examples are needed to delineate the generality and limits of 1,3-allylic stereocontrol in these systems. Nonetheless the results presented here define the elements of a new methodology for the stereocontrolled introduction of fluorine into amino ketones which can be used to access many new examples of fluorinated, densely functionalized compounds.

Acknowledgment We are pleased to acknowledge the support of this work with funds from the National Science Foundation.

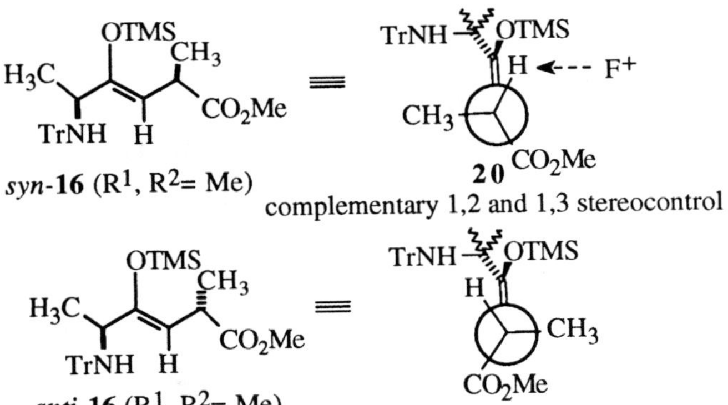

complementary 1,2 and 1,3 stereocontrol

opposing 1,2 and 1,3 stereocontrol

Figure 7. Stereocontrol in *syn* and *anti*-**16**.

References and Notes

. Hudlicky, M., Pavlath, A. E., Eds. "Chemistry of Organic Fluorine Compounds II" ; ACS Monograph 187; American Chemical Society: Washington, DC. **1995**. See chapters by: Filler, R.; Kirk, K. "Biological Properties of Fluorinated Compounds"; Elliott, A. J. "Fluorinated Pharmaceuticals"; Lang, R. W. "Fluorinated Agrochemicals"

. Banks, R. E.; Smart, B. E.; Tatlow, J. C., Eds. "Organofluorine Chemistry: Principles and Commercial Applications", Plenum, New York, **1994**.

. Banks, R. E.; Tatlow, J. C. in ref.1, pp 25-55.

. Banks, R. E. *J. Fluor. Chem.*, **1998**, *87*, 1 is a superb recent overview.

. A good recent listing is found in Lal, G. S.; Pez, G. P.; Syvret, R. G. *Chem. Rev.*, **1996**, *96*, 737.

. Davis, F. A.; Zhou, P.; Murphy, C. K. *Tetrahedron Lett.*, **1993**, *34*, 3971.

. Enders, D.; Potthoff, M.; Raabe, G.; Runsink, J. *Angew.Chem. Int. Ed. Eng.*, **1997**, *36*, 362.

. Davis, F. A.; Kasu, P. V. N.; Sundarababu, G., Paper 285, Organic Division, Asymmetric Synthesis of Fluoroorganic Compounds, 216th Meeting of the American Chemical Society, Boston, MA, August 24, 1998.

. Hoffman, R. V.; Kim H.-O., *Tetrahedron Lett.*, **1992**, *33*, 3579.

0. Hoffman, R. V.; Kim H.-O., *Tetrahedron Lett.*, **1993**, 34, 2051.

1. Hoffman, R. V.; Kim H.-O., *J. Org. Chem.*, **1995**, *60*, 5107.

2. Hoffman, R. V.; Tao. J. *Tetrahedron*, **1997**, *53*, 7119.

3. Reetz, M. T., *Angew.Chem. Int. Ed. Eng.*, **1984**, *23*, 556.

4. Hoffman, R. V.; Tao, J. *J. Org. Chem.*, **1997**, *62*, 6240.

5. Hoffman, R. V.; Tao, J. *J. Org. Chem.*, **1998**, *63*, 3979.

6. Hoffman, R. V.; Tao, J. *Tetrahedron Lett.*. submitted.

7. Lubell, W. D.; Rapoport, H. *J. Am. Chem. Soc.*, **1988**, *110*, 7447.

8. Alexander, C. W.; Liotta, D. C. *Tetrahedron Lett.*, **1996**, *37*, 1961 and references therein.

9. Xie, L.; Isenberger, K. M.; Held, G.; Dahl, L. M. *J. Org. Chem.*, **1997**, *62*, 7516.

0. Garrett, G. S.; Emge, T. J.; Lee, S. C.; Fischer, E. M.; Dyehouse, K.; McIver, J. M. *J. Org. Chem.*, **1991**, *56*, 4823.

1. The semiempirical AM1 calculations were performed on Silicon Graphics workstation (Indigo 2) using the MOPAC program in Insight II (4.0.0) from MSI. The input geometries for the AM1 calculations were obtained by using the Discover force field (Steepest first 1000 steps and then Newton; Derivative = 0.001) within Insight II (4.0.0).

Chapter 5

Asymmetric Friedel–Crafts Reactions with Fluoral Catalyzed by Chiral Binaphthol-Derived Titanium Catalysts

Akihiro Ishii[1] and Koichi Mikami[2]

[1]Chemical Research Center, Central Glass Company Ltd.,
Saitama 350–1151, Japan
[2]Department of Chemical Technology, Tokyo Institute of Technology,
Tokyo 152–8552, Japan

Asymmetric catalysis of Friedel-Crafts reaction with fluoral is established using chiral binaphthol-derived titanium catalysts with or without asymmetric activation to provide a practical synthetic route not only for chiral α-trifluorobenzylalcohols but also for highly enantiopure functionalized β-trifluoroaldols through the sequential diastereoselective reactions of the resultant vinyl ethers or silyl enol ethers with electrophiles.

Asymmetric synthesis of organofluorine compounds is an important issue in pharmaceutical chemistry (1,2) and optoelectronic material science (3,4). In particular, asymmetric catalysis of carbon-carbon bond-forming reactions is the most attractive method, because the carbon skeleton of chiral organofluorine molecules can be constructed at the time of asymmetric induction (5-10). The Friedel-Crafts (F-C) reaction is one of the most fundamental carbon-carbon bond-forming reactions in organic synthesis (11-14). However, its application to catalytic asymmetric synthesis has been quite limited (diastereoselective:15-22, enantioselective:23-25, stereospecific:26,27). Herein, we report the catalytic asymmetric F-C reaction with fluoral catalyzed by chiral binaphthol-derived titanium (BINOL-Ti) catalysts (28,29) and the sequential diastereoselective reactions of the resultant F-C products with electrophiles.

Asymmetric Friedel-Crafts Reactions of Aromatic Compounds

The F-C reactions of aromatic compounds can provide a practical synthetic route for chiral α-trifluorobenzylalcohols of synthetic importance (Scheme 1). In previous asymmetric syntheses of α-trifluorobenzylalcohols, the asymmetric reductions of trifluoromethyl ketone were used as a key step (30-34). In this F-C reaction, the catalytic activity and enantioselectivity of BINOL-Ti catalysts (35-37) were found to be critically influenced by the substituents of BINOL derivatives (Table I). 1) (R)-6,6'-Br$_2$-BINOL-Ti catalyst was the most effective catalyst. This F-C reaction did not proceed easily as compared with the carbonyl-ene reaction (7,8) or the Mukaiyama-aldol reaction (7) with fluoral. Therefore, the role of the electron-withdrawing group at the 6,6'-position of BINOL is very important for increasing the Lewis acidity (runs 1~3). Relatively high enantioselectivity was obtained even when using 1 mol% of (R)-6,6'-Br$_2$-BINOL-Ti catalyst (run 4).   2) Polar solvent was more effective for producing higher *para*-regioselectivity (run 5).   When toluene was used as a solvent, the enantio-enriched adducts of fluoral to toluene were also obtained, along with the expected F-C product with anisole.   3) Interestingly, a lower reaction temperature leads to a decrease in the enantioselectivity of *para*-isomer, presumably because of the oligomeric nature of the BINOL-Ti catalysts at lower temperature (run 6).   4) The steric bulkiness of the alkyl ether portion of the aromatic substrates was essential for producing higher *para*-regioselectivity (run 7).   Interestingly, the bis-adduct with fluoral was not obtained even when using a large excess of fluoral in the reaction of diphenylether (run 8).

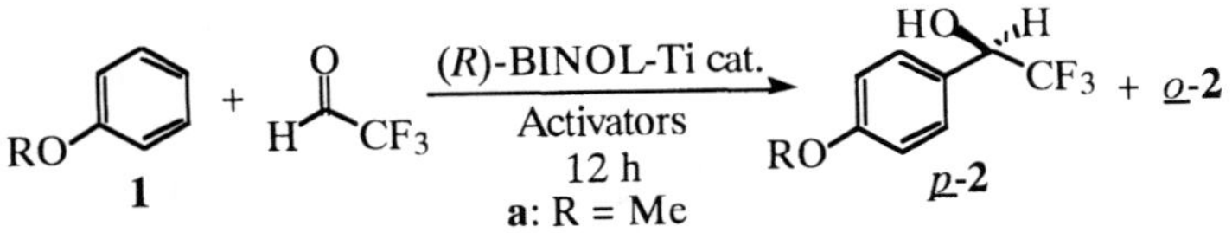

a: R = Me
b: R = *n*-Bu
c: R = Ph

Scheme 1 Asymmetric F-C reactions of aromatic compounds.

Table I The F-C reactions of aromatic compounds with fluoral catalyzed by (R)-BINOLs / Cl$_2$Ti(OPr^i)$_2$ / 4Å MS.

run	**1**	chiral ligand	cat. (mol%)	solvent	temp. (°C)	yield[a] (%)	*p*-**2** : *o*-**2**[a]	ee[b] (%)
1	a	(R)-BINOL	30	CH$_2$Cl$_2$	0	82	4 : 1	73 (R)
2	a	(R)-H$_8$-BINOL[c]	5	CH$_2$Cl$_2$	0	11	4 : 1	22 (R)
3	a	(R)-6,6'-Br$_2$-BINOL	5	CH$_2$Cl$_2$	0	94	4 : 1	84 (R)
4	a	(R)-6,6'-Br$_2$-BINOL	1	CH$_2$Cl$_2$	0	99	4 : 1	72 (R)
5	a	(R)-6,6'-Br$_2$-BINOL	30	toluene	0	99	2 : 1	83 (R)
6	a	(R)-6,6'-Br$_2$-BINOL	5	CH$_2$Cl$_2$	-30	94	4 : 1	79 (R)
7	b	(R)-6,6'-Br$_2$-BINOL	15	CH$_2$Cl$_2$	0	85	8 : 1	83 (R)[d]
8	c	(R)-6,6'-Br$_2$-BINOL	10	CH$_2$Cl$_2$	0	90	3 : 1	54 (R)[d]

[a] Isolated yield. [b] The enantiomeric excess of *p*-**2**. Determined by chiral HPLC. *p*-**2a**: Daicel, CHIRALPAK OD-H, *n*-hexane :*i*-PrOH = 98 : 2, 0.8 ml/min, 254 nm, t_R = 43 min (S), 49 min (R). *p*-**2b**: Daicel, CHIRALPAK AS, *n*-hexane : *i*-PrOH = 99 : 1, 0.8 ml/min, 254 nm, t_R = 40 min (S), 42 min (R). *p*-**2c**: Daicel, CHIRALPAK OD-H, *n*-hexane : *i*-PrOH = 98 : 2, 0.8 ml/min, 254 nm, t_R = 34 min (R), 44 min (S). The absolute configuration of *p*-**2a** was determined by the comparison of optical rotation with the literature values: $[\alpha]^{20}_D$ -35.9° (c 1.35, EtOH) for 80.2% ee of R-isomer (*38*). [c] (R)-octahydrobinaphthol. [d] The absolute configurations of *p*-**2b** and *p*-**2c** were deduced to be R from a similarity.

The representative procedure for run 3: the mixture containing (R)-6,6'-Br$_2$-BINOL (0.05 mmol), Cl$_2$Ti(OPr^i)$_2$ (0.05 mmol), 4Å MS (0.25 g) and dehydrated dichloromethane (0.5 ml) was stirred at room temperature under an argon atomosphere (35-37). After stirring for 1 h, a solution of anisole (1a, 1 mmol) in dehydrated dichloromethane (2.5 ml) was added at 0 °C, and then freshly dehydrated and distilled fluoral (ca. 7 mmol), which was generated by the addition of fluoral hydrate to conc. H$_2$SO$_4$ at 100 °C, was passed to the solution. After stirring for 12 h at the same temperature, dichloromethane and H$_2$O were added to the reaction mixture. Insoluble material was filtered off through a pad of Celite and the aqueous layer was extracted three times with dichloromethane. The combined organic layer was washed with brine, dried over MgSO$_4$, and evaporated under reduced pressure. Chromatographic separation by silica gel (dichloromethane : n-hexane = 3 : 2) gave the product (2a).

The sense of asymmetric induction was the same as observed in BINOL-Ti-catalyzed asymmetric reactions such as carbonyl-ene reaction (35-37,39) and Mukaiyama-aldol reaction (40,41) regardless of the preparative procedure of the catalysts; (R)-BINOL-Ti catalyst produces an (R)-alcohol product. This F-C reaction would not proceed through a six-membered transition state (A) involving a chiral Lewis acid, which has been reported to preferentially produce an ortho-F-C-product in the reaction of phenol (19,24) or 1-naphthol (23). In sharp contrast, the para-isomer was obtained as the major product in our case.

A: Transition state for ortho-F-C-product.

Asymmetric Activation. The catalyst efficiency and enantioselectivity of BINOL-Ti complexes can be further increased through asymmetric activation (42-45). Thus, the (R)-6,6'-Br$_2$-BINOL-Ti(OPr^i)$_2$, prepared from Ti(OPr^j)$_4$ and (R)-6,6'-Br$_2$-BINOL, was activated by the addition of acidic activators such as (R)-5-Cl-BIPOL and (R)-6,6'-Br$_2$-BINOL (Table II). In all runs, chemical yields were obviously improved by the addition of acidic activators, in sharp contrast to the same reaction catalyzed by Yb(OTf)$_3$, in which the addition of (R)-6,6'-Br$_2$-BINOL obviously decreases the catalytic activity: 10 mol% Yb(OTf)$_3$ without (R)-6,6'-Br$_2$-BINOL; 78% (para/ortho = 7), the addition of (R)-6,6'-Br$_2$-BINOL (1 eq); 42% (para/ortho = 13), 2 eq; 45% (para/ortho = 16). The more acidic catalyst was prepared by the coordination of the acidic 6,6'-Br$_2$-BINOL with the (R)-6,6'-Br$_2$-BINOL-Ti(OPr^i)$_2$, in particular (42,43). In combination with a matched chiral activator, the enantioselectivity can be improved up to 90% ee as well as producing a high chemical yield (runs 6, 7).

Table II The F-C reactions of aromatic compounds with fluoral catalyzed by
(R)-6,6'-Br2-BINOL-Ti complex through asymmetric activation.

run	1	cat. (mol%)	additive	temp. (°C)	yield[a] (%)	p-2 : o-2[a]	ee[b] (%)
1	a	10	-	0	66	4 : 1	70 (R)
2	a	5	4Å MS[c]	0	33	3 : 1	62 (R)
3	a	10	pentafluorophenol	0	94	3 : 1	68 (R)
4	a	10	(R)-BINOL	0	97	3 : 1	64 (R)
5	a	10	(R)-5-Cl-BIPOL[d]	0	88	4 : 1	78 (R)
6	a	10	(R)-6,6'-Br2-BINOL	0	89	4 : 1	90 (R)
7	b	10	(R)-6,6'-Br2-BINOL	0	90	8 : 1	90 (R)

[a] Isolated yield. [b] Refers to that of p-2. [c] 0.25 g per 0.05 mmol of (R)-6,6'-Br2-BINOL-Ti(OPr^i)2. [d] 5,5'-dichloro-4,4',6,6'-tetramethyl-2,2'-biphenol.

The representative procedure for run 7: to a solution of Ti(OPr^i)4 (0.1 mmol) in dehydrated dichloromethane (1 ml) was added (R)-6,6'-Br2-BINOL (0.1 mmol) at room temperature under an argon atomosphere. After stirring for 1 h, (R)-6,6'-Br2-BINOL (0.1 mmol) in dehydrated dichloromethane (1 ml) was added to the mixture again. The catalyst solution was prepared by stirring for additional 30 min (42,43). A solution of n-butyl phenyl ether (1b, 1 mmol) in dehydrated dichloromethane (1 ml) was added at 0 °C, and then freshly dehydrated and distilled fluoral (ca. 7 mmol) was passed to the solution. After stirring for 12 h at the same temperature, usual work-up followed by chromatographic separation by silica gel (dichloromethane : n-hexane = 3 : 2) gave the product (2b).

Asymmetric Friedel-Crafts Reactions of Vinyl Ethers

Next, the catalytic asymmetric F-C reaction of alkyl vinyl ethers instead of aromatic compounds as a nucleophile and the sequential diastereoselective oxidation reaction of the F-C product were examined. The catalyst prepared from 6,6'-Br2-BINOL-Ti(OPr^i)2 / 6,6'-Br2-BINOL, which was the most effective catalyst in the F-C reaction of aromatic compounds, was first employed in this reaction (Scheme 2, Table III, run 1). However, only the aldol product (5a) was obtained with high enantioselectivity, presumably because the F-C product would be hydrolyzed by the acidic protons of the 6,6'-Br2-BINOL-Ti(OPr^i)2 / 6,6'-Br2-BINOL complex. Accordingly, the catalysts without proton sources were expected to be effective to this F-C reaction. These catalysts were prepared in isolated form by the previously reported procedure (35-37). Thus, the same reaction was examined using the isolated catalyst prepared from (R)-6,6'-Br2-BINOL and Cl2Ti(OPr^i)2 in the presence of 4Å MS without the addition of Br2-BINOL (run 2). The F-C product (4a) was obtained together with a small amount of the aldol product (5a). The enantiomeric excess of the F-C product was moderate. The product ratio of 4 vs. 5 was found to be critically influenced by the preparative procedure and the ligands of the catalysts. The isolated catalyst derived from (R)-BINOL, which is less acidic than the corresponding 6-Br-analogue, proved to be more effective in terms of both chemical yield and enantioselectivity (run 3). The aldol product was essentially not obtained in this case. Thus, the BINOL-Ti catalyst was superior in this reaction to the 6,6'-Br2-

BINOL-Ti catalyst in sharp contrast to the F-C reaction of aromatic compounds. The higher enantioselectivity was obtained in the reaction of methyl vinyl ether (**3b**) possessing β-methyl substituent irrespective of the geometry (run 4). This F-C reaction with alkyl vinyl ethers was found to more easily proceed as compared with that of aromatic compounds. Therefore, an excess amount of fluoral was not necessary because of the high nucleophilicity of alkyl vinyl ethers.

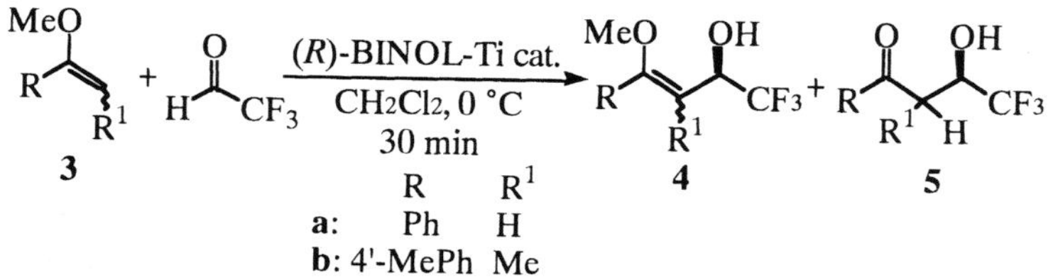

	R	R^1
a:	Ph	H
b:	4'-MePh	Me

Scheme 2 Asymmetric F-C reactions of vinyl ethers.

Table III The F-C reactions of methyl vinyl ethers with fluoral catalyzed by (*R*)-BINOL-Ti catalysts.

run	3	cat.	mol%	yield (%)[a] 4 (E : Z) 5	ee[b] (%)
1	a	(*R*)-6,6'-Br$_2$-BINOL-Ti(OPr^i)$_2$ / (*R*)-6,6'-Br$_2$-BINOL	20	0 53	70 (*R*)[c]
2	a	(*R*)-6,6'-Br$_2$-BINOL / Cl$_2$Ti(OPr^i)$_2$ / 4Å MS (Isolated)	20	48 (1 : 2) 7	58 (*R*)[d]
3	a	(*R*)-BINOL / Cl$_2$Ti(OPr^i)$_2$ / 4Å MS (Isolated)	10	54 (1 : 2) -	72 (*R*)[e]
4	b[f]	(*R*)-BINOL / Cl$_2$Ti(OPr^i)$_2$ / 4Å MS (Isolated)	20	64 (5 : 1) -	85 (*R*)[g]

[a] Isolated yield. [b] The enantiomeric excess of **4**. Determined by chiral HPLC analysis of **5a** or *anti*-**5b** obtained by acidic hydrolysis of **4a** or **4b**. In the hydrolysis of (*E*)-**4b**, *anti*-isomer was obtained as the major diastereomer (*anti* / *syn* = ca. 2). The relative configuration of the major diastereomer was determined to be *anti* on the basis of the chemical shifts of α-methyl carbon in ^{13}C-NMR spectra (CDCl$_3$) of **5b**: *syn*; 11.9 ppm, *anti*; 16.3 ppm (*48*). Chiral HPLC conditions: **5a**; Daicel, CHIRALPAK OD-H, *n*-hexane : *i*-PrOH = 95 : 5, 0.8 ml/min, 254 nm, t_R = 10 min (3*S*), 12 min (3*R*), *anti*-**5b**; Daicel, CHIRALPAK AS, *n*-hexane : *i*-PrOH = 98 : 2, 0.8 ml/min, 254 nm, t_R = 11 min (2*R*, 3*S*), 38 min (2*S*, 3*R*). [c] The enantiomeric excess of **5a**. [d] The enantiomeric excess of *E*, *Z*-mixture (2 : 3) of **4a**. [e] The enantiomeric excess of *E*, *Z*-mixture (1 : 2) of **4a**. [f] *E* : *Z* = 1 : 1. [g] The enantiomeric excess of (*E*)-**4b**.

The representative procedure for run 4: isolated BINOL-Ti catalysts were prepared as previously reported (35-37). The mixture containing (R)-6,6'-Br$_2$-BINOL or (R)-BINOL (1 mmol), Cl$_2$Ti(OPr^j)$_2$ (1 mmol), 4Å MS (5 g) and dehydrated dichloromethane (10 ml) was stirred at room temperature under an argon atomosphere. After 1 h, dehydrated toluene (20 ml) was added to the mixture. Clear catalyst solution was recovered by centrifugal separator followed by Celite filtration. Isolated catalyst was obtained as a reddish brown solid by evaporating under reduced pressure at room temperature. To a solution of isolated (R)-BINOL-Ti catalyst (0.05 mmol) in dehydrated dichloromethane (1.5 ml) was added vinyl ether (3b, 0.25 mmol) at 0 °C under an argon atomosphere, and then freshly dehydrated and distilled fluoral (ca. 0.75 mmol) was passed to the solution. After stirring for 30 min at the same temperature, usual work-up followed by chromatographic separation by silica gel (dichloromethane : n-hexane = 3 : 2) gave the product (4b).

The stereochemical assignments deserve special comments. The geometries of products (4) were determined on the basis of the chemical shifts in ^{1}H-NMR spectra (CDCl$_3$) (Figure 1). In a similar manner to β-methylstyrene (46), (Z)-4a shows the vinylic proton at higher field (4.8 ppm). Likewise, (E)-4b shows the vinylic methyl group at higher field (1.7 ppm). The absolute configuration of (E)-4b was determined to be R by Mosher's method: the chemical shift of vinylic methyl proton in ^{1}H-NMR spectra (CDCl$_3$) of MTPA ester of (E)-4b; 1.60 ppm (S), 1.36 ppm (R); δ_S - δ_R = positive (47).

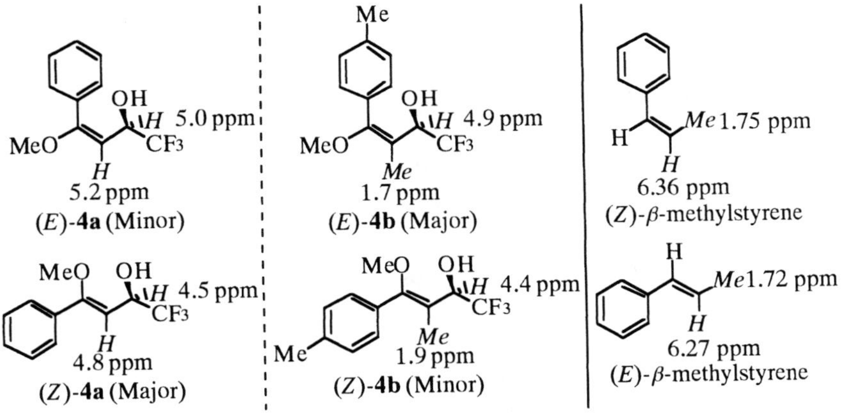

Figure 1 ^{1}H-NMR spectra of 4 and β-methylstyrene.

Sequential Diastereoselective Reactions of Resultant Vinyl Ethers.

Then we focused our attention to sequential diastereoselective oxidation reactions of the F-C products by m-CPBA (Scheme 3). Regardless of the geometry of the F-C products, single diastereomer of α,β-dihydroxy ketone (8) having a chiral quarternary carbon was obtained in high chemical yield and high diastereoselectivity through the acidic hydrolysis of intermediates (6 and 7). The other diastereomer was not observed in the reaction mixture as determined by ^{1}H-NMR spectra (CDCl$_3$); δ 1.74 (q, J = 1.8 Hz, 3H), 2.44 (s, 3H), 3.21 (d, J = 10.5 Hz, 1H), 4.53 (s, 1H), 4.57 (dq, J = 10.5, 6.9 Hz, 1H), 7.30 (d, J = 8.1 Hz, 2H), 7.95 (d, J = 8.1 Hz, 2H). The relative stereochemistry was determined to be syn on the bases of the chemical shifts of α-methyl carbons in ^{13}C-NMR spectra

(CDCl₃) of **8**: *syn*; 23.4 ppm, *anti*; 25.1 ppm (*48*).   Therefore, the diastereoselectivity can be explained as follows, the trifluoromethyl substituent is located outside because of a 1,3-allylic strain (*49*), and then *m*-CPBA should attack from the direction of the hydroxy substituent perpendicular to the olefin π-face through hydrogen bonding (Figure 2) (*50*, Sharpless proposed the formation of hydrogen bond between hydroxy groups and *m*-CPBA: *51*).

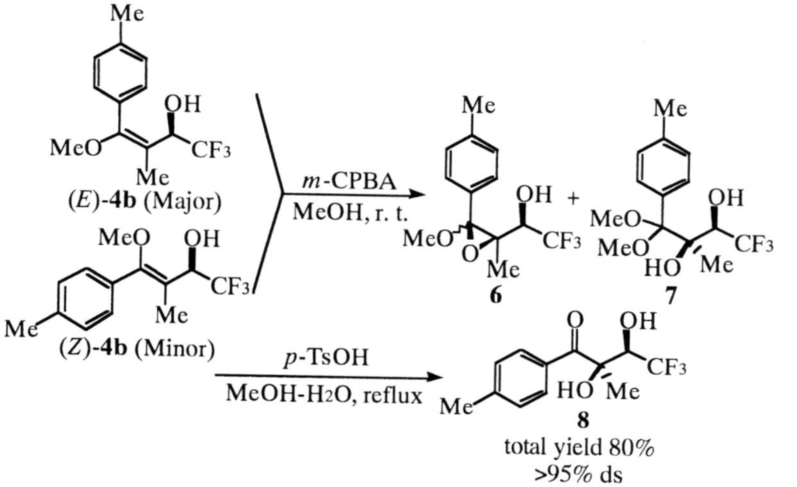

Scheme 3 Diastereoselective oxidation reaction of **4b** with *m*-CPBA.

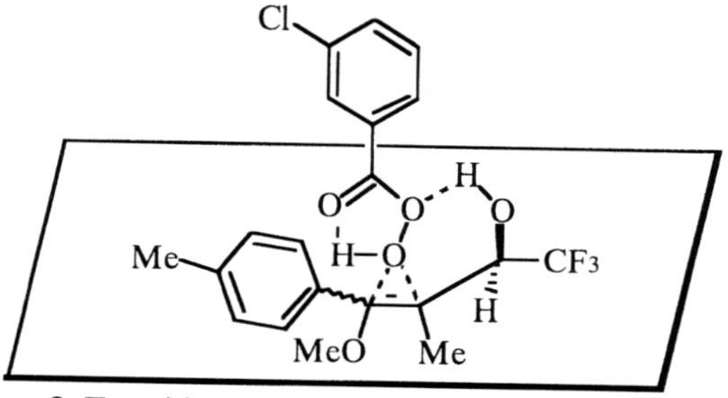

Figure 2 Transition state of diastereoselective oxidation reaction of **4b** with *m*-CPBA.

Asymmetric Friedel-Crafts Reactions of Silyl Enol Ethers: a Possible Mechanism of the Mukaiyama-aldol Reactions

The Mukaiyama-aldol reaction of silyl enol ethers is one of the most important carbon-carbon bond-forming reactions in organic synthesis.   Therefore, its application to catalytic asymmetric synthesis has been investigated in depth for the last decade (*52-56*).   In an analogy to the F-C reaction of vinyl ether, if a catalytic asymmetric F-C reaction proceeds under Mukaiyama conditions to give additional reactive silyl enol ethers, sequential reactions with electrophiles could provide further functionalized products   (chiral *syn*- or *anti*-α,β-dihydroxythioesters as α-protected forms: *57,58*, chiral α-amino-β-hydroxy: *59,60*) possessing adjacent stereogenic centers at the α and β positions in a highly diastereoselective manner (Scheme 4). As a possible mechanism of this reaction, proto- and sila-tropic ene processes have been proposed (prototropic type: *61,62*, silatropic type: *63,64*).

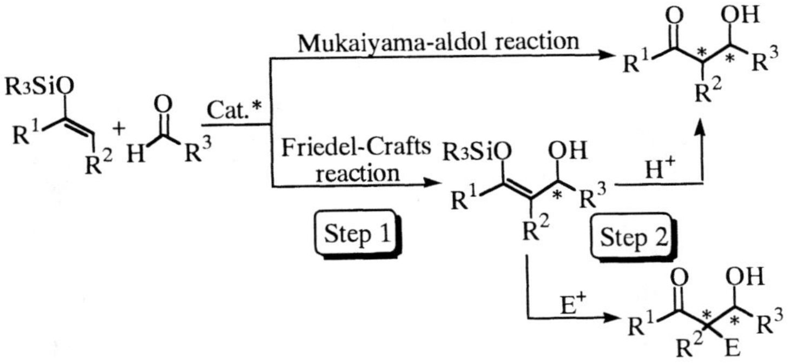

Scheme 4 Possible mechanism of Mukaiyama-aldol reactions.

The reaction of *t*-butyldimethylsilyl enol ether (**9a**) with fluoral was examined using the BINOL-Ti catalyst (*35-37*) prepared from (*R*)-BINOL and Cl₂Ti(OPrⁱ)₂ in the presence of 4Å MS. Significantly, the F-C product (**10a**) was obtained in high yield together with the usual aldol product (**5a**) (Scheme 5, Table IV, runs 1,2). The enantiomeric excess of the major isomer (*Z*)-**10a** was excellent (98% ee). The reaction of silyl enol ethers (**9b, c, d**) (*65*) possessing a methyl substituent at the α position were also observed to give the F-C products (**10b, c, d**) with similarly high enantioselectivity (runs 3~5). The F-C product was also obtained in the reaction between *n*-butyl glyoxlate and *t*-butyldimethylsilyl enol ether (**9c**). The F-C products were not obtained in the reaction of *t*-butyldimethylsilyl enol ethers without aryl substituent such as 1-*t*-butyldimethylsilyloxy-1-cyclohexene and 2-*t*-butyldimethylsilyloxy-1-heptene. In the reaction of trimethylsilyl enol ether (**9e**), the aldol-type product (**5e**) was mainly obtained in the trimethylsilyl ether form (run 6).

The major isomers of the F-C products (**10b, c, d**) were determined to be *Z* on the basis of the chemical shifts of allylic protons in ¹H-NMR spectra (Figure 3) (*46*). Both vinylic proton and allylic proton of the major isomer are observed at higher field than the minor's one in ¹H-NMR spectra (CDCl₃) of **10a**: vinylic proton; 5.08 ppm (major), 5.15 ppm (minor), allylic proton; 4.46 ppm (major), 4.98 ppm (minor). The absolute configuration of (*Z*)-**10c** was determined to be *R* by Mosher's method: the chemical shift of vinylic methyl proton in ¹H-NMR spectra (CDCl₃) of MTPA ester of (*Z*)-**10c**; 1.79 ppm (*S*), 1.55 ppm (*R*); $\delta_S - \delta_R$ = positive (*47*). The steric hindrance of the silyl substituent and the electron-withdrawing nature of the trifluoromethyl substituent may be important in preventing the intra- or inter-molecular nucleophilic attack to the silyl substituent in the zwitterion intermediate (**B**), wherein the aryl substituent probably stabilizes the carbenium ion. It is interesting to note that the enantiomeric excesses of the four products ((*E*)-**10b**, (*Z*)-**10b**, *syn*-**5b**, *anti*-**5b**) were significantly different (66 (*R*), 94 (*R*), 22 (*R*), 62% ee (*R*), respectively). In the zwitterion (**B**), the four possible diastereomers may be formed in different ratios, and the ratio of deprotonation *vs.* desilylation may be critically dependent on the stereochemistries of the diastereomers.

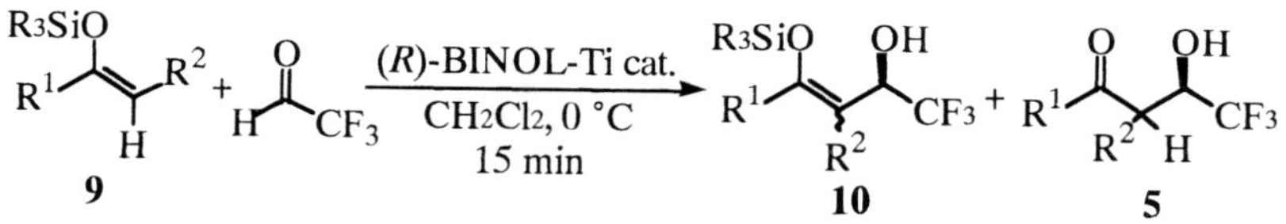

68

R$_3$SiO R^1 R^2 H + O H CF$_3$ →[(R)-BINOL-Ti cat.][CH$_2$Cl$_2$, 0 °C, 15 min] R$_3$SiO R^1 R^2 CF$_3$ + O R^1 R^2 H OH CF$_3$

9 → **10** + **5**

Scheme 5 Asymmetric F-C reactions of silyl enol ethers.

Table IV The F-C reactions of silyl enol ethers with fluoral catalyzed by (R)-BINOL-Ti catalyst.

run	9	SiR$_3$	R^1	R^2	cat. (mol%)	yield (%)[a] **10** (E : Z)	**5** (syn : anti)	ee[b] (%)
1	a	Sit-BuMe$_2$	Ph	H	10	34 (1 : 4)	15	98 (R)
2	a	Sit-BuMe$_2$	Ph	H	5	67 (1 : 5)	14	98 (R)
3	b[c]	Sit-BuMe$_2$	4'-MePh	Me	20	77 (1 : 6)	10 (1 : 1)	94 (R)[d]
4	c[c]	Sit-BuMe$_2$	4'-MeOPh	Me	20	68 (1 : 5)	22 (5 : 3)	95 (R)
5	d[c]	Sit-BuMe$_2$	4'-MeSPh	Me	20	72 (1 : 6)	18 (2 : 1)	95 (R)
6	e[c]	SiMe$_3$	4'-MePh	Me	20	0	27[e]	-

[a] Isolated yield. [b] The enantiomeric excess of (Z)-**10**. Determined by chiral HPLC analysis of **5a** or anti-**5b, c, d** obtained by acidic hydrolysis of (Z)-**10**. In the hydrolyses of (Z)-**10b, c, d**, anti-isomers were obtained as the major diastereomer (anti / syn = ca. 3). The relative configuration of the major diastereomer was determined to be anti on the basis of the chemical shifts of α-methyl carbon in ^{13}C-NMR spectra (CDCl$_3$) of **5c**: anti; 16.5 ppm, see table III b for **5b** (48). Chiral HPLC conditions: anti-**5c**; Daicel, CHIRALPAK AS, n-hexane : i-PrOH = 95 : 5, 0.8 ml/min, 254 nm, t_R = 19 min (2R, 3S), 37 min (2S, 3R), anti-**5d**; Daicel, CHIRALPAK AS, n-hexane : i-PrOH = 98 : 2, 0.8 ml/min, 254 nm, t_R = 28 min (2R, 3S), 64 min (2S, 3R), see table III b for **5a** and anti-**5b**. [c] >95% Z. [d] The enantiomeric excess of (E)-**10b**, syn-**5b** and anti-**5b** were 66 (R), 22 (R) and 62% ee (R), respectively: syn-**5b**; Daicel, CHIRALPAK AS, n-hexane : i-PrOH = 98 : 2, 0.8 ml/min, 254 nm, t_R = 14 min (2R, 3R), 25 min (2S, 3S). The enantiomeric excess of (E)-**10b** was determined by the same chiral HPLC analysis as (Z)-**10b**. Also in this hydrolysis anti-**5b** was obtained as the major diastereomer (anti / syn = ca. 4). [e] The usual aldol product (**5e**) was obtained as the trimethylsilyl ether. The diastereomeric ratio = 1 : 4.

The representative procedure for run 2: to a solution of isolated (R)-BINOL-Ti catalyst (0.05 mmol) in dehydrated dichloromethane (6 ml) was added silyl enol ether (**9a**, 1 mmol) at 0 °C under an argon atmosphere, and then freshly dehydrated and distilled fluoral (ca. 3 mmol) was passed to the solution. After stirring for 15 min at the same temperature, usual work-up followed by chromatographic separation by silica gel (dichloromethane : n-hexane = 3 : 2) gave the product (**10a**).

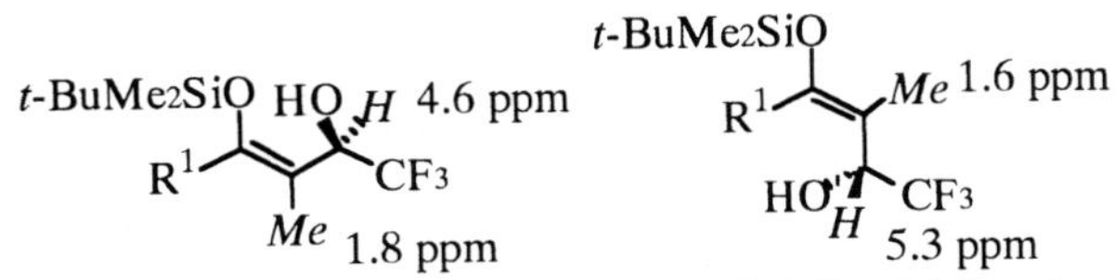

Figure 3 ^{1}H-NMR spectra of **10b~d**.

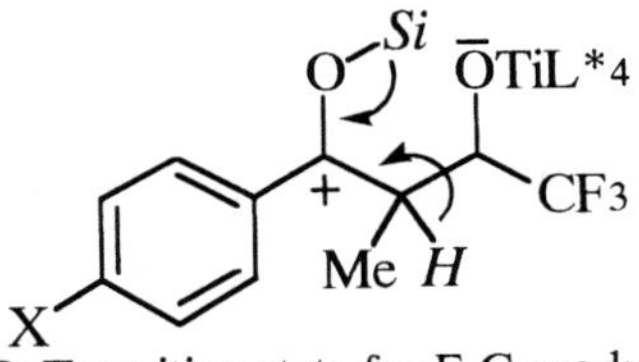

B: Transition state for F-C-product.

Sequential Diastereoselective Reactions of Resultant Silyl Enol Ethers. Next, our attention was focused on the sequential diastereoselective reactions with electrophiles (Scheme 6). The oxidation reaction of the F-C product ((Z)-**10b**) by m-CPBA proceeded to give the syn-diastereomer in its unprotected form in high chemical yield and high diastereoselectivity through the above transition state (Figure 2). These products are of synthetic importance because of similar skeletal features to Merck L-784512 (66) with cyclooxygenase-2 selective inhibitory activity. The protodesilylation reaction of the F-C product ((Z)-**10c**) by TBAF also proceeded stereoselectively to give the anti-diastereomer in quantitative yield in a similar manner to that by m-CPBA oxidation. The diastereomeric excess of **5c** was determined by HPLC analysis; Daicel, CHIRALPAK AS, n-hexane : i-PrOH = 95 : 5, 0.8 ml/min, 254 nm, t_R = 16 min (syn), 37 min (anti).

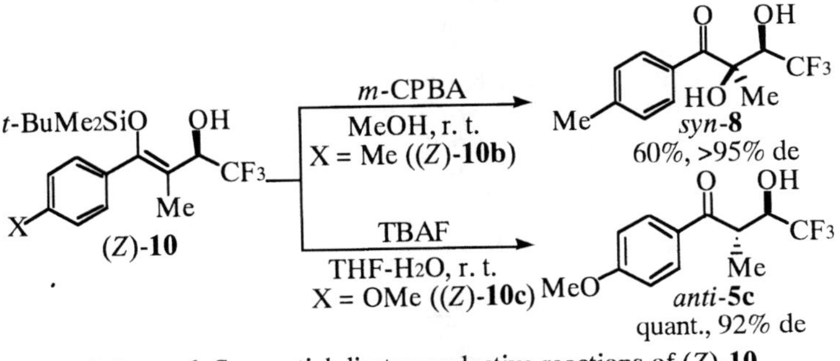

Scheme 6 Sequential diastereoselective reactions of (Z)-**10**.

Conclusion

We have reported the first example of asymmetric catalysis of the Friedel-Crafts reaction with fluoral to provide a practical synthetic route not only for chiral α-trifluorobenzylalcohols but also for highly enantiopure functionalized β-trifluoroaldols through the sequential diastereoselective reactions of the resultant vinyl ethers or silyl enol ethers with electrophiles.

70

Acknowledgment

The authors are grateful to Drs. M. Terada and T. Yajima of Tokyo Institute of Technology. Support from Monbusho (Grant-in-Aid for Scientific Research on Priority Areas, No.09238209: Innovative Synthetic Reactions) is gratefully acknowledged.

Literatures cited

1   *Biomedical Frontiers of Fluorine Chemistry*; Ojima, I.; McCarthy, J. R.; Welch, J. T., Eds.; American Chemical Society: Washington, D. C., 1996.
2   Resnati, G. *Tetrahedron* **1993**, *49*, 9385.
3   Resnati, G.; Soloshonok, V. A., Eds. *Tetrahedron* **1996**, *52*, 1.
4   Olah, G. A.; Chambers, R. D.; Prakash, G. K. S. *Synthetic Fluorine Chemistry*; Wiley: N.Y., 1992.
5   Iseki, K.; Kuroki, Y.; Kobayashi, Y. *Tetrahedron Lett.* **1997**, *38*, 7209.
6   Iseki, K.; Oishi, S.; Sasai, H.; Shibasaki, M. *Tetrahedron Lett.* **1996**, *37*, 9081.
7   Mikami, K.; Yajima, T.; Takasaki, T.; Matsukawa, S.; Terada, M.; Uchimaru, T.; Maruta, M. *Tetrahedron* **1996**, *52*, 85.
8   Mikami, K.; Yajima, T.; Terada, M.; Uchimaru, T. *Tetrahedron Lett.* **1993**, *34*, 7591.
9   Soloshonok, V. A.; Hayashi, T. *Tetrahedron Lett.* **1994**, *35*, 2713.
10 Hayashi, T.; Soloshonok, V. A. *Tetrahedron: Asymmetry* **1994**, *5*, 1091.
11 Smith, M. B. *Organic Synthesis*; McGraw-Hill: N. Y., 1994; pp 1313.
12 Heaney, H. In *Comprehensive Organic Synthesis*; Trost, B. M.; Fleming, I., Eds.; Pergamon Press: Oxford, 1991, Vol. 2; pp 733.
13 Roberts, R. M.; Khalaf, A. A. In *Friedel-Crafts Alkylation Chemistry. A Century of Discovery*; Dekker: N. Y., 1984.
14 Olah, G. A. *Friedel-Crafts Chemistry*; Wiley-Interscience: N. Y., 1973.
15 Costa, P. R. R.; Cabral, L. M.; Alencar, K. G.; Schmidt, L. L.; Vasconcellos, M. L. A. A. *Tetrahedron Lett.* **1997**, *38*, 7021.
16 El Kaim, L.; Guyoton, S.; Meyer, C. *Tetrahedron Lett.* **1996**, *37*, 375.
17 Bigi, F.; Sartori, G.; Maggi, R.; Cantarelli, E.; Galaverna, G. *Tetrahedron: Asymmetry* **1993**, *4*, 2411.
18 Terada, M.; Sayo, N.; Mikami, K. *Synlett* **1995**, 411.
19 Casiraghi, G.; Bigi, F.; Casnati, G.; Sartori, G.; Soncini, P.; Gasparri Fava, G.; Ferrari Belicchi, M. *J. Org. Chem.* **1988**, *53*, 1779.
20 Cox, E. D.; Hameker, L. K.; Li, J.; Yu, P.; Czerwinski, K. M.; Deng, L.; Bennett. D. W.; Cook, J. M. *J. Org. Chem.* **1997**, *62*, 44.
21 Dai, W.-M.; Zhu, H. J.; Hao, X.-J. *Tetrahedron Lett.* **1996**, *37*, 5971.
22 Soe, T.; Kawate, T.; Fukui, N.; Hino, T.; Nakagawa, M. *Heterocycles* **1996**, *42*, 347.
23 Erker, G.; van der Zeijden, A. A. H. *Angew. Chem. Int. Ed. Engl.* **1990**, *29*, 512.
24 Bigi, F.; Casiraghi, G.; Casnati, G.; Sartori, G.; Gasparri Fava, G.; Ferrari Belicchi, M. *J. Org. Chem.* **1985**, *50*, 5018.
25 Kawate, T.; Yamada, H.; Soe, T.; Nakagawa, M. *Tetrahedron: Asymmetry* **1996**, *7*, 1249.
26 Toshimitsu, A.; Hirosawa, C.; Tamao, K. *Synlett* **1996**, 465.
27 Muehldorf, A. V.; Guzman-Perez, A.; Kluge, A. F. *Tetrahedron Lett.* **1994**, *35*, 8755.
28 Mikami, K. *Pure and Appl. Chem.* **1996**, *68*, 639.
29 Mikami, K.; Terada, M.; Narisawa, S.; Nakai, T. *Synlett* **1992**, 255.

30   Corey, E. J.; Bakshi, R. K. *Tetrahedron Lett.* **1990**, *31*, 611.
31   Chong, J. M.; Mar, E. K. *J. Org. Chem.* **1991**, *56*, 893.
32   Corey, E. J.; Cheng, X.-M.; Cimprich, K. A.; Sarshar, S. *Tetrahedron Lett.* **1991**, *32*, 6835.
33   Ramachandran, P. V.; Teodorovic, A. V.; Gong, B.; Brown, H. C. *Tetrahedron: Asymmetry* **1994**, *5*, 1075.
34   Fujisawa, T.; Sugimoto, T.; Shimizu, M. *Tetrahedron: Asymmetry* **1994**, *5*, 1095.
35   Mikami, K.; Terada, M.; Nakai, T. *J. Am. Chem. Soc.* **1990**, *112*, 3949; **1989**, *111*, 1940.
36   Mikami, K.; Terada, M.; Narisawa, S.; Nakai, T. *Org. Synth.* **1993**, *71*, 14.
37   Mikami, K.; Motoyama, Y.; Terada, M. *J. Am. Chem. Soc.* **1994**, *116*, 2812.
38   Ohno, A.; Nakai, J.; Nakamura, K.; Goto, T.; Oka, S. *Bull. Chem. Soc. Jpn.*, **1981**, *54*, 3486.
39   Corey, E. J.; Barnes-Seeman, D.; Lee, T. W.; Goodman, S. N. *Tetrahedron Lett.* **1997**, *37*, 6513.
40   Mikami, K.; Matsukawa, S.; Sawa, E.; Harada, A.; Koga, N. *Tetrahedron Lett.* **1997**, *38*, 1951.
41   Mikami, K.; Matsukawa, S.; Nagashima, M.; Funabashi, H.; Morishima, H. *Tetrahedron Lett.* **1997**, *38*, 579.
42   Mikami, K.; Matsukawa, S. *Nature* **1997**, *385*, 613.
43   Matsukawa, S.; Mikami, K. *Tetrahedron: Asymmetry* **1997**, *8*, 815.
44   Matsukawa, S.; Mikami, K. *Enantiomer* **1996**, *1*, 69.
45   Matsukawa, S.; Mikami, K. *Tetrahedron: Asymmetry* **1995**, *6*, 2571.
46   Rummens, F. H. A.; de Haan, J. W. *Org. Magn. Res.* **1970**, *2*, 351.
47   Dale, J. A.; Mosher, H. S. *J. Am. Chem. Soc.* **1973**, *95*, 512.
48   Heathcock, C. H.; Pirrung M. C.; Sohn, J. E. *J. Org. Chem.* **1979**, *44*, 4294.
49   Hoffmann, R. W. *Chem. Rev.* **1989**, *89*, 1841.
50   Schwesinger, R.; Willaredt, J. In *Houben-Weyl E21*; Helmchen, G.; Hoffmann, R. W.; Mulzer, J.; Schaumann, E. Eds.; Thieme: Stuttgart, 1996, Vol. 8.
51   Sharpless, K. B., Verboeven, T. R. *Aldrichim. Acta* **1979**, *12*, 63.
52   Nelson, S. G. *Tetrahedron: Asymmetry* **1998**, *9*, 357.
53   Gröger, H.; Vogl, E. M.; Shibasaki, M. *Chem. Eur. J.* **1998**, *4*, 1137.
54   Bach, T. *Angew. Chem. Int. Ed. Engl.* **1994**, *33*, 417.
55   Mukaiyama, T. *Org. React.* **1982**, *28*, 203.
56   Evans, D. A.; Nelson, J. V.; Taber, T. R. *Topics in Stereochemistry*; Interscience: N. Y., 1982; Vol. 13.
57   Kobayashi, S.; Horibe, M. *J. Am. Chem. Soc.* **1994**, *116*, 9805.
58   Mukaiyama, T.; Shiina, I.; Uchiro, H.; Kobayashi, S. *Bull. Chem. Soc. Jpn.* **1994**, *67*, 1708.
59   Ito, Y.; Sawamura, M; Hayashi, T. *J. Am. Chem. Soc.* **1986**, *108*, 6405.
60   Ito, Y.; Sawamura, M; Shirakawa, E.; Hayashizaki, K.; Hayashi, T. *Tetrahedron* **1988**, *44*, 5253.
61   Mikami, K.; Matsukawa, S. *J. Am. Chem. Soc.* **1993**, *115*, 7039.
62   Mikami, K.; Matsukawa, S.; Nagashima, M.; Funabashi, H.; Morishima, H. *Tetrahedron Lett.* **1997**, *38*, 579.
63   Mikami, K.; Matsukawa, S. *J. Am. Chem. Soc.* **1994**, *116*, 4077.
64   Mikami, K.; Matsukawa, S.; Sawa, E.; Harada, A.; Koga, N. *Tetrahedron Lett.* **1997**, *38*, 1951.
65   Heathcock, C. H.; Buse, C. T.; Kleschick, W. A.; Pirrung M. C.; Sohn, J. E.; Lampe, J. *J. Org. Chem.* **1980**, *45*, 1066.   See ref. 56.
66   Tan, L.; Chen, C.; Larsen, R. D.; Verhoeven, T. R.; Reider, P. J. *Tetrahedron Lett.* **1998**, *39*, 3961.

SUBSTRATE-CONTROLLED ASYMMETRIC SYNTHESIS

Chapter 6

Biomimetic, Reducing Agent-Free Reductive Amination of Fluorocarbonyl Compounds: Practical Asymmetric Synthesis of Enantiopure Fluoroamines and Amino Acids

Vadim A. Soloshonok[1]

Department of Chemistry, NIRIN, Nagoya, Aichi 462, Japan

The development of a truly practical asymmetric methodology for preparing stereochemically defined biomedicinally interesting fluoro-amines and amino acids is described. The synthetic procedure consists of three simple steps: formation of the *Schiff* base between fluoro-carbonyl compound and enantiomerically pure α-phenylethylamine, its base-catalyzed isomerization to the corresponding *Schiff* base of a fluoro-amino compound and acetophenone, and hydrolysis of the latter to the target free amine or amino acid. The whole procedure represents the first example of asymmetric reductive amination of carbonyl compounds without application of a reducing agent. The key step of the method, a biomimetic azomethine-azomethine isomerization of the *Schiff* bases is a unique case of highly stereoselective transfer of chirality from a less to a more configurationally unstable stereogenic center. Intriguing stereodirecting role of a perfluoroalkyl group in the enantioselective step of [1,3]-PSR is highlighted.

Amino compounds, in which a nitrogen is bound to a stereogenic carbon, make up a large body of naturally occurring and biologically important molecules. Therefore, the development of asymmetric methodology, allowing for preparing stereochemically defined and structurally varied amino compounds, has been one of the major goals of synthetic organic chemistry. Considering the unique consequences of fluorine substitution for hydrogen in organic molecules, such as an opportunity for the rational modification of physical, chemical, steric and thus, biological properties of the parent compounds, the synthesis of selectively fluorinated and enantiomerically pure amino compounds, as biologically relevant targets, might be of particular interest (*1-9*).

For quite some time we have been engaged in the development of asymmetric approaches to fluorinated amino acids *via* alkylation (*10, 11*), *Michael* (*12*) and aldol (*13-15*) addition reactions of a Ni(II) complex of the chiral non-racemic *Schiff* base of glycine with (*S*)-*o*-[*N*-(*N*-benzylprolyl)amino]benzophenone, as well as transition metal-catalyzed stereoselective aldol addition reactions between fluoro-carbonyl compounds and derivatives of isocyanoacetic acid (*13, 16*). While studying the chemistry of fluoro-imines, as intermediates in the synthesis of amino acids, we have found a surprising reaction between per(poly)fluoroalkyl carbonyl compounds and *N*-(benzyl)triphenylphosphazene (**1**). This reaction gave rise to a mixture of the target *N*-benzyl imines of starting carbonyl compounds **2**, and the *N*-benzylidene derivatives of fluorinated amino compounds **3** (Scheme 1) (*17, 18*). Detailed investigation into this unexpected outcome revealed that fluorinated *N*-benzyl imines **2** in the presence of a base, even as weak as triethylamine (TEA), easily undergo virtually irreversible

[1]Current address: Department of Chemistry, University of Arizona, Tucson, AZ 85721.

Scheme 1

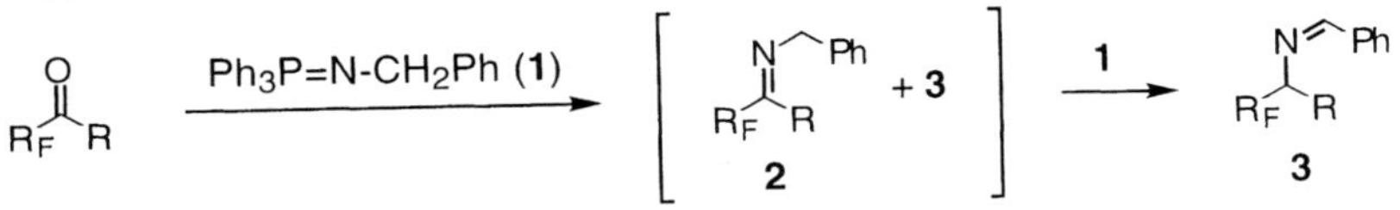

$$R_F = C_nF_{2n+1}, C_n(F_2)_nH; \quad R = H, n\text{-}C_nF_{2n+1}$$

azomethine-azomethine isomerization to give more thermodynamically stable isomers **3**. Since the *N*-benzyl imines **2** were found to be readily prepared by the direct reaction between fluoro-carbonyl compounds and benzylamine, the synthetic potential of this exciting finding for developing conceptually new reductive amination methodology was obvious. Thus, in contrast to the established methods for reductive amination of carbonyl compounds which employ external reducing agents to perform the transformation of a C,N double bond to an amino group, the discovered azomethine-azomethine isomerization, referred by us to as [1,3]-Proton Shift Reaction ([1,3]-PSR), makes use of the intramolecular reduction-oxidation process through a simple transposition of the imine functionality (Scheme 2) providing a formal reduction of the initial C,N double bond with the formation of the corresponding *Schiff* base which could be easily hydrolyzed to afford the targeted amino compound.

Scheme 2

General synthetic application of [1,3]-PSR for preparing various biologically interesting fluorinated amino compounds was demonstrated by an efficient practical synthesis of structurally varied primary and secondary amines (*19, 20*) (Scheme 3), α-amino acids (*21*) (Scheme 4), β-amino acids (*22, 23*) and α-alkyl substituted β-amino acids (*24-26*) (Scheme 5), *via* [1,3]-PSR transamination of the corresponding fluoro-

Scheme 3

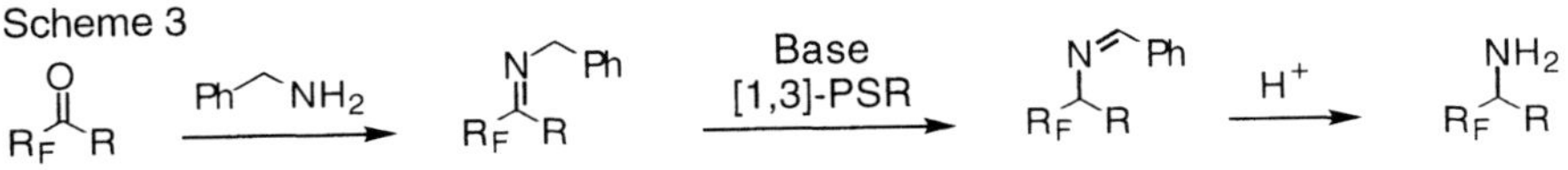

$$R_F = C_nF_{2n+1}, C_n(F_2)_nH, C_6F_5; \quad R = H, n\text{-}Alk, Ar$$

Scheme 4

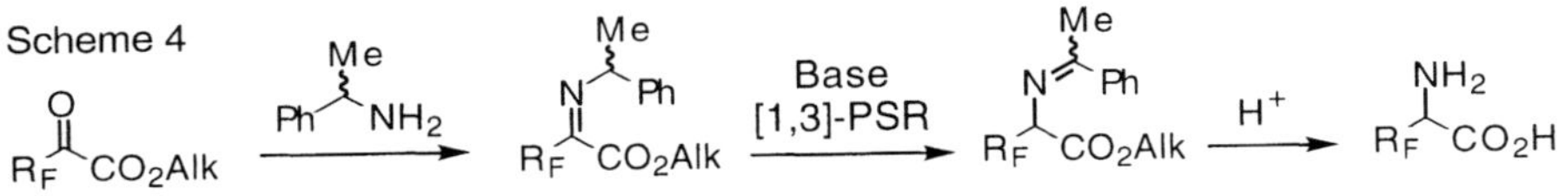

Scheme 5

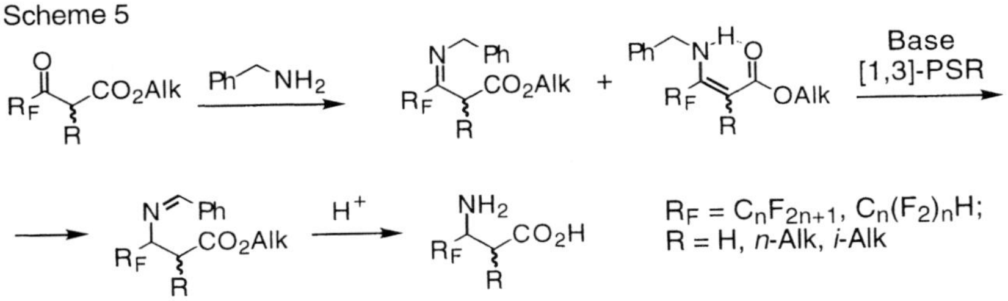

$$R_F = C_nF_{2n+1}, C_n(F_2)_nH; \quad R = H, n\text{-}Alk, i\text{-}Alk$$

aliphatic and fluoro-aromatic aldehydes, fluoroalkyl alkyl and fluoroalkyl aryl ketones, α- and β-keto esters, respectively.

Recently we found that substituting benzylamine with its derivatives containing electron-withdrawing substituent on the phenyl ring or picolylamines, possessing electron-deficient pyridine ring (27, 28), remarkably facilitated the azomethine-azomethine isomerization of the corresponding imines rendering [1,3]-PSR even more general and efficient reducing agent-free reductive amination process. However, for that exciting synthetic potential of the biomimetic [1,3]-proton shift reaction to be realized to the full, the asymmetric version of this reaction, allowing for preparing enantiomerically pure targets, must be developed.

Biological Transamination and Previous Synthetic Models

Biological transamination involves the enzyme-catalyzed isomerization of the imines derived from pyridoxamine and α-keto acids, and from pyridoxal and α-amino acids (Scheme 6). The interconversion of the imines proceeds through the corresponding asymmetric 1,3-azaallylic carbanion **4** with a complete stereoselectivity (29). Obviously, the lure of synthetic application of this biological reaction has attracted a great deal of attention in the past. Numerous investigations into synthetic and mechanistic aspects of hydrocarbon azomethine-azomethine isomerization revealed that the azomethine system of the *Schiff* bases derived from common carbonyl compounds and amines, is rather immobile necessitating strong nucleophilic catalysis, such as NaOMe in MeOH, for the isomerization to occur with the equilibrium normally not favoring targeted products (30).

Scheme 6

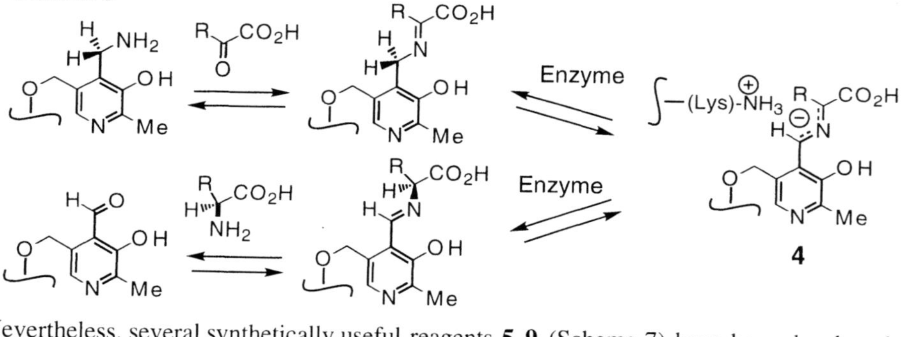

Nevertheless, several synthetically useful reagents **5-9** (Scheme 7) have been developed for highly efficient biomimetic transformation of amines to carbonyl compounds under very mild reaction conditions and with high chemical yields (31-34). However the opposite transformation, preparation of amines from the corresponding carbonyl compounds, turned out to be rather challenging and virtually no synthetically useful generalized solutions have been reported so far (35). As one can see, reagents **5-9** are of

Scheme 7

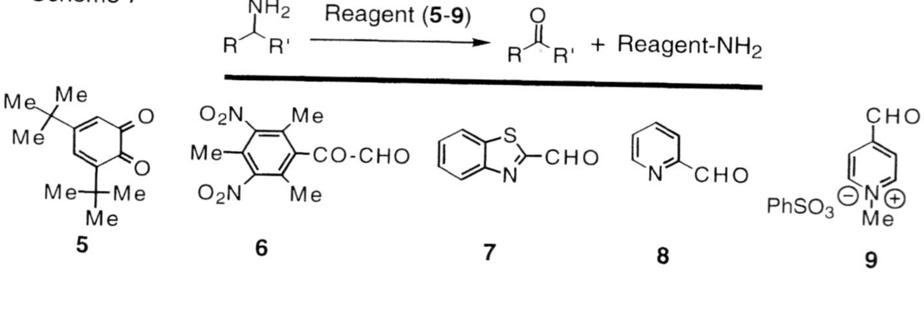

a quite electrophilic nature, that provides an effective stabilization of the developing azaallylic anion and shifts the equilibrium to the target products. In fact, the equilibrium constants of a hydrocarbon base-catalyzed azomethine-azomethine isomerization were found to be adequately correlated by the *Hammett* equation (*36*). In other words, the equilibrium of the isomerization is shifted towards a more C-H acidic tautomer. In terms of stereochemistry, it means that the proton transfer proceeds from a less to a more configurationally unstable stereogenic center and thus, is thermodynamically not possible. Indeed, for the isomerizations of certain chiral hydrocarbon imines it was shown that the isomerization and racemization of both starting and final products occur with comparable rates (*37-39*). Accordingly, the reaction reversibility and racemization of compounds involved, were shown to be the problems which would generally plague the asymmetric biomimetic transamination methodology.

Asymmetric Biomimetic Transamination

The starting chiral fluorinated *Schiff* bases **11a-e** were readily synthesized by the direct condensation between an appropriate ketone **10** and (*S*)-α-phenylethylamine (Scheme 8). An important characteristic of these substrates is that they exist as individual *anti* isomers (by NMR). This property becomes significant when considering the stereochemical outcome of the isomerizations.

Scheme 8

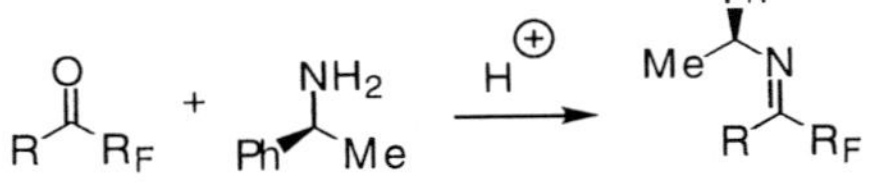

	a	b	c	d	e
R_F	CF_3	CF_3	CF_3	CF_3	C_3F_7
R	Ph	Me	Et	*n*-Oct	Me

As a model reaction, we choose to explore the isomerization of imine **11a**, derived from the α-phenylethylamine and trifluoroacetophenone, since this transformation was expected to be the most challenging due to high C-H acidity of the product in which the benzylic proton is additionally activated by the trifluoromethyl group and *Schiff* base function as well. For the initial studies of the isomerization we tried to apply as mild as possible reaction conditions, since the targeted product **11a** is obviously prone to racemization under the forcing conditions or when strong base is used. Unfortunately, ketimine **11a**, as well as the rest of N-(α-phenyl)ethyl derivatives **11b-e**, were found to be totally inert under the conditions previously established for isomerizations of the N-benzyl analogs (NEt$_3$, rt. or reflux). Thus, no isomerization of **11a** to **12a** was observed in triethylamine (TEA) solution for more than one week. However, at 150 °C the isomerization took place with a slow reaction rate giving rise to the targeted *Schiff* base **12a** of 50% ee in moderate isolated yield (Scheme 9). Further we have found that an addition of DABCO or DBU to the TEA solution allows the isomerization to be completed under milder conditions to give the desired product **12a** with higher chemical yield and enantiomeric purity. Drawing inspiration from these finding, we performed the isomerization in neat DBU. The result was rather impressive: the isomerization was completed at 50 °C in only 1 h giving rise to the product in excellent chemical yield and with markedly improved enantiomeric purity. Lowering of the reaction temperature decreased the isomerization rate however gave the product in higher enantiomeric purity. The best results: 95% chemical yield and 87% enantiomeric purity was obtained in the isomerization conducted in the presence of two equivalents of the DBU. The dramatic influence of the DBU/substrate ratio on the isomerization rate was quite unexpected as it is not consistent with a purely catalytic role of the base in these isomerizations. At this stage we can suggest that DBU, apart from its catalytic role, works as a polar reaction medium facilitating the isomerization. Enantiomeric purity of the product **12a** was determined directly for *Schiff* base **12a** and/or for the N-3,5-dinitrobenzoyl derivative of free amine **14a**, obtained by acidic hydrolysis of **12a**, using chiral stationary phase HPLC analysis. Absolute configuration of (*R*)-**12a** was determined by comparison of the $[\alpha]_D$ values of amine (*R*)-**12a** with those reported in literature (40).

Scheme 9

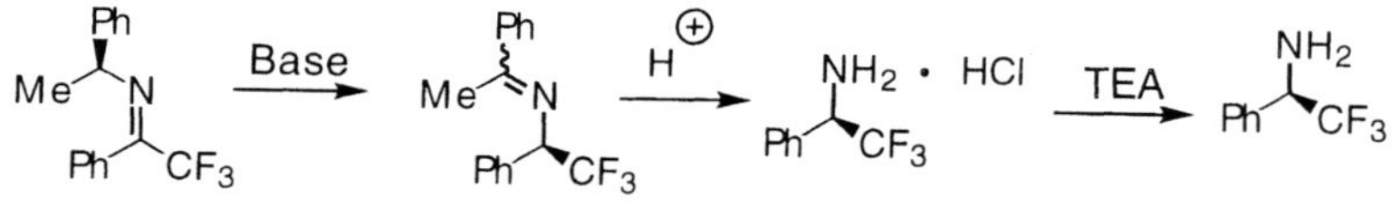

(*S*)-**11a** (*R*)-**12a** (*R*)-**13a** (*R*)-**14a**

Entry	Base (equ.)	*T*, °C	Time, h	Yield, %	ee, %
1	TEA (100)	150	20	64	50
2	TEA/DABCO (0.5)	100	30	76	55
3	TEA/DBU (0.1)	50	130	80	60
4	DBU (1)	50	1	98	77
5	DBU (1)	18	7	95	84
6	DBU (2)	19	4	95	87

Assuming that the asymmetric outcome of the isomerization might be a function of the stereochemical discrimination between substituents at the imine carbon, we designed two types of substrates bearing the trifluoromethyl group *vs* alkyl, and perfluoroalkyl *vs* methyl group, but first we have explored the isomerization of the methyl trifluoromethyl imine (Scheme 10). This isomerization, conducted in the presence of molar or submolar amount of neat DBU occurred smoothly at 60 °C with a reasonable rate to afford the desired *Schiff* base in high chemical yield and enantiomeric purity, both over 90% (entries 1,2). Under the same reaction conditions the isomerization of trifluoromethyl ethyl imine was found to follow a similar pattern of reactivity but gave the corresponding imine with lower stereoselectivity (87% ee) (entries 3,4). The isomerization of *n*-octyl imine **11d**, despite the huge steric bulk of this alkyl substituent, gave the target product **12d** with high chemical yield and in enantiopurity just a bit lower that the values recorded

Scheme 10

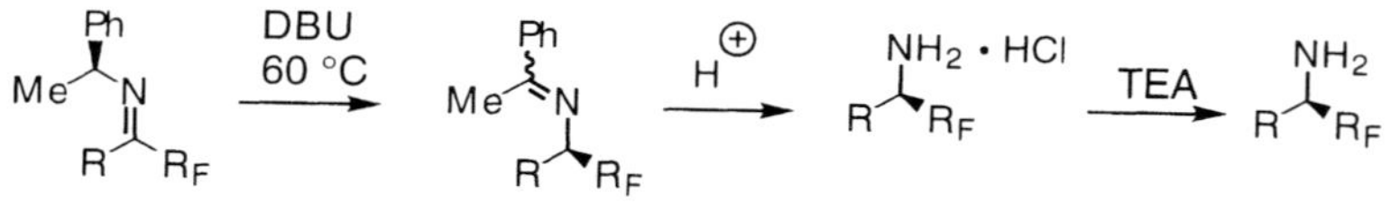

(*S*)-**11b-e** (*R*)-**12b-e** (*R*)-**13b-e** (*R*)-**14b-e**

Entry		R	R_F	DBU (equ.)	Time, h	Yield, %	ee, %
1	**b**	Me	CF_3	1	18	90	93
2	**b**	Me	CF_3	1.5	15	94	93
3	**c**	Et	CF_3	1	18	87	87
4	**c**	Et	CF_3	1.5	15	90	87
5	**d**	*n*-Oct	CF_3	1.5	15	94	84
6	**e**	Me	C_3F_7	1.5	15	74	97

in the transformation of ethyl-bearing ketimine **11c** (entry 5). In contrast, the isomerization of perfluoropropyl methyl imine under the same reaction conditions was accompanied by a formation of some byproducts. However, the product was isolated in appreciated chemical yield and with excellent enantioselectivity of 97% ee. The stereochemical outcome of these isomerizations, conducted under the same reaction conditions (DBU, 60 °C) definitely suggests that the enantioselectivity of the proton transfer is influenced by the steric bulk of the substituents on the imine carbon.

Our next target was the isomerization of highly enaminolizable trifluoromethyl benzyl imine **11f** (Scheme 11). The *Schiff* base **11f**, prepared by the reaction of trifluoromethyl benzyl ketone with α-phenylethylamine, exists as a mixture of ketimine **11f** with the corresponding enamine which could be separated by flash chromatography. We have found that imine **11f**, taken as an individual compound or as a mixture with the enamine, could be completely isomerized in neat DBU at 50 °C to give the desired *Schiff* base **12f** in both high chemical yield and enantiopurity (entry 1). At an elevated temperature, the isomerization proceeded with a higher reaction rate giving the product in lower chemical yield and enantiomeric purity (entry 2). Application of 2 equivalents of the base accelerated the process, allowing us to accomplish the isomerization at room temperature to give the target product of 88% ee in high chemical yield (entry 3).

Scheme 11

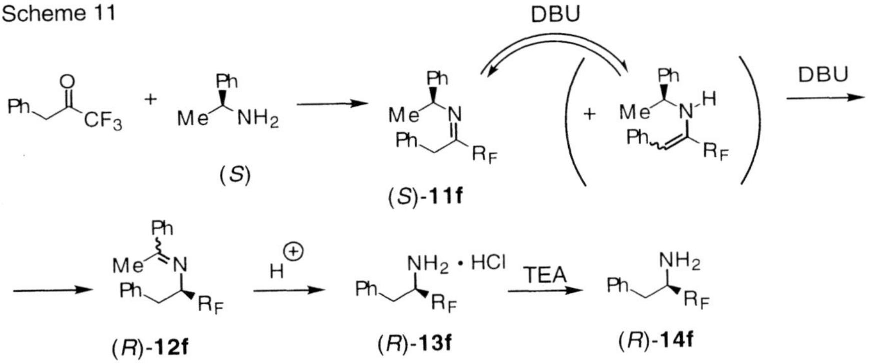

Entry	DBU (equ.)	T, °C	Time, h	Yield, %	ee, %
1	1	50	9	86	88
2	1	80	2	82	85
3	2	21	42	93	88

Finally, we investigated the asymmetric transamination of perfluoroalkyl β-keto carboxylic esters, as a general approach to the stereochemically defined β-amino acids (*41*). An important characteristic of the starting compounds, which can be easily obtained by the direct reaction between the corresponding keto carboxylic esters and phenylethylamine, is that they exist exclusively as (*Z*)-enamines, stabilized by the intramolecular hydrogen bond (Scheme 12). Accordingly, to achieve the target transformation, two sequential 1,3-proton transfers, enamine-azomethine and azomethine-azomethine isomerizations, should take place. Triethylamine was found to be ineffective to assist the isomerizations even under the forced conditions (entry 1). Using DBU, in a catalytic amount, was effective in catalyzing the isomerization however with a poor stereochemical outcome (entry 2). The application of DBU in a molar amount allowed for the reaction to proceed with a reasonable rate at a relatively low temperature giving rise to the desired *Schiff* base in high enantiomeric purity (87% ee) and good chemical yield (entry 3). Further increase of the DBU ratio accelerated the reaction rate and gave the product with better stereochemical outcome (85% yield, 88% ee) (entry 4). Under the same reaction conditions the isomerization of the enamine, bearing a bulkier perfluoropropyl group gave the corresponding Schiff base in excellent (96% ee) enantiopurity albeit with a bit lower chemical yield (entry 5).

Scheme 12

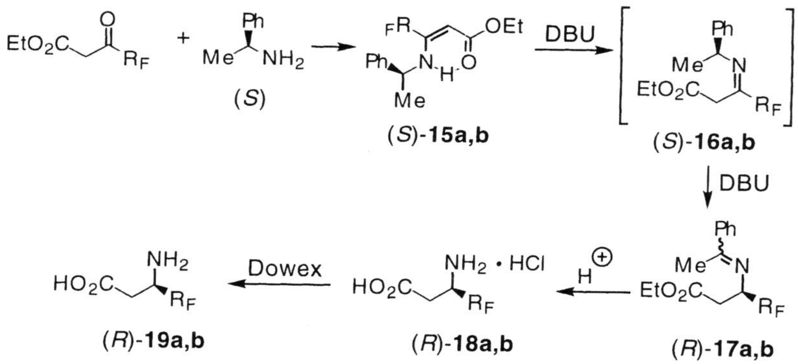

Entry	R_F	Base (equ.)	T, °C	Time, h	Yield, %	ee, %
1	CF_3 (a)	TEA	150	>300	noreaction	
2	CF_3 (a)	DBU (0.1)	125	2	57	50
3	CF_3 (a)	DBU (1)	75	40	78	87
4	CF_3 (a)	DBU (2)	75	24	85	88
5	C_3F_7 (b)	DBU (2)	75	34	57	96

From the data obtained the following important conclusions can be drawn at this stage. First, despite the apparently lower conformational stability of the resultant *Schiff* bases than that of starting imines the isomerization can be conducted with very high enantioselectivity. Second, the application of the DBU in a submolar ratio to the starting imine, as the base and as the solvent, is critical for both the rate and the stereochemical outcome of the reactions. And finally, on the basis of the stereochemical data obtained, we would suggest that the asymmetric outcome of the isomerizations could be described as a function of stereochemical discrimination between the substituents at the imine carbon. Four asymmetric aza-allylic carbanions **A, B, C,** and **D** could be drown as possible intermediates in the isomerizations under study (Scheme 13). Asymmetry to these intermediates can be imparted by ion pairing on only one side of the anion with the conjugate acid of DBU. In other words, the proton abstraction and the following collapse to the new covalent state occur on the same face of the anion. Among these transition states **A** and **D** account for the (R) absolute configuration of the products. Considering geometrical homogeneity of the starting imines, provided by steric demand of the perfluoroalkyl group, and the nature of non-bonding steric interactions in the aza-allylic carbanions, we would suggest that the intermediate **A** might be strongly favored relative to **D** and other possible structures.

Scheme 13

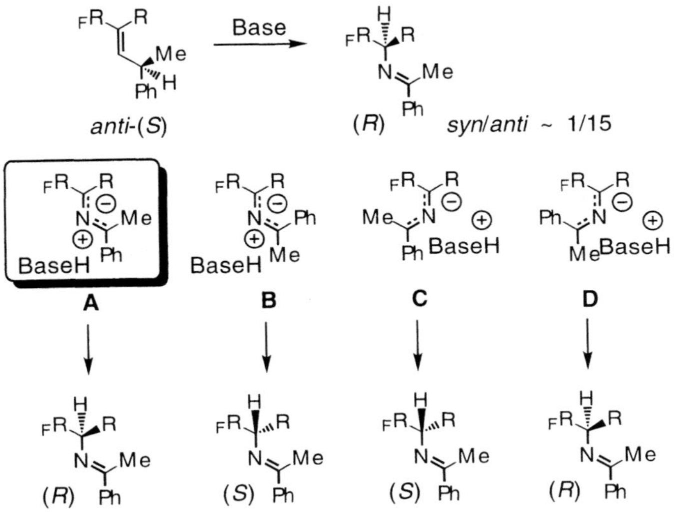

FR R Me Ph H **Base** → FR R N Me Ph

anti-(S)   *(R)*   *syn/anti ~ 1/15*

A   B   C   D

FR R N Me BaseH Ph
FR R N Ph BaseH Me
Me N Ph BaseH
Ph N Me BaseH

↓   ↓   ↓   ↓

FR R N Me (R) Ph
FR R N Me (S) Ph
FR R N Me (S) Ph
FR R N Me (R) Ph

Conclusion

In summary, this study has disclosed a unique example of highly enantioselective [1,3]-proton transfer from a less to a more configurationally unstable stereogenic center. A wide range of synthetic applications of this asymmetric, reducing agent-free transamination for preparing biologically interesting fluoro-amino compounds are readily envisaged. The extreme simplicity of the experimental procedure and the ready availability of all starting materials, combined with substrate generality, high chemical yields and enantiomeric purity of the products, render this biomimetic approach synthetically useful alternative to the methods employing external reducing agents. This study also highlights very intriguing stereodirecting features of perfluoroalkyl group influencing stereochemical result of the isomerizations. Thus this group provides geometric homogeneity of the starting imines and plays the role of enantio-directing substituent in these reactions.

Acknowledgments

The author's deepest gratitude goes to The Science and Technology Agency (STA) for the award of Fellowship Program, which is managed by the Research Development Corporation of Japan (JRDC) in cooperation with the Japan International Science and Technology Exchange Center (JISTEC), allowing continuation of an active research in the field of enantiocontrolled fluoro-organic synthesis.

Literature Cited

1. *Biochemistry Involving Carbon-Fluorine Bonds;* Filler, R., Ed.; ACS Symp. Ser. 28; American Chemical Society: Washington, D.C., 1976.
2. *Fluorinated Carbohydrates,* Taylor, N. F., Ed., ACS Symp. Ser. 374; American Chemical Society: Washington, D.C., 1988.
3. *Organofluorine Compounds in Medicinal Chemistry and Biomedical Applications;* Filler, R.; Kobayashi, Y.; Yagupolskii, L. M., Eds.; Elsevier: Amsterdam, 1993.
4. *Fluorine-containing Amino Acids. Synthesis and Properties;* Kukhar, V. P.; Soloshonok, V. A., Eds.; John Wiley & Sons: Chichester, 1994.

5. *Enantiocontrolled Synthesis of Fluoro-Organic Compounds;* Hayashi, T.; Soloshonok, V. A., Eds.; Tetrahedron: Asymmetry. Special Issue; *Tetrahedron: Asymmetry* **1994**, *5*, issue N 6.
6. Iseki, K.; Kobayashi, Y. In *Reviews on Heteroatom Chemistry*; Oae, S., Ed.; MYU: Tokyo, 1995.
7. *Fluoroorganic Chemistry: Synthetic Challenges and Biomedical Rewards;* Resnati, G.; Soloshonok, V. A., Eds.; Tetrahedron Symposium in Print, **58**; *Tetrahedron* **1996**, *52*, issue N 1.
8. *Biomedical Frontiers of Fluorine Chemistry*; Ojima, I.; McCarthy, J. R.; Welch, J. T., Eds.; ACS Books, American Chemical Society: Washington, D. C., 1996.
9. *Enantiocontrolled Synthesis of Fluoro-Organic Compounds: Stereochemical Challenges and Biomedicinal Targets*, Soloshonok, V. A. Ed.; John Wiley & Sons Ltd., scheduled to appear in 1999.
10. Kukhar, V. P.; Belokon, Y. N.; Svistunova, N. Y.; Soloshonok, V. A.; Rozhenko, A. B.; Kuzmina, N. A. *Synthesis* **1993**, 117.
11. Kukhar, V. P.; Resnati, G.; Soloshonok, V. A. in *Fluorine-containing Amino Acids. Synthesis and Properties;* Kukhar, V. P.; Soloshonok, V. A., Eds.; John Wiley & Sons: Chichester, 1994.
12. Soloshonok, V. A.; Avilov, D. V.; Kukhar', V. P.; Meervelt, L. V.; Mischenko, N. *Tetrahedron Letters* **1997**, *38*, 4903.
13. Soloshonok, V. A. *"Practical Synthesis of Enantiopure Fluoro-Amino Acids of Biological Interest by Asymmetric Aldol Reactions."* In *Biomedical Frontiers of Fluorine Chemistry*; Ojima, I.; McCarthy, J. R.; Welch, J. T., Eds.; ACS Books, American Chemical Society: Washington, D. C., 1996; Chapter 2.
14. Soloshonok, V. A. *"Asymmetric Aldol Reactions of Fluoro-Carbonyl Compounds".* In *Enantiocontrolled Synthesis of Fluoro-Organic Compounds: Stereochemical Challenges and Biomedicinal Targets*, Soloshonok, V. A. Ed., Wiley: Chichester, scheduled to appear in 1999.
15. Soloshonok, V. A.; Avilov, D. V.; Kukhar', V. P.; Meervelt, L. V.; Mischenko, N. *Tetrahedron Letters* **1997**, *38*, 4671.
16. Soloshonok, V. A.; Kacharov, A. D.; Avilov, D. V.; Ishikawa, K.; Nagashima, N.; Hayashi, T. *J. Org. Chem.* **1997**, *62*, 3470.
17. Soloshonok, V. A.; Gerus, I. I.; Yagupolskii, Y. L.; Kukhar, V. P. *Zh. Org. Khim.* **1988**, *24*, 993; *Chem. Abstr. 110*: 134824v.
18. Soloshonok, V. A.; Yagupolskii, Y. L.; Kukhar, V. P. *Zh. Org. Khim.* **1988**, *24*, 1638; *Chem. Abstr. 110*: 154827b.
19. Soloshonok, V. A.; Kirilenko, A. G.: Kukhar, V. P.; Resnati, G. *Tetrahedron Letters* **1994**, *35*, 3119.
20. Ono, T.; Kukhar V. P.; Soloshonok, V. A. *J. Org. Chem.* **1996**, *61*, 6563.
21. Soloshonok, V. A.; Kukhar', V. P. *Tetrahedron* **1997**, *53*, 8307.
22. Soloshonok, V. A.; Kirilenko, A. G.; Kukhar', V. P.; Resnati, G. *Tetrahedron Letters* **1993**, *34*, 3621.
23. Soloshonok, V. A.; Kukhar', V. P. *Tetrahedron* **1996**, *52*, 6953.
24. Soloshonok, V. A.; Kirilenko, A. G.; Fokina, N. A.; Galushko, S. V.; Kukhar, V. P.; Svedas, V. K.; Resnati, G. *Tetrahedron: Asymmetry* **1994**, *5*, 1225.
25. Soloshonok, V. A.; Soloshonok, I. V.; Kukhar, V. P.; Svedas, V. K. *J. Org. Chem.* **1998**, *63*, 1878.
26. Soloshonok, V. A.; Kirilenko, A. G.; Fokina, N. A.; Shishkina, I. P.; Galushko, S. V.; Kukhar, V. P.; Svedas, V. K.; Kozlova, E. V. *Tetrahedron: Asymmetry* **1994**, *5*, 1119.
27. Soloshonok, V. A.; Ono, T. *Synlett*, **1996**, 919.
28. Soloshonok, V. A.; Ono, T. *Tetrahedron*, **1996**, *52*, 14701.
29. *Pyridoxal Catalysis: Enzymes and Model Systems*; Snell, E.E., Braunstein, A.E., Severin, E.S., Torchinsky, Yu.M., Eds.; Interscience: New York, 1968.

30. Layer, R.W. *Chem. Rev.* **1963**, 489.
31. Corey, E. J.; Achiwa, K. *J. Amer. Chem. Soc.* **1969**, *91*, 1429.
32. Caló, V.; Lopez, L.; Todesco, P. E. *J. Chem. Soc. Perlin Trans. I* **1972**, 1652.
33. Babler, J. H.; Invergo, B. J. *J. Org. Chem.* **1981**, *46*, 1937.
34. Buckley, T. F.; Rapoport, H. *J. Amer. Chem. Soc.* **1982**, *104*, 4446.
35. Cainelli, G.; Giacomini, D.; Trerè, A.; Pilo Boyl, P. *J. Org. Chem.* **1996**, *61*, 5134.
36. Smith, P.A.S.; Dang, C.V. *J. Org. Chem.* **1976**, *41*, 2013.
37. Guthrie, R.D.; Jaeger, D.A.; Meister, W.; Cram, D.J. *J. Am. Chem. Soc.* **1971**, *93*, 5137.
38. Jaeger, D.A.; Cram, D.J. *J. Am. Chem. Soc.* **1971**, *93*, 5153.
39. Jaeger, D.A.; Broadhurst, M.D.; Cram, D.J. *J. Am. Chem. Soc.* **1979**, *101*, 717.
40. Soloshonok, V.A.; Ono, T. *J. Org. Chem.* **1997**, *62*, 3030-3031.
41. Soloshonok, V.A.; Ono T.; Soloshonok, I.V. *J. Org. Chem.* **1997**, *62*, 7538-7539.

Chapter 7

Stereoselective and Enantioselective Synthesis of New Fluoroalkyl Peptidomimetic Units: Amino Alcohols and Isoserines

Ahmed Abouabdellah, Jean-Pierre Bégué, Danièle Bonnet-Delpon, Andrei Kornilov, Isabelle Rodrigues, and Truong Thi Thanh Nga

CNRS, URA 1843, Centre d'Etudes Pharmaceutiques, Rue J. B. Clément, 92296 Châtenay-Malabry, France

A concise and stereoselective synthesis of *syn* and *anti* fluoroalkyl β-amino alcohols 1 has been performed in three steps from the cheap ethyl trifluoroacetate. Stereoselectivity arises from stereocontrol in the β-amino ketone reduction. The salen-mediated chiral epoxidation of the 1-trifluoromethyl enol ether 4a led to the epoxy ether in a good enantiomeric excess. Reaction with dimethylaluminum amide, followed by a reduction step provided the non-racemic *anti* amino alcohols 12a and 13a. A stereoselective route to new fluoroalkyl isoserines has been found through the β-lactam chemistry. Cycloaddition of fluoroalkyl imines with benzyloxy ketene provided stereoselectively *cis* fluoroalkyl azetidinones, and corresponding N-Boc isoserinates after ring opening. From chiral imines, non-racemic *syn* and *anti* methyl trifluoromethyl isoserinates could be prepared. These new peptidomimetic units have been used for the design of fluoroalkyl analogues of bioactive compounds.

β-Amino alcohols are important targets which have found use in the treatment of a wide variety of human disorders,[1] as peptidomimetic units and as chiral auxiliaries in organic synthesis.[2] Since the selective introduction of fluorine atoms into a molecule is accompanied by change in physical, chemical and biological properties, fluoroalkyl amino alcohols aroused increasing interest.

The fluorinated moiety can be present either α to the hydroxyl group, or α to the amino group. Regioisomers 1 are much more known than regioisomers 2, because of their use as essential key unit for the synthesis of protease inhibitors. On the contrary, the first reports regarding amino alcohols 2 have been published only very recently.[3-5]

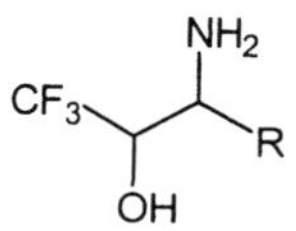

1

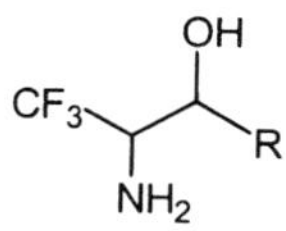

2

© 2000 American Chemical Society

α-Fluoroalkyl β-amino alcohols 1

Fluoroalkyl β-amino alcohols are precursors of the corresponding fluoroalkyl peptidyl ketones which have been shown to be effective inhibitors of proteolytic enzymes,[6] such as serine proteases,[7] (chymotrypsin,[8] elastases,[9,10] trypsin,[11] thrombin,[12]) aspartyl proteases[13,14] or cysteine proteases.[15] In some cases fluoroalkyl β-amino alcohols are themselves inhibitors of the same enzymes.[9,14] This interest in these fluoroalkyl β-amino alcohols 1 aroused efforts for stereoselective and enantioselective synthetic methods.[10a,b,14,16-19] A number of approaches have been reported during these last years, either through the addition of trifluoromethyl equivalent anion to protected aldehyde[16] or through the building block approach.[11,17-19] However the main problem of all these approaches is the diastereoselectivity.

We developed a new easy and versatile diastereoselective synthesis of these amino alcohols. Our synthetic plan was based on the ring opening of trifluoromethyl epoxy ethers **3**, which are easily available in two steps from the cheap ethyl trifluoroacetate, and are useful building blocks. The first step, the Wittig olefination of this ester, is possible because of the increased electrophilicity of the ester carbonyl by the fluorinated moiety (Scheme 1).[20,21]

The second step, the epoxidation of the enol ether **4** has been performed with meta-chloroperbenzoic acid.[21,22] The electron withdrawing character of the CF_3 group stabilizes epoxy ethers towards proteolysis. This stability allows investigations on the oxirane ring opening with various nucleophiles.[18,19,22-24]

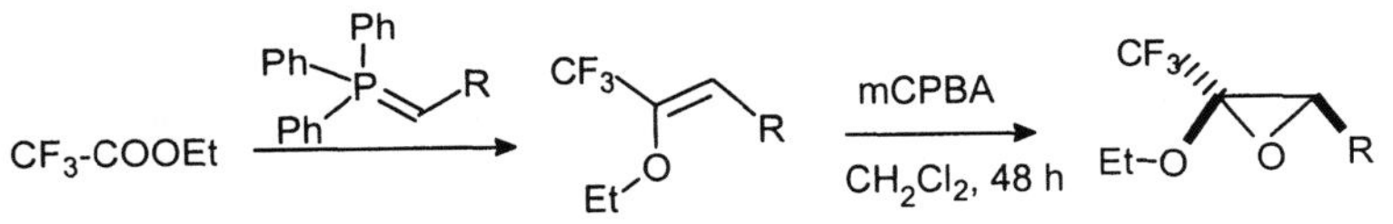

Scheme 1: Preparation of 1-trifluoromethyl epoxy ethers.

An oxirane ring opening of these epoxy ethers by a nitrogen-nucleophile could be a good access to α-amino ketones. Unfortunately, primary amines reacted at high temperature providing N-monosubstituted ketones which, after enolization by 1-4 prototropy, undergo a rapid degradation leading to an unidentified mixture. Probably, the degradation of the unstable produced N-monosubstituted amino ketone is faster than the reaction of epoxy ethers with amines. On the contrary, various secondary amines reacted cleanly.[18a] The addition occurred, as expected, regioselectively on the less hindered site leading to N-disubstituted amino ketones which are stable enough to be further reduced. Reduction with sodium borohydride provided selectively the *syn* amino alcohols.[18a] The stereocontrol of the reduction follows the Felkin-Anh transition state model, where the amino group is the bulkier group. As an example, reaction of **3** ($R = CH_2-CH_2-C_6H_5$) with dibenzylamine, followed by reduction, provided exclusively the *syn* amino alcohol **5**, after debenzylation. However, when the R substituent is the bulky isopropyl group, its steric hindrance competes with the steric hindrance of the amino group, leading to only 75 % of the *syn* isomer (Scheme 2).[18,19,23]

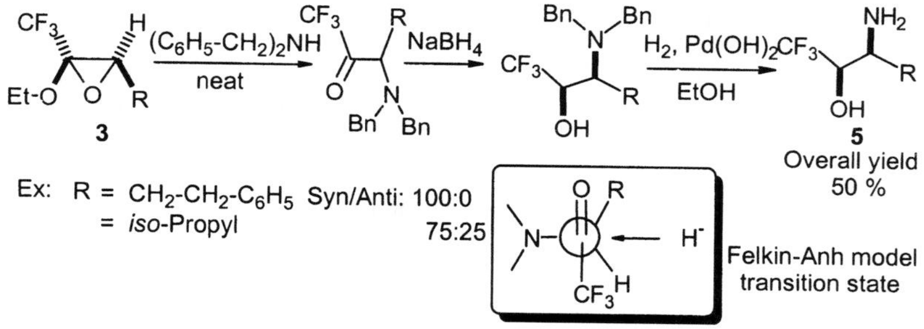

Scheme 2: Reaction of epoxy ethers **3** with secondary amines and reduction.

In order to prepare the *anti* amino alcohols, it was necessary to reverse the control of the reduction of amino ketones, and so to introduce a strong chelation in the transition state. Aluminum amides, as nitrogen-containing agents for the ring opening, appeared to be good candidates for this purpose: first, as Lewis acids, they can facilitate the reaction under mild conditions; second, they can favour the chelation control in the reduction step. Furthermore the aluminum takes the place of the mobile hydrogen involved in prototropy and renders possible the reaction with a primary amine. However, a competitive reaction of the epoxy ether with Lewis acid could also be expected.[26]

Dimethylaluminum amides[25] were prepared in dichloromethane at 0 °C from trimethyl aluminum (Me_3Al) and a primary amine. Epoxy ethers **3** reacted slowly at room temperature with two equivalents of the dimethyl aluminum benzylamide. After 16 h, reduction step was performed *in situ* at - 78 °C with $NaBH_4$ in the presence of ethanol. The use of ethanol as solvent is very important since, in methanol, the reaction failed leading to an inseparable mixture. In ethanol, the *N*-monosubstituted amino alcohols *anti* **6** and *syn* **7** were obtained in good yields (Scheme 3). No trace of products resulting from ring opening by a methyl group of the aluminum reagent could be detected. Furthermore, nucleophilic substitution occurred at C_β, whatever the R group is, unlike reactions with $EtAlCl_2$ and Me_3Al which most often occurred with a $C\alpha$-O bond cleavage.[26] As expected, the diastereoselectivity *anti/syn* is high, ranging from 97/3 (R = C_6H_5) to 73/27 (R = CH_2-C_6H_{13}) (Scheme 3). The *anti* configuration of the major isomer has been determined by NMR data of the corresponding oxazolidinone[15,27] (Scheme 4).

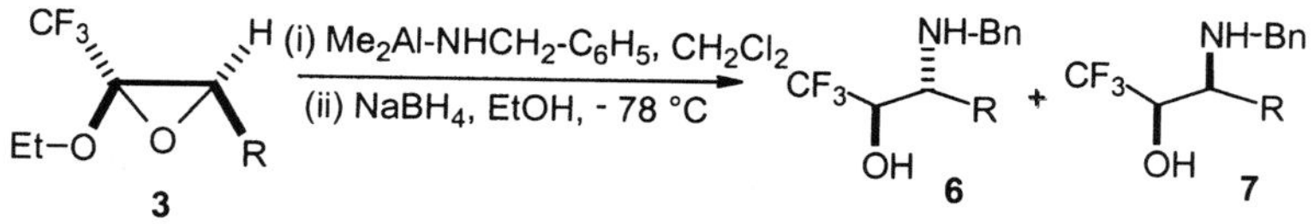

a	R = C_6H_5	75 %	*anti:syn*	97:3
b	R = CH_2-CH_2-C_6H_5	66 %	*anti:syn*	80:20
c	R = CH_2-Cyclohexyl	71 %	*anti:syn*	73:27
d	R = n-Hexyl	62 %	*anti:syn*	80:20

Scheme 3: Reaction of epoxy ethers **3** with dimethylaluminum amide and reduction.

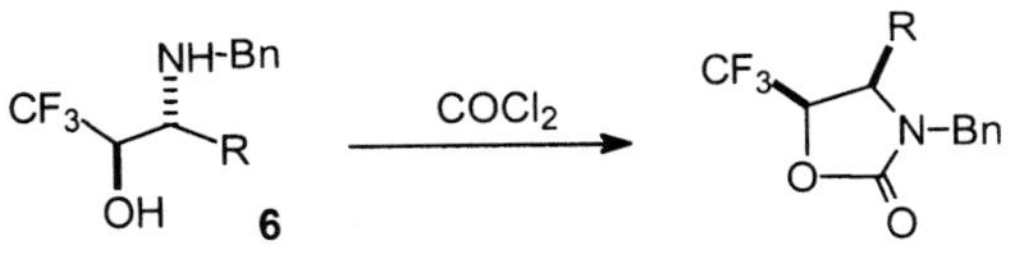

Scheme 4: *Syn* configuration: formation of oxazolidinone.

The selective formation of *anti* diastereoisomers confirms that the aluminum atom allows a chelation control in the reduction reaction. The chelated intermediate is quite stable in dichloromethane. Surprisingly, its ^{13}C NMR spectrum, before addition of $NaBH_4$/ethanol, exhibited no signal corresponding to a ketonic group but a quadruplet at 88 ppm ($^2J_{CF}$ = 26 Hz), indicating a hemiacetal that was supposed to be the intermediate **A** (Figure 1). However, the reactive intermediate is likely the complex **B**, produced immediately on addition of ethanol. Hydride addition takes place on the less hindered face leading to *anti* amino alcohols **6** (Figure 1).

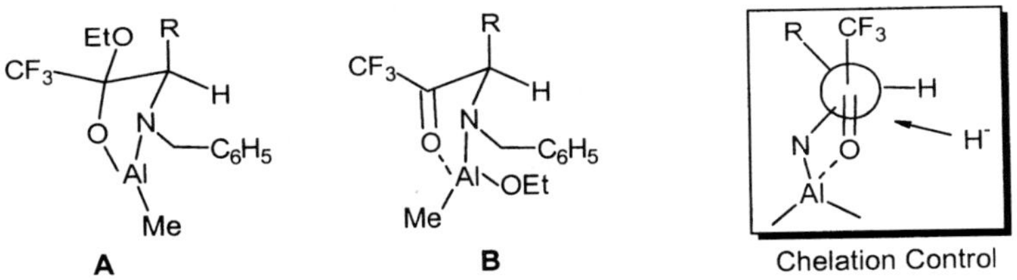

Figure 1 : Stereochemistry of the reduction of amino ketones

So, we were able to prepare selectively *syn* and *anti* trifluoromethyl amino alcohols. The next step was a search for a chiral approach to these compounds. Two approaches have been investigated to obtain chiral *anti* amino alcohols: first we performed the reaction of epoxy ethers **3** with the chiral dimethylaluminum amide, prepared from the *(R)*-phenethylamine and Me_3Al (Scheme 5). From **3a**, the reaction was effective leading, after reduction to the *anti* diastereoisomers **8a** and **9a** stereoselectively (Scheme 5). However, the chiral amine induced no selectivity: *anti* amino alcohols **8a** and **9a** were obtained in a 50/50 mixture. Their separation was performed by crystallisation of the mandelate salts. Although this access to homochiral *anti* amino alcohols is somehow tedious, it is general since oxirane ring opening is efficient whatever the R substituent, and since epoxy ethers, substituted with various fluoroalkyl groups, are available.[22]

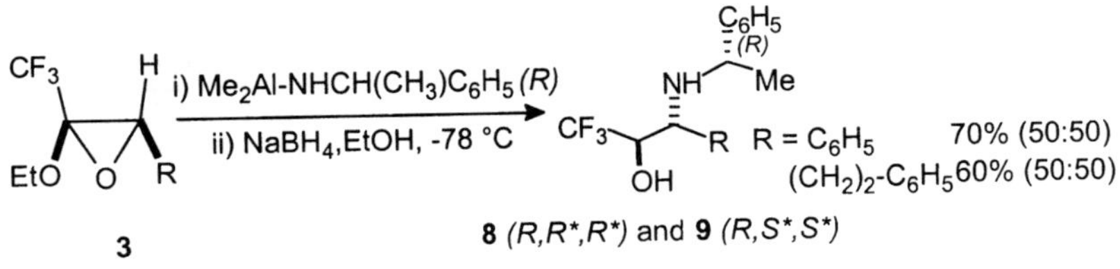

Scheme 5 : Reaction of epoxy ethers with a chiral amide.

A second approach, based on the ring opening of homochiral epoxy ethers **10** and **11** has been investigated. Since the preparation of epoxy ethers from enol ethers bearing a chiral auxiliary was disappointing, we turned to the salen-mediated asymmetric epoxidation, largely developed by Jacobsen[28,29] and by Katsuki.[30] Until now chiral epoxidation of CF_3-substituted double bonds (alkenes, enol ethers) had not been reported. Success of the reaction could, a priori, be limited by the poor The reaction of the enol ether **4a** with the (*R,R*) Mn-salen catalyst, bleach, and 4-phenyl pyridine N-oxide as co-oxidant, under the accurate conditions of pH reported by Jacobsen,[29] has been monitored by GC with an internal standard. The reaction was very slow compared to that of non-fluorinated enol ethers.[31] However epoxy ether **10a** was quite stable in the reaction medium in contrast to non-fluorinated epoxy ethers which could not be isolated in this reaction.[31,32] We noticed that efficiency of the catalyst decreased with the reaction time, and that reaction rate slowed down after some hours. Thus, despite the relative stability of epoxy ether **10a**, degradation partially occurred (Figure 2). The best compromise reaction time was 16 h, with about 50-60 % of conversion. The ee (> 80 %) of the resulting epoxy ether **10a** could be determined by ^{1}H NMR in the presence of the chiral shift reagent Eu(hfc)$_3$. The enantiomeric excess has been confirmed by separation of enantiomers on gas chromatography chiral column. The same reaction performed with the (*S,S*) Mn-salen catalyst led to the epoxy ether **11a**. Enol ether **4b** (CH$_2$-CH$_2$-C$_6$H$_5$) also reacted under the same conditions, but the degradation of the produced epoxy ether **10b** was faster than the disappearance of starting material (Figure 2).

Figure 2 : Chiral epoxidation of enol ethers **4a** and **4b** (% enol ether: ◆, % epoxy ether: ■).

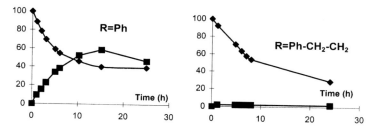

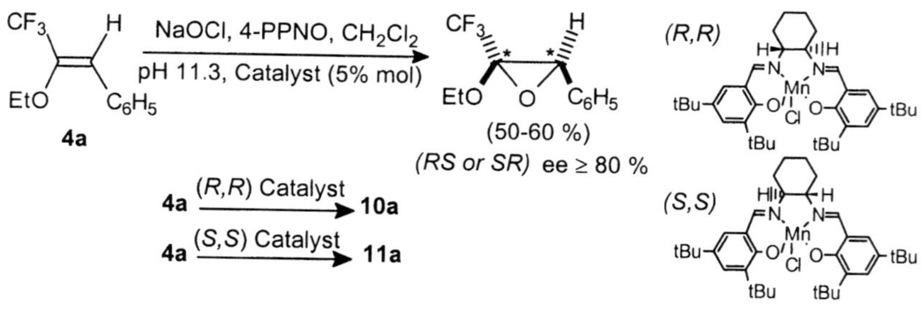

Scheme 6 : Chiral epoxidation of enol ether **4a**.

The preparation of an epoxy ether in high optical purity is not the sole condition for the success of this approach, seeing that α-amino ketones easily undergo enolization and consequently racemization. The stability of complex A of hemiketal form of amino ketone (fig. 1) should prevent from enolization.

Reaction of **10a** and **11a** with the aluminum amide prepared from Me₃Al and the (R)-phenethylamine and subsequent reduction step occurred with the same excellent *anti/syn* diastereoselection as precedently, leading respectively to *anti* amino alcohols **8a** and **9a**. These amino alcohols were obtained in an excellent purity from **10a** (**8a/9a** = 93/7) and from **11a** (**9a/8a** = 90/10). The stereoisomeric excess is the same as the enantiomeric excess of starting epoxy ethers. No racemization occurred in the reaction: ring opening does not involve a carbenium ion, and no enolization occurs from intermediate **A** or **B**. Both enantiomeric amino alcohols **12a** and **13a** were obtained by debenzylation with palladium hydroxide; unfortunately, we have not been able to assign the absolute configuration of the asymmetric carbons of **8a** and **9a** (Scheme 7).

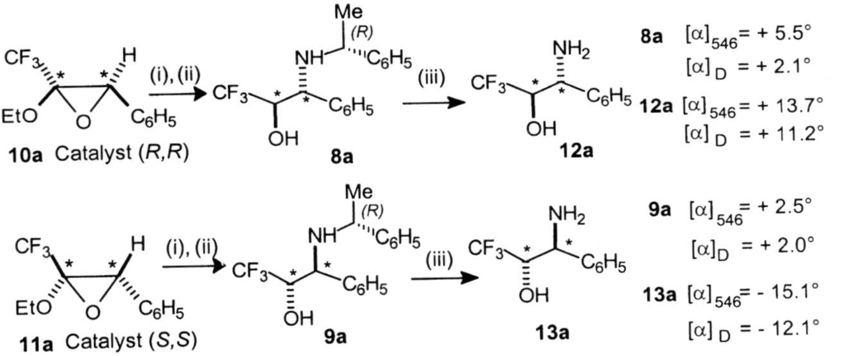

i) Me₂Al-NHCH(CH₃)-C₆H₅ (R), CH₂Cl₂, rt ii) NaBH₄, EtOH, - 78 °C. (iii) Pd(OH)₂/C; H₂

Scheme 7 : Reaction of chiral epoxy ethers with (R) phenethyl amine.

β-Amino β-fluoroalkyl alcohols 2

Due to the lack of easy access, properties of amino alcohols **2** as peptidomimetic units have never been explored although specific features brought by the fluorinated moiety can be expected: for example the presence of the fluoroalkyl group can increase the stability towards non specific proteolysis[33] and strongly weakens the basicity of the amine function, the latter factor reducing the energy requirements for desolvation.[34] Our interest in these amino alcohols was targeted on the unknown and important trifluoromethyl isoserine **14** and **15**. Norstatine, statine, and their analogues have been largely used as peptidomimetic unit in peptide-based inhibitors of aspartyl proteases such as renin and HIV-1 proteases.[35] Fluorinated analogues of these non-proteogenic α-hydroxy β-amino acids could be of great interest, as isosteres of norstatine. Morever, β-substituted isoserine unit is an essential component of taxol derivatives.

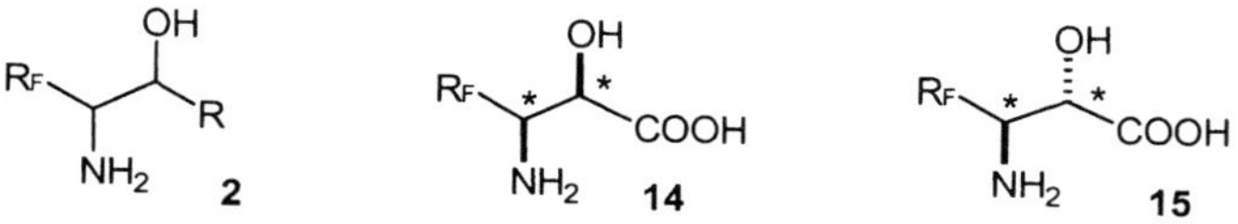

Our approach was based on the ring opening of fluoroalkyl β-lactams. These β-lactams can be obtained either through a [2+2] ketene-imine cycloaddition with a fluoroacetaldimine, or by the cyclocondensation of an alkoxy or silyloxy ester enolate on this same fluoroacetaldimine. The preparation of fluoroalkyl β-lactams **16** (R_F = CF_3, CF_2H and CF_2Cl) and the preparation of homochiral β-lactams and isoserines **14** and **15**, have been have investigated with success by these two approaches.

The Staudinger reaction of ketenes to aldimines is well known to provide *cis*-β-lactams.[36] However, it had never been studied in the case of fluoroalkyl imines. The trifluoroacetaldimine **17a** has been prepared by the usual route from the corresponding ethyl trifluoromethyl hemiketal and *p*-methoxyaniline.[37,38] Reaction of the benzyloxyketene, generated from α-benzyloxyacetyl chloride, with imine **17a**, performed at 45 °C in CH_2Cl_2, provided the expected *cis* azetidinone **16a** in 65 % yield (Scheme 8). In azetidinone **16a**, the $^3J_{H-3,H-4}$ coupling constant of 5 Hz indicates the *cis* relative configuration. Coupling constants in parent non-fluorinated β-lactams have been reported to be 2 Hz for the *trans* isomer and 5-6 Hz for the *cis* one.[39] Condensation of the aldimine **17b** with the same ketene led to the *cis* β-lactam **16b** in 72 % yield. The *cis* β-lactam **16c** was obtained in a satisfactory yield (55 %) from the aldimine **17c** (Scheme 8).

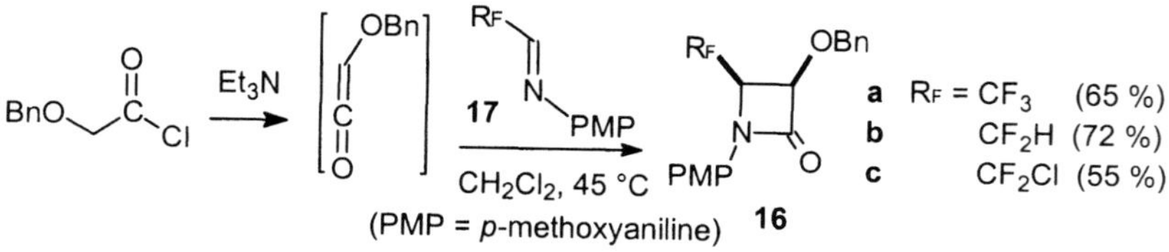

Scheme 8 : Preparation of the *cis*-fluoroalkyl azetidinones **16**.

Classical, but tricky steps of deprotection, protection and azetidinone ring opening have been optimized for the preparation of fluoroalkyl isoserinates, and of properly protected β-lactams **21** for a further coupling (Scheme 9). The best order is first the removal of the 4-methoxyphenyl group with ceric ammonium nitrate (CAN) (65-87 %). The reaction had to be carefully monitored and stopped as soon as azetidinones **16** reacted, because of the fast degradation of the deprotected azetidinones in the presence of excess of CAN. These resulting azetidinones were converted into *N*-Boc derivatives **18a-c**. In the case of **16c**, a low temperature of reaction (- 50 °C) for the protection with Boc, was absolutely required to obtain **18c** in good yield. Only at this stage, the azide-catalyzed ring opening[40] by methanol was performed providing esters **19a-c**. Further debenzylation by catalytic hydrogenation followed by the Boc cleavage led to methyl isoserinates **20a-c**. In these three series, CF_3, CF_2H, CF_2Cl, each step was performed in fairly good yields (Scheme 9). Protected azetidinones **21a-c** were prepared by reductive cleavage of benzyl group in azetidinones **18** and protection of the hydroxyl group by reaction with ethyl vinyl ether (EVE) in presence of *p*-toluene sulfonic acid (Scheme 9).

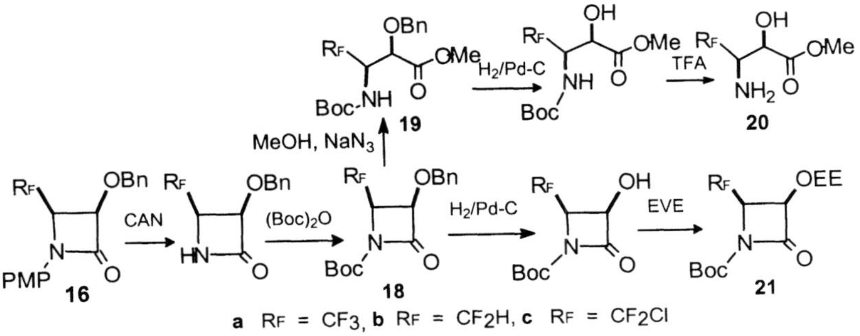

Scheme 9 : Preparation of methyl isoserines **20** and protected azetidinones **21**.

Azetidinones **18** can been used to prepare tri- and tetrapeptide isosteres **22**. As example, cleavage of the azetidinone ring has been performed with esters of simple amino acids such as tryptophan, phenylalanine or esters of dipeptides. Interestingly, these very simple compounds exhibit an inhibition of HIV-1 protease with an IC_{50} of about micromolar range. So without any recognising element on the N-terminal site and no accurate design on the C-terminal part the activity is significant.

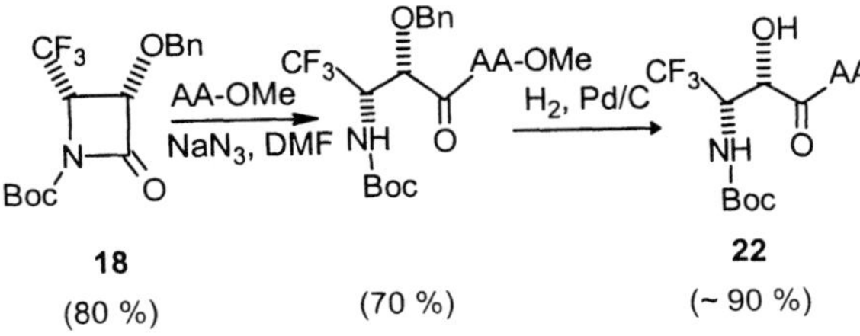

AA-OMe	IC_{50} (μM)
Leu-OMe	50
Val-Phe-OMe	50
Phe-Phe-OMe	1
Tryp-OMe	3
Tyr-OMe	7
Phe-OMe	9

HIV-1 Protease Inhibition

Scheme 10 : Trifluoromethyl isoserine derivatives, HIV-1 protease inhibition.

Azetidinones **21** are suitable synthons for coupling with alcohols, such as baccatin III, in view of the preparation of fluorinated docetaxel analogues. β-Lactam **21a** has been coupled with the properly protected baccatin III, in the group of Prof. Ojima providing an analogue of docetaxel, where the 3'-phenyl group has been replaced by a CF_3 group. The compound presents a higher cytotoxicity *in vitro* towards human tumoral cell lines than docetaxel.[41,42] As already observed in the baccatin series, the reaction occurred with a good kinetic resolution. Using 2 equivalents of the racemic β-lactam **21a**, only the *(3R,4S)*-isomer reacted, leading to the CF_3 analogue of docetaxel, with the natural configuration of the isoserine chain. Coupling with azetidinones **21b** (CF$_2$H) and **21c** (CF$_2$Cl) are under investigation in Stony Brook.

In order to prepare chiral 4-CF$_3$ azetidinones, we first studied the cyclocondensation of chiral ester enolates with trifluoroacetaldimine **17a**, since the lithium ester enolate-imine cyclocondensation has been demonstrated to occur in high yields and with high enantiomeric purity.[36b,43,46] However, the ester cyclocondensation of enolate of benzyloxy esters with the trifluoroacetaldimine **17a** failed.[47] We finally succeeded in the preparation of the *cis* β-lactam **23** by using the more reactive triisopropylsilyloxy esters **24**, as reported by Ojima[46] (Scheme 11). β-Lactams obtained in this reaction have in most cases the *cis* configuration.[44,45] Results were disappointing when reaction was performed with chiral esters **24b,c** : the azetidinone **23** was not obtained from **24b**, and yield from **24c** was very low.

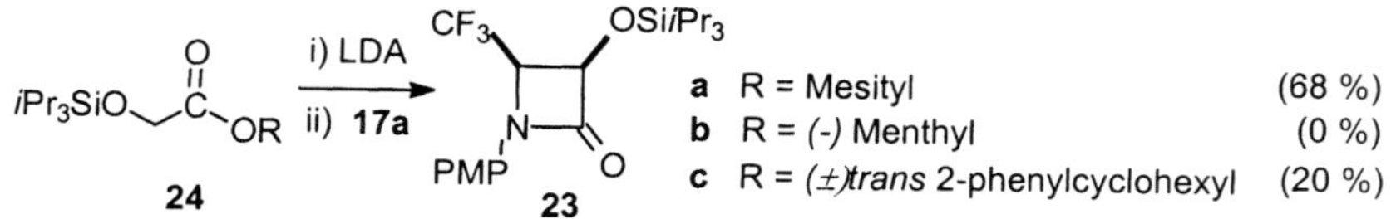

a	R = Mesityl	(68 %)
b	R = *(-)* Menthyl	(0 %)
c	R = *(±)trans* 2-phenylcyclohexyl	(20 %)

Scheme 11 : Cyclocondensation route to azetidinone **23**.

Although the asymmetric Staudinger reaction, controlled by a chiral *N*-substituent on imines, is seldom diastereoselective, because of the relatively long distance between chiral and reaction centers,[47] we turned all the same towards the [2+2] ketene-imine cycloaddition with the chiral trifluoroacetaldimine **25**, prepared from trifluoroacetaldehyde hemiketal and the (*S*)-phenethylamine. The reaction of benzyloxyketene with imine **25** was efficient leading to the mixture of azetidinones **26** and **27** in 90 % yield. High *cis/trans* stereoselectivity was observed, with only 3-5 % of *trans* azetidinones formed. As expected, the chirality transfer was low with a diastereoisomeric excess of only 10 %. Fortunately, it was possible to easily separate the two diastereoisomers by crystallization in ethanol of the crude mixture. Stereoisomer **26** crystallized in ethanol and was obtained in an excellent diasteroisomeric purity (> 99 %). Stereoisomer **27** could be isolated in 95 % diastereoisomeric purity after SiO$_2$ chromatography and crystallization (Scheme 12).

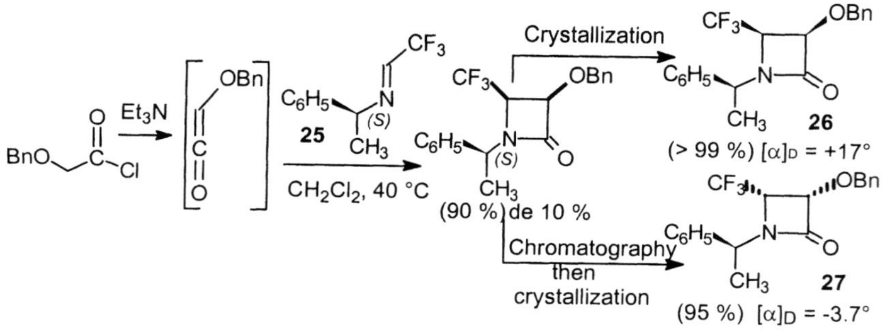

Scheme 12 : Non racemic trifluoromethyl azetidinones **26** and **27**.

Azetidinones **26** and **27** underwent the acidic methanolysis leading to methyl isoserinates **28** and **29**. The X-ray diffraction diagram of crystals of **28** indicated unambiguously the configuration *(2R,3R)* for this isomer.[47] Catalytic debenzylation of isoserinates **28** and **29** in the presence of (Boc)$_2$O provided the two pure enantiomers of methyl *syn* N-Boc isoserinates **30** and **31** (Scheme 13).

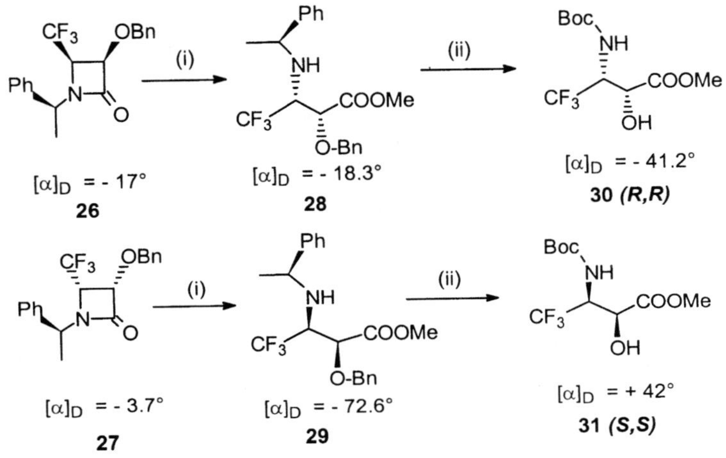

(i) HCl, MeOH; (ii) H$_2$, Pd(OH)$_2$/C, (Boc)$_2$O

Scheme 13 : Homochiral *syn* methyl 3-trifluoromethyl isoserinates **30** and **31**.

Pure enantiomers of protected *anti* methyl isoserinates **34** and **35** could be prepared in good yields through isomerisation of the *syn* β-lactams **26** and **27**, under basic conditions, to *trans* β-lactams **32** and **33**, followed by acidic methanolysis, and reductive cleavage (Scheme 14).

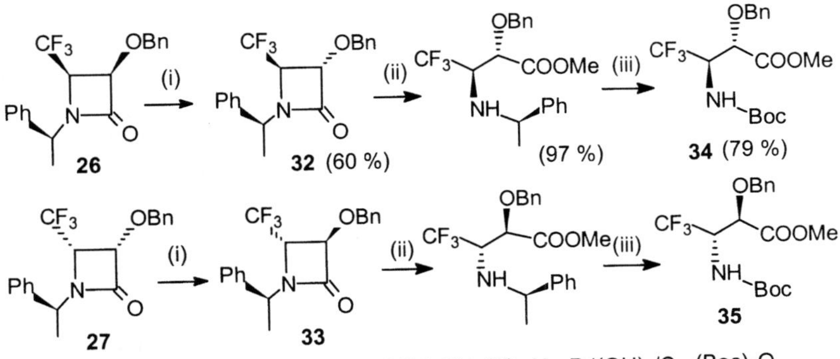

(i) t-BuOK, CH$_3$CN; (ii) HCl/MeOH; (iii) H$_2$, Pd(OH)$_2$/C, (Boc)$_2$O

Scheme 14 : *Anti* methyl 3-trifluoromethyl isoserinates **34** and **35**.

Alkylation of the azetidinone **26** is also efficient. The lithium enolate of **26** could be generated with the use of lithium hexamethyldisilyl amide, and various electrophiles reacted to provide the substituted azetidinones **36**. During this reaction the relative configuration of CF_3 and benzyloxy groups was conserved. The ratio of the *trans* minor isomer was about only 0-20 %. This stereochemistry results from the α-entrance of the electrophile by the less congested face, and is only slightly dependent on the bulkiness of the electrophile. These new substituted β-lactams **36** offer an access to new fluoroalkyl analogues of peptidomimetic units, and of taxotere side chain.

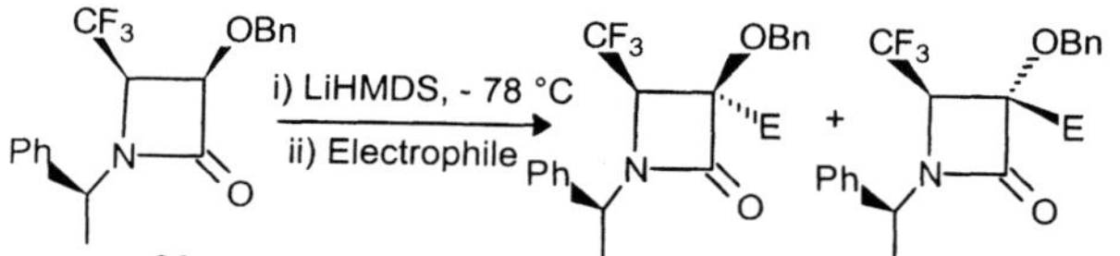

Electrophile	% *Cis:Trans*
H_3O^+	(100) 90:10
Me-S-S-Me	(61) 100:0
ICH_3	(80) 90:10
CH_3-CH_2-CHO	(74) 80:20

Scheme 15 : Alkylation of β-lactam **26**.

Conclusion

This study provides a stereoselective route to *syn* and *anti* β-amino trifluoromethyl alcohols and to *syn* and *anti* trifluoromethyl isoserines. These new preparations allow an access to enantiomerically pure fluoroalkyl amino alcohols which can be used as peptidomimetic units for the design of fluoroalkyl analogues of bioactive compounds. These methods can be extended to different patterns of substitution (Rf and R).

Acknowledgment: We thank Drs L. Allain, F. Benayoud, N. Fischer-Durand, H. Sdassi, F. Chorki, and C. Richard for their contributions and Michèle Ourévitch for NMR experiments. The authors thank Fondation pour la Recherche Médicale (SIDACTION) and MESRT for postdoctoral fellowships (A.A. and A.K.), CNRS and MRT for PhD grant (N.T.T.T. and I.R). We thank also Dr Masamichi Maruta (Central Glass Company) for his generous gift of chemicals, Mrs Sabine Halut (Laboratoire des Métaux de Transition, Université P. & M. Curie) for X-ray diffraction structure. We are pleased to thank Professor Iwao Ojima for fruitful discussions. This study was partially supported by Biomed 2 and INTAS (Bioactive Fluoroorganic Compounds: Asymmetric synthesis, Biotransformations, Molecular Recognition) European Community programs.

References:

1. Grayson, M. Ed. *Kirk-Othmer Encycl. Chem. Techno.* **1982**, *17*, 311-345.
2. Tomioka, K. *Synthesis* **1990**, 541; Noyori, R.; Kitamura, M. *Angew. Chem. Int. Ed. Engl.* **1991**, *30*, 49-68.

3. Marti, R.E.; Heinzer, J.; Seebach, D. *Liebigs Ann.* **1995**, 1193-1215.

4 Bravo, P.; Farina, A.; Kukhar, V.; Markosky, A.; Meille, S.; Soloshonok, V.; Sorochinsky; A.; Viani, F.; Zanda, M.; Zappala, C. *J. Org. Chem.* **1997**, *62*, 3424-3425.

5. Tanaka, K.; Ishiguro, Y.; Mitsuhashi, K. *Bull. Chem. Soc. Jpn* **1993**, 66, 661-663. Uneyama, K.; Hao, J.; Amii, H. *Tetrahedron Lett.* **1998**, *39*, 4079-4082.

6. Imperiali, B. *Synthetic Peptides in Biotechnology*, A. R. Liss, Inc. **1988**, 97-129.

7. Imperiali, B.; Abeles, R.B. *Biochemistry* **1985**, *24*, 1813-1817; Imperiali, B.; Abeles, R.B. *Biochemistry* **1986**, *25*, 3760-3767.

8. Brady, K.; Liang, T.C.; Abeles, R.H. *Biochemistry* **1989**, *28*, 9066-9070; Liang, T.C.; Abeles, R.H. *Biochemistry* **1987**, *26*, 7603-7608; Brady, K.; Wei, A.; Ringe, D.; Abeles, R.H. *Biochemistry* **1990**, *29*, 7600-7607; Brady, K.; Abeles, R.H. *Biochemistry* **1990**, *29*, 7608-7617.

9. (a) Dunlap, R.P.; Stone, P.J.; Abeles, R.H. *Biochem. Biophys. Res. Commun.* **1987**, *145*, 509-513; (b) Stein, R.L.; Strimpler, A.M.; Edwards, P.D., Lewis, J.J.; Mauger, R.C.; Schwartz, J.A.; Stein, M.M.; Trainor, D.A.; Wildonger, R.A.; Zottola, M.A. *Biochemistry* **1987**, *26*, 2682-2689; (c) Govardhan, C.P.; Abeles, R.H. *Arch. Biochem. Biophys.* **1990**, *280*, 137-146; (d) Warner, P.; Green, R.C.; Gomes, B.; Strimpler, A.M. *J. Med. Chem.* **1994**, *37*, 3090-3099; (e) Berstein, P.R.; Gomes, B.C.; Kosmider, B.J.; Vacek, E.P.; Williams, J.C. *J. Med. Chem.* **1995**, *38*, 212-215; (f) Veale, C.A.; Damewood, J.R.; Steelman, G.B.; Bryant, C., Gomes, B.; Williams, *J. Med. Chem.* **1995**, *38*, 86-97; (g) Brown, F.J.; Andisik, D.W.; Berstein, P.R.; Bryant, C.B.; Ceccarelli, C.; Damewood, J.R.; Edwards, P.D.; Earley, R.A.; Feeney, S.; Green, R.C.; Gomes, B.; Kosmider, B.J.; Krell, R.D.; Shaw, A.; Steelman, G.B.; Thomas, R.M.; Vacek, E.P.; Veale, C.A.; Tuthill, P.A.; Warner, P.; Williams, J.C.; Wolanin, D.J.; Woolson, S.A. *J. Med. Chem.* **1994**, *37*, 1259-1261.

10. (a) Peet, N.P.; Burkhart, J.P.; Angelastro, M.R.; Giroux, E.L.; Mehdi, S.; Bey, P.; Kolb, M.; Neises, B.; Schirlin, D. *J. Med. Chem.* **1990**, *33*, 394-407; (b) Skiles, J.W.; Fuchs, V.; Miao, C.; Sorcek, R.; Grozinger, K.G.; Mauldin, S.C.; Vitous, J.; Mui, P.W.; Jacober, S.; Chow, G.; Matteo, M.; Skoog, M.; Weldon, S.M.; Possanza, G.; Keirns, J.; Letts, G.; Rosenthal, A.S. *J. Med. Chem.* **1992**, *35*, 641-662; (c) Angelastro, M.R., Baugh, L.E.; Bey, P.; Burkhart, J.P.; Chen, T.M.; Durham, S.L.; Hare, C.M.; Huber, E.W.; Janusz, M.J.; Koehl, J.R.; Marquart, A.L.; Mehdi, S.; Peet, N.P. *J. Med. Chem.* **1994**, *37*, 4538-4554.

11. Ueda, T.; Kam, C.M.; Powers, J.C. *Biochem. J.* **1990**, *265*, 539-545.

12. Neises, B.; Ganzhorn, A. Eur. Pat. Appl. EP 503,203, **1992**; *Chem. Abstract* **1993**, *118*, 148063y.

13. (a) Sham, H.L.; Stein, H.; Rempel, C.A.; Cohen, J.; Plattner, J.J. *FEBS Lett.* **1987**, *220*, 299-301; (b) Tarnus, C.; Jung, M.J.; Rémy, J.M.; Baltzer, S.; Schirlin, D. *FEBS Lett.* **1989**, *249*, 47-50; (c) Thaisrivongs, S.; Pals, D.T.; Turner, S.R. in *Selective Fluorination in Organic and Bioorganic Chemistry*, J. T. Welch Ed., ACS Symposium Series 456, Washington D.C. **1991**, pp 164-173.

14. Patel, D.V.; Rielly-Gauvin, K.; Ryono, D.E. *Tetrahedron Lett.* **1988**, *29*, 4665-4668; Patel, D.V.; Rielly-Gauvin, K.; Ryono, D.E.; Free, C.A.; Smith, S.A.; Petrillo Jr, E.W. *J. Med. Chem.* **1993**, *36*, 2431-2447.

15. (a) Giordano, C.; Gallina, C.; Consalvi, V.; Scandurra, R. *Eur. J. Med. Chem.* **1989**, *24*, 357-362; (b) Smith, R.A.; Copp, L.J.; Donnelly, S.L.; Spencer, R.W.; Krantz, A. *Biochemistry* **1988**, *27*, 6568-6573.

16. (a) Bergeson, S.H.; Schwartz, J.A.; Stein, M.M.; Wildonger, R.A.; Edwards, P.D.; Shaw, A; Trainor, D.A.; Wolanin, D.J. U.S. Pat. 4,910,190; **1990**; (b) *Chem. Abstr.* **1991**, *114*, 120085m. (c) Edwards, P.D. *Tetrahedron Lett.* **1992**, *33*, 4279-4282. (d) Kolb, M.; Neises, B. *Tetrahedron Lett.* **1986**, *27*, 4437-4440; (d) Walter, M.W.; Adlington, R.M.; Baldwin, J.E.; Schofield C.J. *J. Org. Chem.* **1998**, *63*, 5179-5179.

17. Imperiali, B.; Abeles, R.H. *Tetrahedron Lett.* **1986**, *27*, 135-138; Kolb, M.; Barth, J.; Neises, B. *Tetrahedron Lett.* **1986**, *27*, 1579-1582; Kolb, M.; Neises, B.; Gerhart, F. *Liebigs Ann. Chem.* **1990**, 1-6.

18. (a) Bégué, J.P.; Bonnet-Delpon, D.; Sdassi, H. *Tetrahedron Lett.* **1992**, *33*, 1879-1882. (b) Abouabdellah, A.; Bégué, J.P.; Bonnet-Delpon, D.; Rodriguez, I.; Kornilov, A.; Richard, C. *J. Org. Chem.* **1998**, *63*, 6529.

19. Bégué, J.P.; Bonnet-Delpon, D.; Fischer-Durand N.; Reboud-Raveaux, M.; Amour, A. *Tetrahedron: Asymmetry* **1994**, *5*, 1099-1110.

20. Bégué, J.P.; Mesureur, D. *J. Fluorine Chem.* **1988**, *39*, 271; Bégué, J.P.; Bonnet-Delpon, D.; Née, G.; Wu, S.W. *J. Org. Chem.* **1992**, *57*, 3807.

21. Bégué, J.P.; Bonnet-Delpon, D.; Kornilov, A. *Organic Synth.* **1998**, *75*, 153-160.

22. Bégué, J.P.; Benayoud, F.; Bonnet-Delpon, D.; Fischer-Durand, N.; Sdassi, H. *Synthesis*, **1993**, 1083-1085.

23. Bégué, J.P.; Bonnet-Delpon, D. in *Biomedical Frontiers of Fluorine Chemistry*, ACS symposium Series, vol. 639, Ojima, I.; McCarthy, J.R.; Welch, J.T. Eds, Washington DC, Chapter 4, **1996**, pp 59-72.

24. Bégué, J.P.; Bonnet-Delpon, D.; Kornilov, A. *Synthesis* **1996**, 529-531.

25. (a) Levin, J.I.; Turos, S.M.; Weinreb S.M. *Synth. Commun.* **1982**, *12*, 989-993; (b) Overman, L.E. Flippin, L.A. *Tetrahedron Lett.* **1981**, *22*, 195-198.

26. Bégué, J.P.; Bonnet-Delpon, D.; Benayoud, F. *J. Org. Chem.* **1995**, *60*, 5029-5036.

27. Sham, H.L.; Trempel, C.A.; Stein, H.; Cohen, J.J. *J. Chem. Soc. Chem. Comm.* **1990**, 904-905.

28. Jacobsen, E.N.; Zhang, W.; Muci, A.R.; Ecker, J.R.; Deng, L. *J. Am. Chem. Soc.* **1991**, *113*, 7063-7064.

29. Brandies, B.D.; Jacobsen, E.N. *J. Org. Chem.* **1994**, *59*, 4378-4380.

30. Katsuki, T. *J. Synth. Org. Chem. Jpn* **1995**, *53*, 940-951.

31. Fukuda, T.; Katsuki, T. *Tetrahedron Lett.* **1996**, *37*, 4389-4392.

32 Adam, W.; Fell, R.T.; Mock-Knoblauch, C.; Saha-Möller, C. *Tetrahedron Lett.* **1996**, *37*, 4389-4392.

33 Ojima, I.; Kato, K.; Jameison, F.A.; Conway, J. *Bioorganic Med. Chem. Lett.* **1992**, *2*, 219-222.

34. Schirlin, D.; Tarnus, C.; Baltzer, S.; Rémy, J.M. *Bioorg. Med. Chem. Lett.* **1992**, *2*, 651-654.

35. Rich, D.H. *J. Med. Chem.* **1985**, *28*, 263. Moore, M.L.; Dreyer, G.B. *Perspect. Drug Dis. Design* **1993**, *1*, 85-108. Gait, M.J.; Karn, J. *TIBTECH*, **1995**, *13*, 430-437.

36. (a) Georg, G.I.; Ravikumar, V.T. In *The Organic Chemistry of β-Lactams*; Georg, G.I. Ed.; VCH Publishers: New York, 1992, pp 295-368 and references cited therein. (b) Ojima, I. In *The Organic Chemistry of β-Lactams*; Georg, G.I. Ed.; VCH Publishers: New York, 1992, pp 197-255.

37 Guanti, G.; Banfi, L.; Narisano, E; Scolastico, C.; Bosone, E. *Synthesis* **1985**, 609-611.

38. Kaneko, S.; Yamazaki, T.; Kitazume, T. *J. Org. Chem.* **1993**, *58*, 2302-2312.

39. Browne, M.; Burnett, D.A.; Caplen, M.A.; Chen, L.Y.; Clader, J.W.; Domalsky, M.; Dugar, S.; Pushpavanam, P.; Sher, R.; Vaccaro, W.; Zhao, H. *Tetrahedron Lett.* **1995**, *36*, 2555-2558.

40. Palomo, C.; Aizpurua, J.M.; Galarza, R.; Mielgo, A. *J. Chem. Soc. Chem. Comm.* **1996**, 633-634.

41. Ojima, I.; Slater, J.C. Pera, P.; Veith, J.M.; Abouabdellah, A.; Bégué, J.P.; Bernacki, R.J. *Bioorganic Med. Chem. Lett.* **1997**, *7*, 133-138.

42. Ojima, I.; Kuduk, S.D.; Slater, J.C.; Gimi, R.H.; Sun, C.M.; Chabravarty, S.; Ourevitch, M.; Abouabdellah, A.; Bonnet-Delpon, D.; Bégué, J.P.; Veith, J.M.; Pera, P.; Bernacki, R.J. chapter 17 in "*Biomedical Frontiers of Fluorine Chemistry*", Ojima, I.; McCarthy, J.R.; Welch, J.T. Ed.; A.C.S.: Washington, D.C. 1996, pp 228-245.

43. For reviews, Hart, D.J.; Ha, D.C. *Chem. Rev.* **1989**, *89*, 1447; Ojima, I. *Account Chem. Res.* **1995**, *28*, 383-389 and references cited herein.

44. (a) Ojima, I.; Habus, I. *Tetrahedron Lett.* **1990**, *46*, 3841. (b) Georg, G.I.; Kant, J.; Gill, H.S. *J. Am. Chem. Soc.* **1987**, *109*, 1129.

45. (a) Ojima, I.; Habus, I.; Zhao, M.; Zucco, M.; Park, Y.H.; Sun, C.M.; Brigaud, T. *Tetrahedron* **1992**, *48*, 6985-7012. (b) Ojima, I.; Habus, I.; Zhao, M.; Georg, G.I.; Jayasinghe, L.R. *J. Org. Chem.* **1991**, *56*, 1681-1683.

46. Ojima, I.; Park, Y.H.; Sun, C.M.; Brigaud, T.; Zhao, M. *Tetrahedron Lett.* **1992**, *33*, 5737-5740.

47. Abouabdellah, A.; Bégué, J.P.; Bonnet-Delpon, D. *Synlett* **1996**, 399-400. Abouabdellah, A.; Bégué, J.P.; Bonnet-Delpon, D.; T. T. Thanh Nga *J. Org. Chem.* **1997**, *62*, 8826-8833; *ibidem* **1998**, *63*, 5294.

48. Hashimoto, Y.; Kai, A.; Saigo, K. *Tetrahedron Lett.* **1995**, *48*, 8821-8824. Georg, G.I.; Wu, Z. *Tetrahedron Lett.* **1994**, *47*, 381-384.

Chapter 8

Asymmetric Synthesis of Fluoroalkyl Amino Compounds via Chiral Sulfoxides

Pierfrancesco Bravo[1], Luca Bruché[2], Marcello Crucianelli[2], Fiorenza Viani[1], and Matteo Zanda[2]

[1]CNR-Centro di Studio sulle Sostanze Organiche Naturali, via Mancinelli 7, I–20131 Milano, Italy
[2]Dipartimento di Chimica, Politecnico di Milano, via Mancinelli 7, I–20131 Milano, Italy

A wide range of biologically interesting, structurally diverse, chiral and stereochemically defined fluoro-amino compounds have been synthesized in nonracemic form, exploiting the manifold reactivity of the stereogenic sulfinyl group, as a primary source of chirality. Fluorinated and fluoroalkylated analogs of chiral amines, β-amino alcohols, α-amino acids, β-hydroxy α-amino acids, β-amino α-hydroxy acids, γ-amino β-hydroxy acids, *ephedra* and tetrahydroisoquinoline alkaloids were stereoselectively obtained from easily available and nonexotic starting materials, such as fluorinated acetic or pyruvic esters, a cheap and easy-to-handle source of fluorine. Some new asymmetric reactions disclosed during the latest five years are described.

Because of the profound stereoelectronic modifications brought about by incorporation of fluorine into organic molecules, the reactivity of fluoro-organic compounds is intriguing, exciting and often unpredictable (*1*). Additional interest, fascination and scientific challenge stem from the introduction of the stereochemical factor (*2,3*). Finally, on considering the well known importance of fluorinated fine chemicals and materials in many fields of science, such as engineering and medicine, and therefore of life (*4-6*), one can safely state that the chemistry of chiral fluoro-organic compounds holds strong promises for the future and is therefore worth of intensive scientific investigation. In this scenario fluorinated chiral amino compounds can play a major role. In fact, chiral compounds incorporating amino functions have proven very useful building blocks in pharmaceutical field, as witnessed by the large number of drugs belonging to the family of amino compounds, in agrochemical field and in material science, for example liquid crystals and polymers. Fluorinated counterparts feature a great potential, but the serious synthetic difficulties connected with their stereocontrolled preparation have strongly limited the availability, study and use of chiral fluoro-amino compounds (*3*). Scope of this review is to present and discuss the stereocontrolled preparation of structurally varied chiral nonracemic fluoro-amino compounds using a stereogenic sulfinyl group as chiral auxiliary (*7,8*). Most synthetic routes were carried out using original methodologies as key steps, as a result of the studies carried out by our group during the latest five years (*9*).

The first section of this chapter describes the preparation and several synthetic applications of α-fluoroalkyl β-sulfinyl enamines and imines; the second deals with the chemistry of di- and trifluoropyruvaldehyde *N,S*-ketals, stereochemically stable synthetic equivalents of β-di and β-trifluoro α-amino aldehydes, which can be prepared from the corresponding β-sulfinyl enamines; the third overviews the preparation of chiral sulfinimines of trifluoropyruvate and their use to prepare a library of α-trifluoromethyl (Tfm) α-amino acids; the fourth section is mainly dedicated to the asymmetric synthesis of monofluorinated amino compounds, using a miscellany of methods such as *Mitsunobu*-like azidation of β-hydroxy sulfoxides, ring opening of fluoroalkyl epoxides with nitrogen-centered nucleophiles and 1,3-dipolar cycloadditions with chiral fluorinated dipolarophiles.

Preparation and Use of α-Fluoroalkyl β-Sulfinyl Enamines and Imines.

Our general strategy to prepare nonracemic fluoro amino compounds using the sulfinyl group as chiral auxiliary is summarized in Scheme 1. α-Fluoroalkyl β-sulfinyl imines **C** (and their enamine tautomers, which prevail when PG = Cbz or H) can be prepared by direct assembling of methyl aryl sulfoxide **A** with the fluoroalkylimine framework **B** by C-C bond forming (path 1), or alternatively using the β-ketosulfoxide precursor **D**, prepared from **A** and fluoroacetic esters **E**, by C=N bond forming (path 2). Three entries to the saturated reduced counterparts of **C**, namely the β-amino sulfoxides **F**, have been developed: reduction of the C=N bond of **C** (path 3) provides polyfluorinated α-unsubstituted derivatives **F** (R^1 = H); condensation between aryl alkyl sulfoxides **G** and fluorinated imines **H** represents a general entry to polyfluorinated β-amino sulfoxides **F** (path 4); *Mitsunobu*-like amination of monofluorinated β-hydroxy sulfoxides **J** (R$_F$ = CH$_2$F), obtained by reduction of keto derivatives **D**, leads to the corresponding β-amino sulfoxides **F** (path 5). The condensation between sulfoxides **G** and fluoro imines **H** (path 4), as well as a detailed review of the powerful methodology to transform stereoselectively the resulting β-amino sulfoxides **F** into the corresponding sulfur-free β-amino alcohols (the non-oxidative *Pummerer* reaction), are covered in the chapter entitled "Stereoselective Synthesis of β-Fluoroalkyl β-Amino Alcohol Units".

Two efficient entries to the title compounds have been developed (Scheme 2).

Route A to synthesize fluorinated β-sulfinyl imines was first disclosed in our laboratories (*10*), then developed jointly with the group of *Fustero* (*11*). Addition of lithiated aryl methyl sulfoxides **1** to fluoroacetimidoyl chlorides **2**, prepared one-pot from fluoroacetic acids and amines by the method of *Uneyama* (*12,13*), produced high yields of the corresponding fluorinated α-sulfinyl *N*-alkyl/aryl imines. The method is efficient and features a rather general scope, being suitable also for preparing some fluorine-free derivatives.

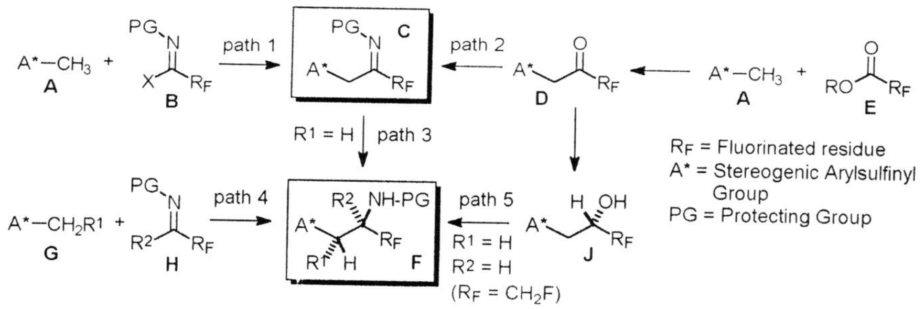

Scheme 1. General strategy to prepare nonracemic fluoro amino compounds via chiral sulfoxide chemistry.

Route B is also versatile and efficient (*14,15*), allowing for the preparation of *N*-aryl, *N*-alkoxycarbonyl and also *N*-unsubstituted polyfluoroalkyl enamines and imines in good to excellent yields. Starting material in this case are fluorinated β-sulfinyl ketones **7**, which were submitted to *Staudinger* (aza-*Wittig*) reaction (*16,17*) with iminophosphoranes **8**. Monofluorinated ketones **7** were found to be poorly reactive when submitted to the *Staudinger* reaction, therefore low yields (10-30%) of the corresponding products were generally obtained. No reaction was achieved with nonfluorinated β-ketosulfoxides.

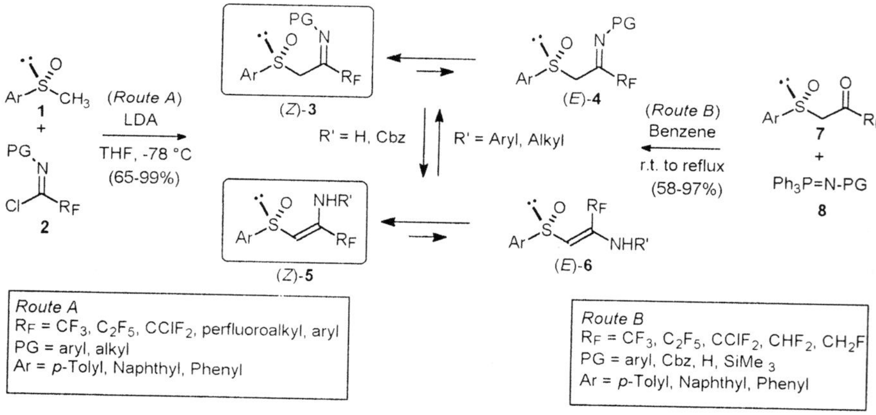

Scheme 2. Synthetic approaches to β-sulfinyl imines, and their tautomerism.

α-(Polyfluoroalkyl) β-sulfinyl enamines and imines (Scheme 2) are ideal building blocks for the asymmetric synthesis of fluoro-amino compounds, because despite the fact that they may exist as complex mixtures of imine/enamine tautomers (**3,4/5,6**, respectively) and geometric isomers E/Z (**4,6/3,5**, respectively), an almost total tautomeric/geometric homogeneity and a great chemical stability were always observed. The nitrogen ligand PG determines the tautomery: when PG = Cbz or H, the enamine is dominant, whereas the imine form largely prevails when PG = aryl or alkyl. Spectroscopic experiments and X-ray diffraction studies demonstrated that β-sulfinyl enamines exist exclusively as Z-isomers **5**, with the exception of the difluoromethyl derivative (R_F = CHF$_2$) existing as a thermodynamic mixture of Z/E enamines **5/6** in 3:2 ratio, separable by standard chromatographic methods. Also β-sulfinyl imines exist almost exclusively as Z isomers **3**, as shown by *Fustero* (*11*) in collaboration with us, by means of NMR experiments and X-ray diffraction studies, supported by *ab initio* calculations (HF/6-31G*). Fluorosubstitution is very important with respect to the chemical stability of these compounds: in fact, the stability of the C=N bond toward aqueous hydrolysis dramatically drops in the case of monofluorinated and fluorine-free derivatives.

α-(Fluoroalkyl) β-sulfinyl enamines and imines can be stereoselectively reduced with hydride reducing agents to the corresponding amines, which are very useful intermediates. The C=C bond of enamino sulfoxides is not very easy to reduce, therefore long reaction times were generally required and a few hydride reagents proved to be effective (*15*). However, K- or L-Selectride reduced N-unsubstituted enamines **9** (Scheme 3) with very good stereoselectivity, providing the corresponding *anti* β-sulfinyl amines, which were almost quantitatively transformed into the NH-Cbz derivatives **10** with K$_2$CO$_3$/ClCOOBn. Conversion of NH-Cbz α-(fluoroalkyl) β-sulfinyl amines **10** to the corresponding (R)-fluoroalaninols **11**, was accomplished by means of the "non-oxidative" *Pummerer* reaction (NOP reaction) (*18-20*) (see chapter entitled "Stereoselective Synthesis of β-Fluoroalkyl β-Amino Alcohol Units"). This methodology allows for a mild, high yielding and one-pot displacement of the sulfinyl group by a hydroxy one from fluorinated and nonfluorinated β-sulfinyl amines protected as NH-acyl derivatives, whereas N-unsubstituted (NH$_2$) and NH-aryl β-sulfinyl amines do not undergo efficiently the same transformation. According to the NOP protocol, Tfm NH-Cbz derivative **10a** was transformed into **11a** upon treatment with trifluoroacetic anhydride (TFAA)/*sym*-collidine, followed by aqueous K$_2$CO$_3$ and finally NaBH$_4$. Trifluoroalanine (R)-**12a** was obtained by oxidation of the corresponding alaninol **11a**, according to the *Sharpless* protocol (catalytic RuO$_4$) (*21*). Several fluoro-amino derivatives selectively labeled with deuterium were obtained according to the same strategy (*15*) (Scheme 3). Reduction of the α-Tfm β-sulfinyl enamine **9a** with NaBD$_4$ provided with fair stereocontrol the monodeuterated β-sulfinyl amine **13**. Exchange of **9** with D$_2$O to give the deuterated enamine **14**, followed by treatment with NaBD$_4$ afforded with moderate stereocontrol the trideuterated β-sulfinyl amine **15**, which was transformed, in excellent yields, into the trideuterated trifluoroalaninol (R)-**16** by means of the NOP reaction.

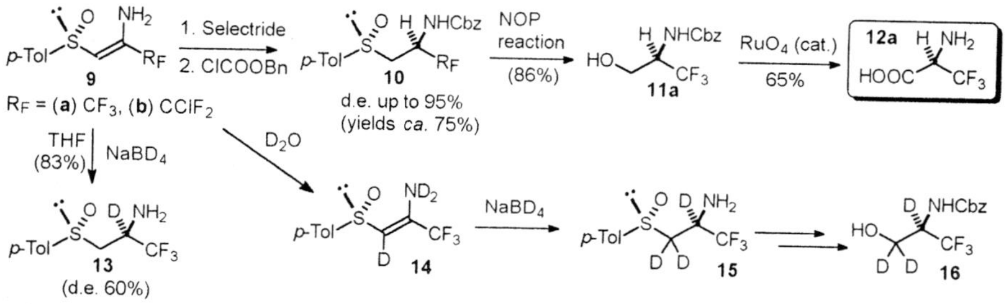

Scheme 3. Stereoselective synthesis of deuterium labeled fluoro amino compounds.

Much more smooth and stereoselective was the reduction of the C=N bond of β-sulfinyl imines **17**, studied and developed in collaboration with the group of *Fustero* (*10,22*). β-(p-Tolyl)sulfinyl N-(p-methoxyphenyl) (PMP) imines **17a** (Scheme 4) were reduced at r.t. with NaBH₄, producing the corresponding β-(p-tolyl)sulfinyl amines **18a** with 60% d.e. and quantitative yields. More usefully, reduction of β-(1-naphthyl)sulfinyl imines **17b** with NBu₄BH₄ in methanol at -78 °C provided quantitative yields of the β-(1-naphthyl)sulfinyl amines **18b** with diastereoselection up to 99:1 in the case of R_F = CF₃ or CClF₂. Reduction of β-(2-naphthyl)sulfinyl imines was also investigated, but the stereocontrol was lower under the same conditions.

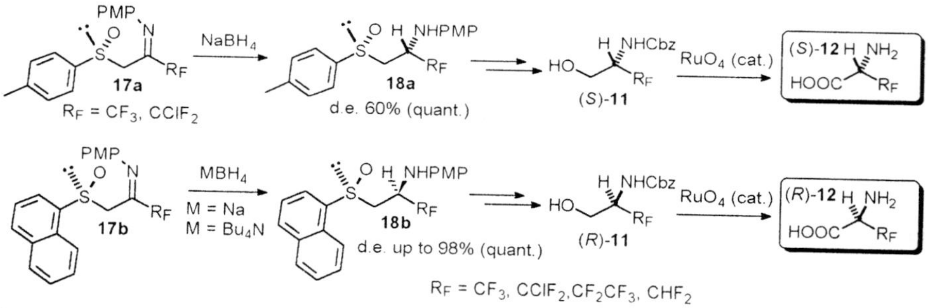

Scheme 4. Stereoselective reduction of β-sulfinyl imines, and elaboration into both enantiomers of 3,3,3-trifluoroalanine.

Adducts **18a,b** were elaborated according to the usual procedure, namely oxidative cleavage of the PMP group, protection as NH-Cbz derivatives, and NOP reaction, which delivered several enantiopure fluoroalaninols **11** in good overall yields. *Sharpless* oxidation (*21*) afforded a wide range of enantiopure β-fluoro alanines having both (R) and (S) configurations. The excellent stereocontrol achieved in the reduction of naphthyl derivatives **17b** has been rationalized by *Fustero* on the basis of semiempirical and *ab initio* calculations (HF/6-31G*), which evidenced a π-π stacking interaction between the N-PMP and the 1-naphthyl group. This should produce a very efficient shielding of one of the C=N diastereofaces (Scheme 5), therefore the hydride attacks selectively the other one. The drop of stereocontrol experimentally observed for

2-naphthyl derivatives was accordingly interpreted in terms of a less efficient π-π stacking effect, which is also supported by the calculations.

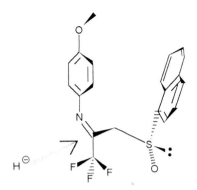

Scheme 5. Model for the highly stereocontrolled reduction of β-naphthylsulfinyl imines.

α-Difluoromethyl (Dfm) α-amino acids are man made analogs of the natural compounds, which feature a great biomedicinal interest (23,24). These compounds are even more promising than the more popular α-Tfm α-amino acids (24,25), but unfortunately they are much more difficult to prepare, particularly in nonracemic form. A convenient synthesis of enantiopure (R)-α-Dfm alanine **22** and (S)-α-Dfm-serine **23** was carried out starting from α-Dfm β-sulfinyl NH-Cbz enamines **19** (26) (Scheme 6). The major drawback of this strategy is represented by the modest diastereocontrol featured by the key hydrocyanation reaction, in spite of the fact that a variety of reagents and conditions were examined. However, almost quantitative yields of diastereomeric nitriles **20,21** (R_F = CF₃, CClF₂, CHF₂) and 30-40% d.e. were obtained by treatment of **19** with KCN, or with NaH followed by (EtO)₂POCN or Me₃SiCN. The major difluoromethylated diastereomer **20** was transformed into the final product (R)-**22** by reductive desulfenylation and hydrolysis of the NHCbz and CN moieties. The minor diastereomer **21** was submitted to the NOP reaction, and then to hydrolysis, providing enantiopure (S)-**23**. It is worth noting that, due to the availability of many starting NH-Cbz α-(fluoroalkyl) β-sulfinyl enamines **19**, even in deuterium labeled form like **14** (Scheme 3), this method could be used to prepare a wide range of enantiopure α-amino acids bearing diverse α-fluoroalkyl residues.

The interesting β-amino nitro derivatives **24** (Scheme 6), potential precursors of α-fluoroalkyl β-diamino compounds, like **25**, were also obtained from β-sulfinyl enamines **19** and nitromethane (15). Unfortunately, even in this case a low stereocontrol was obtained.

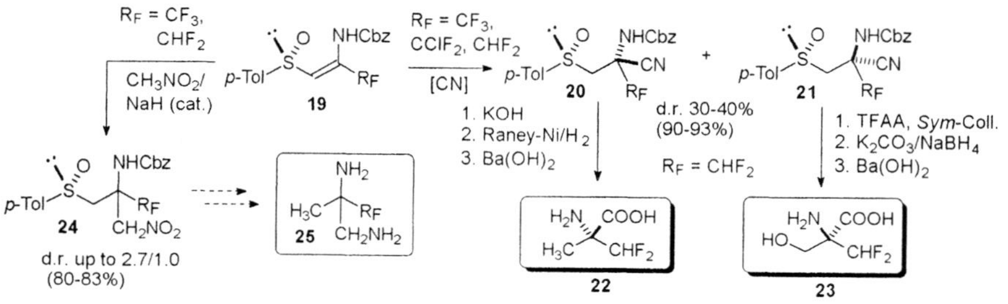

Scheme 6. Synthesis of α-fluoroalkyl amino compounds from β-sulfinyl enamines.

The great synthetic potentialities of fluorinated β-sulfinyl imines as chiral building blocks are well exemplified by the synthesis of 1-Tfm-analogs of 1,2,3,4-tetrahydroisoquinoline alkaloids, having a quaternary stereocenter at C-1 (**27**) (Scheme 7). The starting compound **26**, readily prepared by condensation of lithium methyl p-tolyl sulfoxide (R)-**1** with the corresponding acetimidoyl chloride **2** (see *Route A*, Scheme 2), underwent stereoselective *Pictet-Spengler* cyclization (**28,29**) under catalysis of TFA, providing the cyclic derivative **27**.

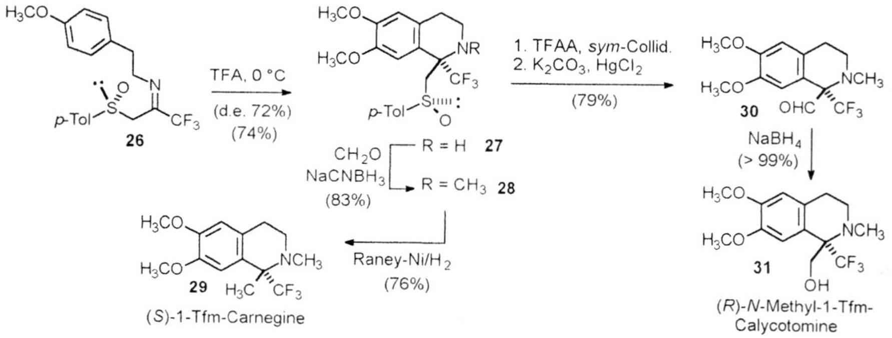

Scheme 7. Total synthesis of 1-Tfm tetrahydroisoquinoline alkaloids using β-sulfinyl imines.

Such a good stereocontrol can be understood if considering that the imine **26** is homogenously (Z)-configurated, therefore the electronrich 3,4-dimethoxyphenyl group and the stereogenic p-tolylsulfinyl group should be spatially close to each other. Thus the sulfinyl auxiliary should be able to exert a strong stereodirecting effect on the ring closure. N-Methylation to **28** and reductive desulfenylation smoothly provided (S)-1-Tfm carnegine **29**. The *Pummerer* reaction (**30,31**) performed on **28** and subsequent hydrolysis of the thioacetal intermediate afforded the highly stable quaternary α-amino aldehyde **30**, whose reduction delivered (R)-N-methyl 1-Tfm calycotomine **31** in very good overall yield.

Fluoropyruvaldehyde *N,S*-Ketals: Nonracemic Synthetic Equivalents of α-Fluoroalkyl α-Amino Aldehydes.

One of the most direct approaches to non-racemic β-fluoroalkyl β-amino alcohols **K** would be the addition of an appropriate nucleophile R⁻ to a chiral α-amino aldehyde **L** (Scheme 8). However, the preparation of β-fluoro α-amino aldehydes is very troublesome, given the lack of readily available precursors and appropriate synthetic methodologies (*32*).

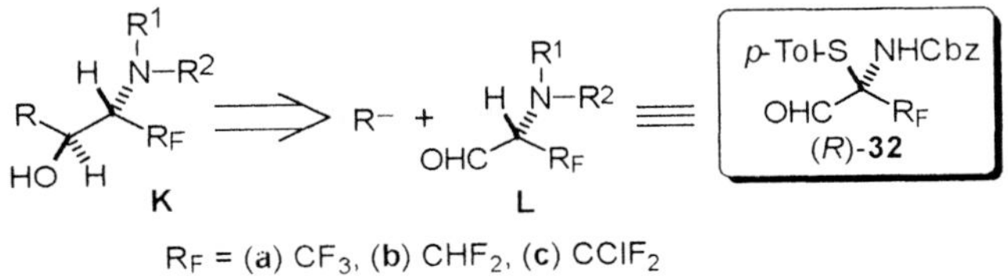

R_F = (a) CF$_3$, (b) CHF$_2$, (c) CClF$_2$

Scheme 8. Retrosynthetic strategy to prepare β-fluoroalkyl β-amino alcohols from pyruvaldehyde *N,S*-ketals.

Furthermore, β-fluoro α-amino aldehydes are expected to have high proclivity toward racemization even in a very weakly basic environment, the α-proton being strongly acidic due to the presence of fluorine (*32,33*).

Trifluoro, difluoro and chlorodifluoropyruvaldehyde *N,S*-ketals (*R*)-**32a-c** (Scheme 8) represent a new class of readily available, versatile and stereochemically stable equivalents of chiral nonracemic β-fluoro-α-aminoaldehydes **L**. Substrates (*R*)-**32** are non-enolizable, and therefore optically stable, because the *p*-tolylthio substituent creates a quaternary α-stereogenic center. In a later stage of the synthetic sequence, the *p*-tolylthio residue can be stereoselectively substituted by a hydrogen atom, providing the targeted γ-fluoro β-amino alcohol frameworks **K**. The *p*-tolylthio group plays the key-role of removable auxiliary function, and acts as the main stereocontrolling group as well.

Fluoropyruvaldehyde *N,S*-ketals (*R*)-**32** can be prepared in good yields (up to 88%) and e.e.'s (up to 82%) from α-(fluoroalkyl)-β-sulfinylenamines (*R*)-*Z*-**19a-c** (Scheme 9), through an unprecedented self-immolative tandem sequential process (*34,35*), consisting in a Pummerer reaction promoted by TFAA (*step 1*) that provides the intermediate α-trifluoroacetoxy *N*-Cbz imine **36**, followed by a 1,2-migration of the *p*-tolylthio group (*step 2*), triggered by addition of silica gel or aqueous base.

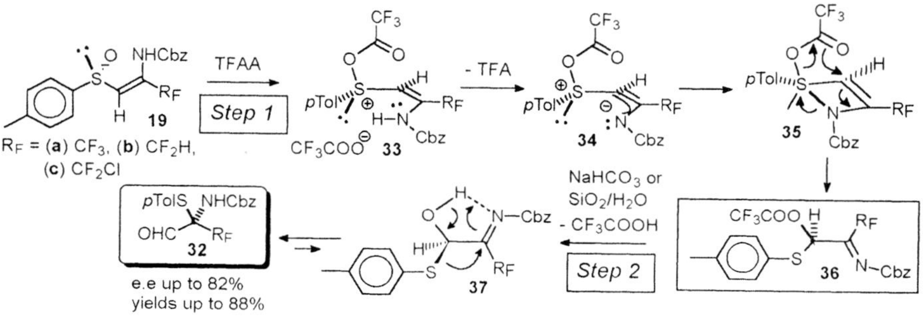

Scheme 9. Mechanism of the enantioselective tandem rearrangement leading to pyruvaldehyde N,S-ketals.

Each transfer of stereogenic center, from sulfur to the α-carbon and then to the β-carbon, occurs with an average degree of enantioselectivity up to 95.5:4.5. Such process occurs for sulfinyl enamines (R)-**19** having diverse α-fluoroalkyl residues, like CF_3, CF_2Cl, CF_2H. The NHCbz provides a neighboring group participation in the *step 1* of the tandem process, preventing from TFAA-promoted racemization of the sulfinyl stereocenter, as well as allowing an efficient transfer of the stereochemical information from the sulfinyl group to the α-stereogenic center. This is the reason why *cis* geometry between the sulfinyl and the amino groups of the starting enamine (R)-**19** is necessary for achieving high level of stereocontrol. The intramolecular mechanism proposed (Scheme 9) is strongly supported by crossover and dilution experiments. In particular, the enantiospecifity of the process progressively improves when decreasing the concentration of the starting material **19**, as demonstrated by means of chemometric studies (*36*). The trivalent trifluoroacetoxysulfonium salt **33** should form by action of TFAA on the sulfinyl enamine (R)-Z-**19**. Loss of TFA from **33** produces the chiral ylide **34**. Formation of a new bond between the positive sulfur and the negatively charged nitrogen atom immediately locks the conformation of **34**, leading to the formation of the highly reactive four membered sulfurane **35**. The latter undergoes a fast and highly stereospecific migration of the trifluoroacetoxy group from the sulfur to the *Si* face of the C-1 position, producing the imine (R)-**36**. In contrast, the sulfur and the nitrogen atoms of the *trans* difluoromethyl enamine **19b** are far away and cannot interact. For this reason, a fast and extended racemization of the sulfinyl stereocenter occurs, producing the corresponding imine **36b** with very low e.e. Formation of the final pyruvaldehyde (R)-**32** from (R)-**36** (*step 2*) is triggered by addition of SiO_2 or aqueous NaHCO_3, which cleave the trifluoroacetyl group. The resulting transient α-hydroxy α-p-tolylthio imine (R)-**37** undergoes a fast suprafacial 1,2-migration of the p-tolylthio group, through the five-membered cyclic conformation having hydrogen bonding between the hydroxy and the imino functions. *Step 2* can be formally considered an α-hydroxy-imine/α-amino-aldehyde rearrangement, which is known to be an equilibrium process (*37*). In this case, the equilibrium is strongly shifted toward the α-amino-aldehyde form (R)-**32** even at low temperature, probably because of the strong stabilization of the sp^3 N,S-ketal carbon induced by the electron-withdrawing fluoroalkyl group.

Addition of Grignard reagents to fluoropyruvaldehydes (R)-**32** occurs with moderate to excellent *anti*-stereocontrol (*38*,*39*), depending on the nature of the organomagnesium halides, providing the nonracemic β-*p*-tolylthio β-benzyloxycarbonylamino secondary carbinols **39** (Scheme 10). The highest stereoselectivity (36:1) was achieved with vinylmagnesium bromide, the lowest one with ethylmagnesium bromide (4:1), in both cases with **32a**. A weak loss of enantiomeric purity was observed only in the case of difluoromethyl derivatives.

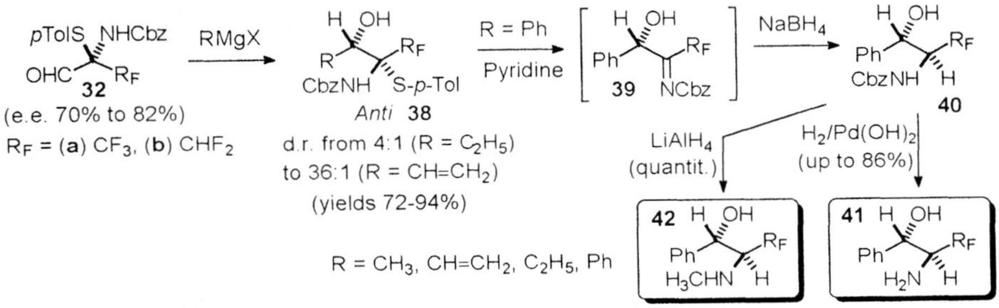

Scheme 10. Diastereoselective elaboration of pyruvaldehyde N,S-ketals into fluorinated ephedra-alkaloids.

The stereochemical outcome of these reactions can be rationalized by means of a chelated Cram's cyclic model **M** (*40*) (Scheme 11), where the N-Cbz group is the chelating ligand and the p-tolylthio residue acts as the stereocontrolling "large" group.

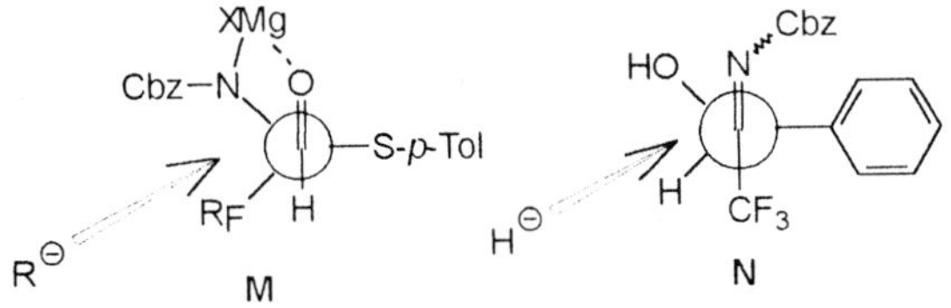

Scheme 11. Models for the stereoselective transformations involving pyruvaldehyde N,S-ketals and derivatives.

Adducts **38** can be used as versatile intermediates to obtain biologically interesting sulfur-free γ-tri- and γ-difluorinated β-amino alcohols. In fact, reductive displacement of the 2-*p*-tolylthio substituent of **38** efficiently takes places by means of the NaBH$_4$/pyridine system (Scheme 10). The first step of the process should be the elimination of *p*-thiocresol from the N,S-ketalic center C-2, promoted by pyridine, that produces the transient N-Cbz imines (S)-**39**. Then, the imines **39** undergo reduction by NaBH$_4$, providing the sulfur-free NH-Cbz β-amino alcohols **40** with high levels of *anti*-stereoselectivity (d.r. 12:1 for R_F = CF$_3$, 3:1 for R_F = CHF$_2$). The high *anti* selectivity featured by the Tfm-ketal **38a** could be explained by means of the Felkin-Anh model **N** (*41*,*42*) (Scheme 11): the staggered position should be occupied by the phenyl ring, and attack of the hydride should therefore occur from the *Re* face of the imine, with the hydroxyl acting as the "medium" group and, obviously, the hydrogen as the "small" group. Further stabilization of this conformation might derive from hydrogen bonding involving the hydroxyl and the iminic nitrogen, or the carbonyl oxygen of Cbz. Cleavage and reduction of the NHCbz moiety of **40** provided tri- and difluoro analogs of, respectively, norephedrine **41** and ephedrine **42**. Encouraging preliminary experiments suggest that intermediates **38** could be used for the synthesis of β-hydroxy α-fluoroalkyl α-amino acids, since the *p*-tolylthio group can be displaced by a masked carboxyl, namely the cyano group, under conditions similar to those used with NaBH₄.

An efficient approach to nonracemic (70% e.e.) γ-Tfm GABOB (3*R*,4*S*)-**43**, starting from pyruvaldehyde **32a** is shown in Scheme 12 (*43*).

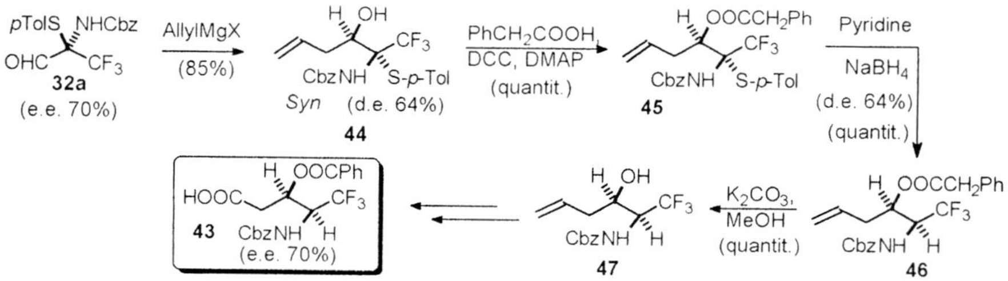

Scheme 12. Synthesis of a nonracemic γ-Tfm-GABOB from pyruvaldehyde *N*,*S*-ketals.

Addition of allylmagnesium chloride to the aldehyde **32a** occurred with fair *syn* selectivity, producing the homoallylic alcohol **44**. This inversion of facial selectivity in comparison with the other reactions of **32** (Scheme 10), and also with a wide range of other *N*-monoprotected α-amino aldehydes, with Grignard reagents (*44,45*), including allylic ones (*46*), have been rationalized with the *syn*-selective chair-like transition state **P** (Scheme 13), involving a sterically controlled approach of the allylic nucleophile to the *Re* face of the carbonyl (Felkin-Anh) (*41,42*).

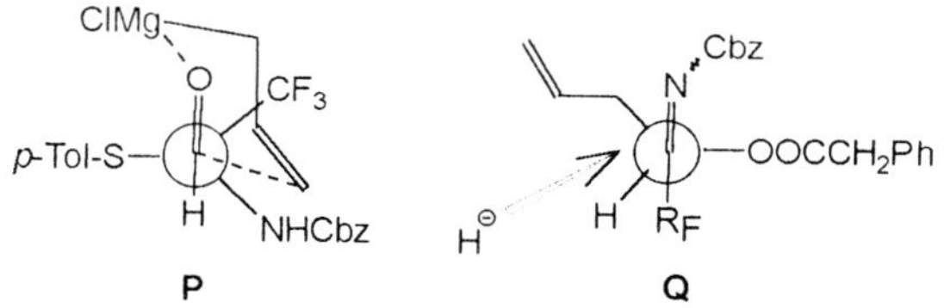

Scheme 13. Models for the asymmetric elaborations leading to the γ-Tfm-GABOB.

Although the reductive desulfenylation of compound **44** with NaBH₄/pyridine occurred without stereocontrol, the reaction applied to the phenylacetic ester **45** provided the corresponding sulfur free derivative **46** with reasonable *syn* selectivity. The rationale might be as follows: pyridine catalyzes elimination of *p*-thiocresol from **45** forming the corresponding transient imine, which is reduced by NaBH₄ mainly through the Felkin-Anh model **Q** (Scheme 13), in which the phenylacetoxy group plays the role of electronegative, "large" substituent. The target γ-Tfm-GABOB $(3R,4S)$-**43** can be readily obtained from **46**, with overall conservation of the original enantiomeric purity (70%).

Sulfinimines of Trifluoropyruvate: Toward a Library of Nonracemic α-Tfm α-Amino Acids.

α-Trifluoromethyl α-amino acids are synthetic analogues of naturally occurring amino acids (AAs), which are gaining a preminent position in the man-made area of fluorine containing amino acids (24,25). The great interest in α-TFM α-amino acids arises from the fact that some of them exhibit anticancer, antibacterial and antihypertensive properties, ability to behave as potent suicide inhibitors of pyridoxalphosphate-dependent enzymes (transaminases, decarboxylases), and that some peptides containing these compounds have increased metabolic stability and biological activity. On the other hand, chiral sulfinimines RC(R')=NS(O)R'' have recently become very popular intermediates in the asymmetric synthesis of amino compounds (47). The breakthrough was probably represented by the reports of *Davis* (48), who disclosed a very convenient one-pot method for their preparation: the reaction of LiN[Si(CH$_3$)$_3$]$_2$ with menthyl sulfinate, followed by addition of the appropriate carbonyl compound RC(O)R'. However, this method in our hands proved to be not suitable for preparing sulfinimines having a strongly electrophilic iminic carbon, like those derived from trifluoropyruvate. Therefore, we have developed an alternative method (49) for preparing such peculiar sulfinimines, exploiting a new chiral *Staudinger* reagent (16,17), the N-sulfinyl iminophosphorane **51** (50) (Scheme 14), prepared in enantiopure form by a *Mitsunobu*-type reaction (51) of the sulfinamide (S)-**50**. *Staudinger* reaction between **51** and trifluoropyruvic esters smoothly produced the intermediate sulfinimines **52**, which can be reacted *in situ* with a wide range of Grignard reagents. In Scheme 14 is shown the synthesis of (R)-α-Tfm phenylalanine **55**, with recovery and regeneration of the chiral auxiliary, which allows for a considerable atom economy. Addition of benzylmagnesium chloride to the sulfinimines **52** provided the benzyl sulfinamides **53** with moderate d.e. After chromatographic separation, the sulfinyl auxiliary was removed with TFA/L-menthol **49** (recycled from menthyl sulfinate **48**) according to the *Davis* protocol (52), providing the target α-Tfm phenylalaninate **54** and menthyl sulfinate **48**, which can be regenerated in diastereopure form (53,54) and recycled for the preparation of the iminophosphorane **51**.

Many other nonracemic α-Tfm amino acids can be prepared with high enantiopurity according to the same procedure, thus providing the first library of these compounds (Scheme 15). Addition of the appropriate Grignard reagent to the sulfinimine of ethyl trifluoropyruvate **52b** occurred with variable stereocontrol, producing the sulfenamides **53**, isolated in diastereopure form by standard chromatographic separation. The best diastereoselectivity (up to 87/13) was obtained with bulky Grignard reagents, like *iso*-propyl and *iso*-butyl derivatives. Methanolysis (55) of the major sulfinamides **53** provided methyl sulfinate and the corresponding amino esters, easily hydrolyzed to α-Tfm alanine **56**, butyrine **57**, leucine **58**, valine **59** and the *n*-butyl derivative **60**.

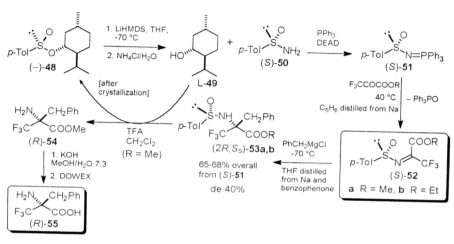

Scheme 14. Synthesis of (R)-α-Tfm-phenylalanine with regeneration and recycle of the sulfin chiral auxiliary.

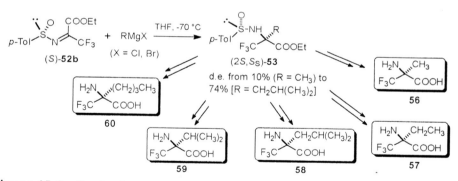

Scheme 15. Synthesis of a library of nonracemic α-Tfm-amino acids from sulfinimines of trifluoropyruvate.

Current efforts are directed toward the use of this methodology for preparing nonracemic α-Tfm and α-Dfm amino acids to be incorporated into oligopeptidic structures, having improved lipophilicity, biological availability, hydrolytic stability and biological activity, for example new fluorinated analogs of the RGD sequence (*56*).

Syntheses of Monofluoro-Amino Compounds.

The above mentioned strategies have been applied mainly to the synthesis of difluoromethyl or other perfluoroalkyl amino compounds. However, also monofluoro amino compounds have been obtained in our laboratories using taylored synthetic protocols and the sulfinyl group as auxiliary.
The potent antibacterial agent 3-fluoro-D-alanine **66** (*57*) was prepared starting from the α-sulfinyl α'-fluoroacetone **61** (*58*) (Scheme 16). Stereoselective reduction with DIBAH (*59,60*) produced the alcohol **62**, which was submitted to a *Mitsunobu*-type azidation (*61,62*), providing the β-sulfinyl azide **63**. Unfortunately, this interesting reaction could not be applied to compounds having higher degree of fluorination, because in those cases elimination to give the corresponding γ-fluoro α,β-unsaturated sulfoxides was the main event.

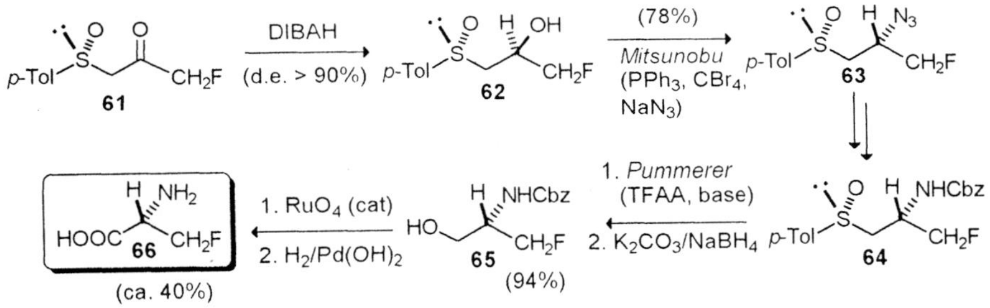

Scheme 16. Stereoselective synthesis of 3-fluoro-D-alanine.

Chemoselective reduction and *N*-protection as Cbz derivative to give **64**, followed by NOP reaction afforded the fluoro alaninol **65**, readily transformed into the target amino acid **66**.
Several enantiopure compounds having a quaternary monofluoromethyl or Tfm-substituted carbinolic center C-2, like 2-hydroxy 3-amino acids and 3-deoxy 3-amino glycerols (Scheme 17), were obtained by amine-promoted ring opening of oxiranes (*63-66*), prepared by stereocontrolled addition of diazomethane to the corresponding α-sulfinyl α'-fluoro acetones **67** (d.e. up to 96% when R = H, otherwise d.e. up to 86%) (*67,68*). Two distinct synthetic pathways were followed, depending on the nature of the fluoroalkyl residue. In the first one (R$_F$ = CH$_2$F) the sulfinyl oxiranes **68** were transformed via *Pummerer* reaction (*30,31*) into the corresponding sulfur-free aldehyde **69**. This compound is the starting material for preparing the fluoro amino alcohol **70** and the acid **71**, using a regioselective opening of the oxirane ring with benzylamine or ammonia, respectively, as key-step (*69*). The second pathway (R$_F$ = CF$_3$) relies on the regioselective ring opening of the trifluoro sulfinyl oxirane **68** with dibenzylamine, followed by *Pummerer* reaction and the usual cleavage of the protecting groups (*70*).

112

By this route the corresponding Tfm derivatives **73,74** could be prepared. Alternatively, oxirane ring-opening and reductive desulfinylation provided both enantiopure α-monofluoromethyl and α-Tfm β-amino alcohols **75**.

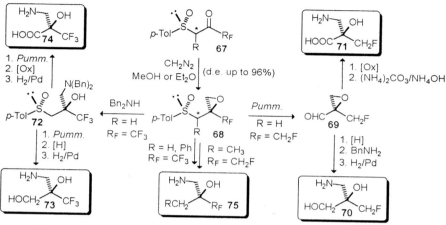

Scheme 17. Approach to enantiopure quaternary fluoro amino compounds via sulfinyl epoxide chemistry.

Ring opening of the sulfur-free oxiranes **76**, prepared from **68** by *Pummerer* reaction, reduction of the intermediate aldehyde and subsequent *O*-benzylation, by action of nucleobases, for example thymine and fluorouracil, provided an efficient entry to fluoronucleosides **77** (*71*) (Scheme 18).

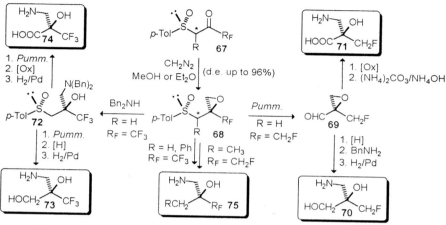

Scheme 18. Synthesis of monofluorinated nucleosides.

Some monofluoro amino derivatives have been prepared with excellent stereocontrol exploiting the potential of inter- and intra-molecular 1,3-dipolar cycloadditions (*72*) with fluorinated dipolarophiles. An intermolecular application is shown in Scheme 19 (*73,74*). Nitrile oxides **78** were reacted with fluorinated chiral enol ethers **79**: 4,5-dihydroisoxazoles **80** were obtained with total regio- and stereo-selectivity. By subsequent elimination of methanol and reductive opening of the isoxazole ring **81**, fluoroiminoalcohol **82** was obtained in fair yield.

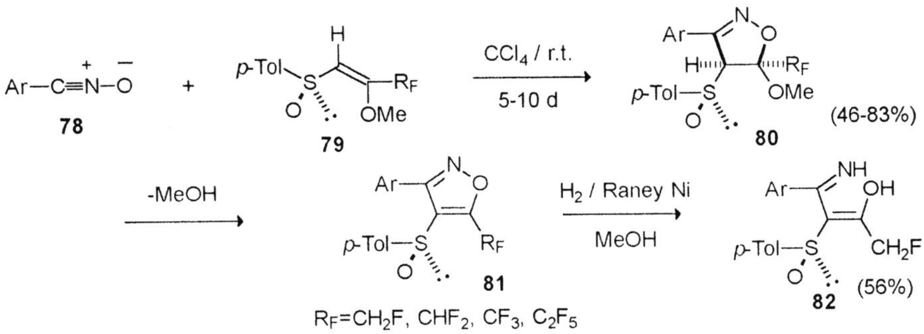

Scheme 19. 1,3-Dipolar cycloadditions involving fluorinated sulfinyl dipolarophiles.

An intramolecular application is illustrated in Scheme 20 (75,76): *Pummerer* reaction of the β-allyloxy-sulfoxides **83** produced the corresponding aldehydes **84**, which were transformed one-pot into the transient nitrones **85**. Under moderate heating, a fully regio- and diastereoselective intramolecular cycloaddition took place, leading to 6-fluoromethyl furo[3,4-c]isoxazolidines **86**. Hydrogenolysis of the N-O bond of **86** (R = Ph) provided the corresponding furo-amino alcohol **87**.

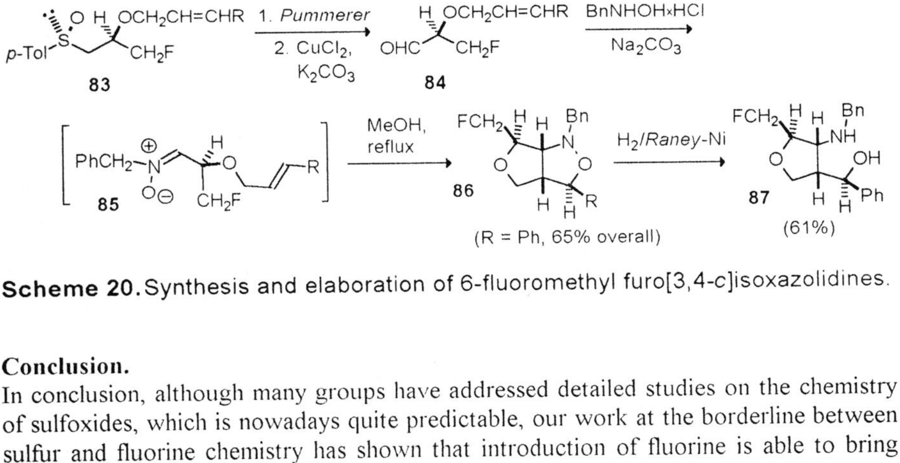

Scheme 20. Synthesis and elaboration of 6-fluoromethyl furo[3,4-c]isoxazolidines.

Conclusion.
In conclusion, although many groups have addressed detailed studies on the chemistry of sulfoxides, which is nowadays quite predictable, our work at the borderline between sulfur and fluorine chemistry has shown that introduction of fluorine is able to bring about intriguing and useful modifications of the normal reactivity, which have been exploited to prepare a wide range of biologically interesting compounds.

Acknowledgements. The following scientists are acknowledged for their fundamental contribution to the studies reviewed herein: Dr. Alessandro Volonterio, Dr. Vadim A. Soloshonok, Prof. Stefano V. Meille, Dr. Antonio Navarro and Prof. Santos Fustero-Lardies. Thanks are due also to all the authors of our papers cited in the references below.

114

Literature Cited.

1. Schlosser, M. *Angew. Chem., Int. Ed. Engl.* **1998**, *110*, 1496.
2. Resnati, G. *Tetrahedron* **1993**, *49*, 9385.
3. *Enantiocontrolled Synthesis of Fluoro-Organic Compounds: Stereochemical Challenges and Biomedicinal Targets*; Soloshonok, V. A., Ed.; Wiley: Chichester, 1999.
4. Welch, J. T.; Eswarakrishnan, S. *Fluorine in Bioorganic Chemistry*; Wiley: New York, 1991.
5. Banks, R. E.; Tatlow, J. C.; Smart, B. E. *Organofluorine Chemistry: Principles and Commercial Applications*; Plenum Press: New York, 1994.
6. *Biomedical Frontiers of Fluorine Chemistry*; Ojima, I.; McCarthy, J.R.; Welch, J.T., Eds.; ACS Books, American Chemical Society: Washington, D.C., 1996.
7. Walker, A. J. *Tetrahedron: Asymmetry* **1992**, *3*, 961.
8. Carreño, M. C. *Chem. Rev. (Washington, D. C.)* **1995**, *95*, 1717.
9. Bravo, P.; Zanda, M. "Asymmetric Synthesis of Fluoro-Organic Compounds via Chiral Sulfoxide Chemistry" see ref. 3.
10. Bravo, P.; Cavicchio, G.; Crucianelli, M.; Markovsky, A. L.; Volonterio, A.; Zanda, M. *Synlett* **1996**, 887.
11. Fustero, S.; Navarro, A.; Pina, B.; Asensio, A.; Bravo, P.; Crucianelli, M.; Volonterio, A.; Zanda, M. *J. Org. Chem.* **1998**, *63*, 6210.
12. Uneyama, K.; Kobayashi, M. *J. Org. Chem.* **1994**, *59*, 3003.
13. Uneyama, K.; Tamura, K.; Mizukami, H.; Maeda, K.; Watanabe, H. *J. Org. Chem.* **1993**, *58*, 32.
14. Bravo, P.; Crucianelli, M.; Zanda, M. *Tetrahedron Lett.* **1995**, *36*, 3043.
15. Arnone, A.; Bravo, P.; Capelli, S.; Fronza, G.; Meille, S. V.; Zanda, M.; Cavicchio, G.; Crucianelli, M. *J. Org. Chem.* **1996**, *61*, 3375; corrigenda: *J. Org. Chem.* **1996**, *61*, 9635.
16. Staudinger, H.; Meyer, J. *Helv. Chim. Acta* **1919**, *2*, 635.
17. Johnson, A. W.; Kasha, W. C.; Starzewsky, K. A. O.; Dixon, D. A. *Ylides and Imines of Phosphorus*; John Wiley: New York, 1993, chapt. 13.
18. Arnone, A.; Bravo, P.; Bruché, L.; Crucianelli, M.; Vichi, L.; Zanda, M. *Tetrahedron Lett.* **1995**, *36*, 7301.
19. Bravo, P.; Zanda, M.; Zappalà, C. *Tetrahedron Lett.* **1996**, *37*, 6005.
20. Bravo, P.; Farina, A.; Kukhar, V. P.; Markovsky, A. L.; Meille, S. V.; Soloshonok, V. A.; Sorochinsky, A. E.; Viani, F.; Zanda, M.; Zappalà, C. *J. Org. Chem.* **1997**, *62*, 3424.
21. Carlsen, P. H. J.; Katsuki, T.; Martin, V. S.; Sharpless, K. B. *J. Org. Chem.* **1981**, *46*, 3936.
22. Fustero, S.; Garcia, M.; Salavert, E.; García, J.; Navarro, A.; Bravo, P.; Crucianelli, M.; Volonterio, A.; Zanda, M. 12th European Symposium on Fluorine Chemistry, Berlin (Germany), August 29-September 2, 1998. Abstract N. PI-5.
23. Osipov, S. N.; Golubev, A. S.; Sewald, N.; Michel, T.; Kolomiets, A. F.; Fokin, A. V.; Burger, K. *J. Org. Chem.* **1996**, *61*, 7521.
24. *Fluorine-Containing Amino Acids: Synthesis and Properties*; Kukhar, V. P.; Soloshonok, V. A. Eds.; Wiley: Chichester, 1994.

25. Koksch, B.; Sewald, N.; Jakubke, H.-D.; Burger, K. "Synthesis and Incorporation of α-Trifluoromethyl-Substituted Amino Acids into Peptides". In the book of ref. 6.

26. Bravo, P.; Capelli, S.; Meille, S.V.; Seresini, P.; Volonterio, A.; Zanda, M. *Tetrahedron: Asymmetry* **1996**, *7*, 2321.

27. Bravo, P.; Crucianelli, M.; Farina, A.; Meille, S. V.; Volonterio, A.; Zanda, M. *Eur. J. Org. Chem.* **1998**, 435.

28. Pictet, A.; Spengler, T. *Ber. Dtsch. Chem. Ges.* **1911**, *44*, 2030.

29. Cox, A. D.; Cook, J. M. *Chem. Rev.* **1995**, *95*, 1797.

30. Pummerer, R. *Ber.* **1909**, *42*, 2282.

31. De Lucchi, O.; Miotti, U.; Modena, G. *Organic Reactions*; Paquette, L. A. Ed.; John Wiley and Sons: New York, 1991, Vol. 40.

32. Konno, T.; Yamazaki, T.; Kitazume, T. *Tetrahedron* **1996**, *52*, 199.

33. Angert, H.; Czerwonka, R.; Reissig, H.-U. *Liebigs Ann.* **1997**, 2215.

34. Bravo, P.; Crucianelli, M.; Fronza, G.; Zanda, M. *Synlett* **1996**, 249.

35. Volonterio, A.; Zanda, M.; Bravo, P.; Fronza, G.; Cavicchio, G.; Crucianelli, M. *J. Org. Chem.* **1997**, *62*, 8031.

36. Bjørsvik, H.-R.; Bravo, P.; Crucianelli, M.; Volonterio, A.; Zanda, M. *Tetrahedron: Asymmetry* **1997**, *8*, 2817.

37. Compain, P.; Goré, J.; Vatèle, J.-M. *Tetrahedron* **1996**, *52*, 6647 and references cited therein.

38. Volonterio, A.; Bravo, P.; Capelli, S.; Meille, S. V.; Zanda, M. *Tetrahedron Lett.* **1997**, *38*, 1847.

39. Volonterio, A.; Vergani, B.; Crucianelli, M.; Zanda, M.; Bravo, P. *J. Org. Chem.* **1998**, *63*, 7236.

40. Cram, D. J.; Wilson, D. R. *J. Am. Chem. Soc.* **1963**, *85*, 1245.

41. Chérest, M.; Felkin, H.; Prudent, N. *Tetrahedron Lett.* **1968**, *9*, 2199.

42. Anh, N. G.; Eisenstein, O. *Tetrahedron Lett.* **1976**, *17*, 155.

43. Bravo, P.; Corradi, E.; Pesenti, C.; Vergani, B.; Viani, F.; Volonterio, A.; Zanda, M. *Tetrahedron: Asymmetry* **1998**, *9*, 3731.

44. Devant, R. M.; Radunz, H.-E. In *Houben-Weyl: Methods in Organic Synthesis*; Müller, E., Ed.; Thieme Verlag: Stuttgart, 1995; Vol. E21b, pp 1236-1239.

45. Jurczak, J.; Golebiowski, A. *Chem. Rev.* **1989**, *89*, 149-164.

46. Roush, W. R. In *Comprehensive Organic Synthesis Vol. 2*; Heathcock, C. H. Ed.; Pergamon Press: Oxford, 1991, pp 1-54.

47. Davis, F. A.; Zhou, P.; Chen, B.-C. *Chem. Soc. Rev.* **1998**, *27*, 13.

48. Davis, F. A.; Reddy, R. E.; Szewczyk, J. M.; Reddy, G. V.; Portonovo, P. S.; Zhang, H.; Fanelli, D.; Reddy, R. T.; Zhou, P.; Carroll, P. J. *J. Org. Chem.* **1997**, *62*, 2555.

49. Bravo, P.; Crucianelli, M.; Vergani, B.; Zanda, M. *Tetrahedron Lett.* **1998**, *39*, in press.

50. Senning, A.; Kelly, P. *Naturwissenschaften* **1968**, *55*, 543. *Chem. Abstr.* 70:47555v.

51. Bittner, S.; Assaf, Y.; Krief, P.; Pomerantz, M.; Ziemnicka, B. T.; Smith, C. G. *J. Org. Chem.* **1985**, *50*, 1712.

52. Davis, F. A.; Reddy, R. E.; Szewczyk, J. M. *J. Org. Chem.* **1995**, *60*, 7037.

53. Mioskowski, C.; Solladié, G. *Tetrahedron* **1980**, *36*, 227.

116

54. Hulce, M.; Mallamo, J. P.; Frye, L. L.; Kogan, T. P.; Posner, G. H. *Org. Synth.* **1984**, *64*, 196.
55. Mikolajczyk, M.; Drabowicz, J.; Bujnicki, B. *J. Chem. Soc., Chem. Commun.* **1976**, 568.
56. Dal Pozzo, A.; Muzi, L.; Moroni, M.; Rondanin, R.; de Castiglione, R.; Bravo, P.; Zanda, M. *Tetrahedron* **1998**, *54*, 6019.
57. Kollonitsch, J. *Suicide Substrate Enzyme Inactivators of Enzymes Dependent on Pyridoxal-Phosphate*, in *Biomedicinal Aspects of Fluorine Chemistry*; Filler, R.; Kobayashi, Y. eds.; Elsevier: Amsterdam, 1993.
58. Bravo, P.; Cavicchio, G.; Crucianelli, M.; Poggiali, A.; Zanda, M. *Tetrahedron: Asymmetry* **1997**, *8*, 2811.
59. Bravo, P.; Resnati, G. *Tetrahedron Lett.* **1987**, *28*, 4865.
60. Solladié, G.; Greck, C.; Demailly, G. *Tetrahedron Lett.* **1985**, *26*, 435.
61. Mitsunobu, O. *Synthesis* **1981**, 1-28.
62. Shuto, S.; Ono, S.; Hase, Y.; Kamiyama, N.; Takada, H.; Yamasihita, K.; Matsuda, A. *J. Org. Chem.* **1996**, *61*, 915.
63. Arnone, A.; Bravo, P.; Frigerio, M.; Salani, G.; Viani, F.; Zappalà, C.; Cavicchio, G.; Crucianelli, M. *Tetrahedron* **1994**, 8289.
64. Bravo, P.; Frigerio, M.; Fronza, G.; Soloshonok, V.; Viani, F.; Cavicchio, G.; Fabrizi, G.; Lamba, D. *Can. J. Chem.* **1994**, 1769.
65. Arnone, A.; Bravo, P.; Frigerio, M.; Viani, F.; Soloshonok, V. A. *Tetrahedron* **1998**, *54*, 11825.
66. Arnone, A.; Bravo, P.; Frigerio, M.; Viani, F.; Soloshonok, V. A. *Tetrahedron* **1998**, *54*, 11841.
67. Arnone, A.; Bravo, P.; Cavicchio, G.; Frigerio, M.; Marchetti, V.; Viani, F.; Zappalà, C. *Tetrahedron Lett.* **1992**, 5609.
68. Arnone, A.; Bravo, P.; Frigerio, M.; Meille, S. V.; Romita, V.; Viani, F.; Zappalà, C.; Soloshonok, V. A.; Shishkin, O. V.; Struchkov, Y. T. *Gazz. Chim. Ital.* **1997**, *127*, 819.
69. Arnone, A.; Bravo, P.; Frigerio, M.; Salani, G.; Viani, F. *Tetrahedron* **1994**, 13485.
70. Bravo, P.; Farina, A.; Frigerio, M.; Meille, S. V.; Viani, F.; Soloshonok, V. *Tetrahedron: Asymmetry* **1994**, 987.
71. Bravo, P.; Frigerio, M.; Soloshonok, V.; Viani, F. *Tetrahedron Lett.* **1993**, 7771.
72. Carruthers, W. *Cycloaddition Reactions in Organic Synthesis*; Pergamon: Oxford, 1990.
73. Bravo, P.; Bruché, L.; Merli, A.; Fronza, G. *Gazz. Chim. Ital.* **1994**, *124*, 275.
74. Bravo, P.; Bruché, L.; Crucianelli, M.; Farina, A.; Meille, S. V.; Merli, A.; Seresini, P. *J. Chem. Res. (S)* **1996**, 348.
75. Arnone, A.; Bandiera, P.; Bravo, P.; Bruché, L.; Zanda, M. *Gazz. Chim. Ital.* **1996**, *126*, 773.
76. Bandiera, P.; Bravo, P.; Bruché, L.; Zanda, M.; Arnone, A. *Synth. Commun.* **1998**, *28*, 2665.

Chapter 9

Studies Toward the Syntheses of Functionalized, Fluorinated Allyl Alcohols

P. V. Ramachandran, M. Venkat Ram Reddy, and Michael T. Rudd

H. C. Brown and R. B. Wetherill Laboratories of Chemistry,
Purdue University, West Lafayette, IN 47907–1393

This chapter reviews our study of Morita-Baylis-Hillman (MBH) and vinylmetalation reactions for the synthesis of achiral and chiral, functionalized, fluorinated allyl alcohols. Vinylmetalation of fluorocarbonyls with aluminum and copper reagents for the synthesis of unsubstituted, and β-substituted fluorinated allyl alcohols is summarized. An exploratory study of terpenyl alcohols as chiral auxiliaries in these reactions is also discussed.

Carbon-carbon bond forming reactions constitute the core of organic synthesis, and novel, efficient methods are always desirable. One such reaction that has attracted much attention in recent days is the Morita-Baylis-Hillman (MBH) reaction; the reaction of activated olefins with reactive carbonyls or imines in the presence of a catalytic amount of trialkyl(aryl)phosphine or amine (1-3). Of all the amines tested, 1,4-diazabicyclo[2.2.2]octane (Dabco) has been found to be superior (Scheme 1). This reaction, which provides multifunctional molecules has been accommodated in certain undergraduate curriculum (4), and does not demand any sophisticated techniques or instrumentation.

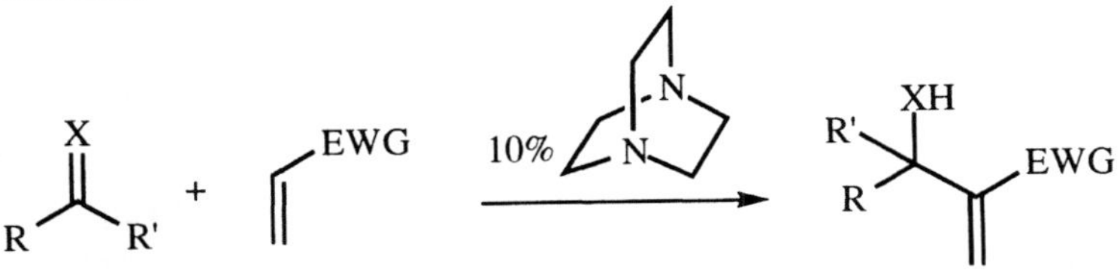

$X = O, NR''$ $EWG = CHO, CN, COOMe, COCH_3, SO_2Ph, etc.$

Scheme 1. General scheme for MBH reaction

Various electron-withdrawing groups, such as aldehyde, nitrile, esters, ketones, sulfones, etc. have been utilized to activate the vinyl moiety. The generally accepted mechanism (1) shown in Scheme 2 involves the Michael addition of the amine catalyst to the alkene, followed by an aldol type addition to the carbonyl or imine compound. Subsequent elimination releases the catalyst, and the cycle continues.

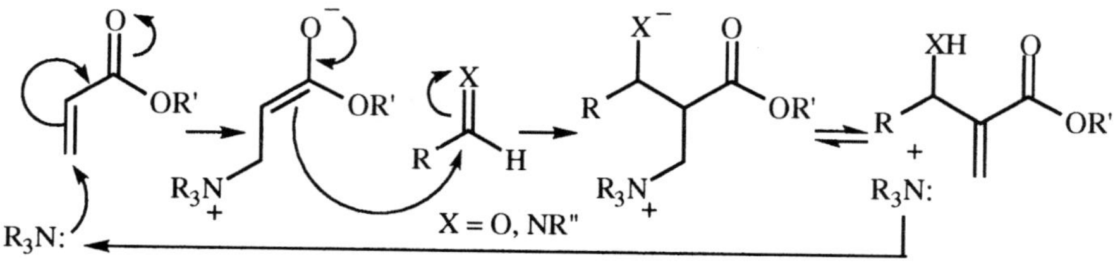

Scheme 2. Mechanism for MBH reaction

The product allyl alcohols have been employed in the synthesis of various biologically active compounds, such as the anti-tumor agent Sarkomycin ester (5) and α-methylene-γ-butyrolactones (6). These lactones constitute 10% of all natural products, including medicinals for cardio-vascular diseases (7). In materials chemistry, the Morita-Baylis-Hillman products have been applied to prepare novel side chains of liquid crystal polymers (8).

There are several drawbacks for this otherwise simple reaction. One of the shortcomings is its impractical slow rate, often requiring two or more weeks for completion. Also, the yields are inconsistent. In addition, the reaction is not applicable to β-substituted alkenes and is limited to only a few classes of activated ketones, such as α-keto esters (9, 10), perfluoroalkyl ketones (11), and non-enolizable 1,2-diketones (12). For example, only 7% conversion occurs when acetone is reacted with n-butyl acrylate at 120 °C in 4-6 d and aralkyl ketones fail to react even under high pressure (2).

A convenient alternative to the MBH reaction which avoids the slow rate is the vinylmetalation reaction (13-15). 1,2-Addition of a vinyl group to carbonyl compounds and imines via a vinylmetal intermediate is known as the vinylmetalation reaction. This methodology allows the reactions of ketones, β-substituted, and β,β-disubstituted olefins also to be included in the scheme. We carried out a systematic study of both MBH and vinylmetalation reactions for the syntheses of functionalized, fluorine-containing allyl alcohols from fluoro-aldehydes and –ketones. We have also explored the asymmetric version of these reactions. This review discusses our successes and failures, with pointers for the future.

Morita-Baylis-Hillman Reaction of Fluoro-carbonyls

MBH Reaction of Fluorinated Benzaldehydes. Although MBH reaction with benzaldehyde and several of its substituted derivatives has been studied (1), the reactions of the corresponding fluorinated benzaldehydes are not known. We studied the effect of fluorine subsitution of benzaldehydes in MBH reaction by reacting ethyl acrylate, acrylonitrile, acrolein and methyl vinyl ketone with several fluorinated benzaldehydes. The reactions of mono (2'-, 3'-, and 4'-), di- (2',3'-,2',4'-, 2',5'-, 2',6'-, 3',4'-, 3',5'-), and 2',3',4',5',6'-pentafluorobenzaldehydes were complete within 4 d and the product allylic alcohols were obtained in good yields (Scheme 3).

MBH Reaction of Perfluoro-aliphatic Aldehydes. Fluoral is known to polymerize in the presence of amines (Scheme 4) (16). However, we tested whether the polymerization is favored in the presence of activated olefins.

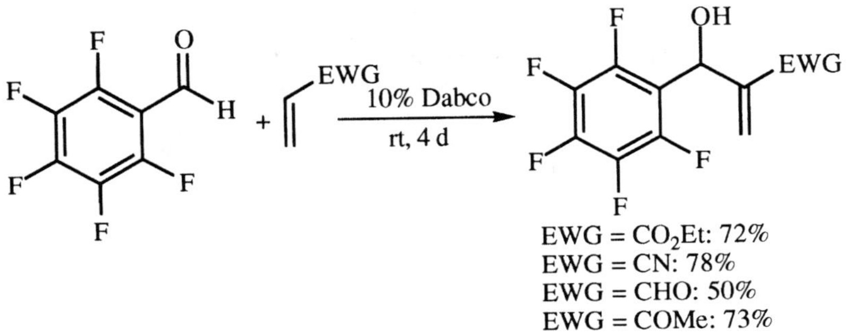

EWG = CO$_2$Et: 72%
EWG = CN: 78%
EWG = CHO: 50%
EWG = COMe: 73%

Scheme 3. MBH reaction of pentafluorobenzaldehyde

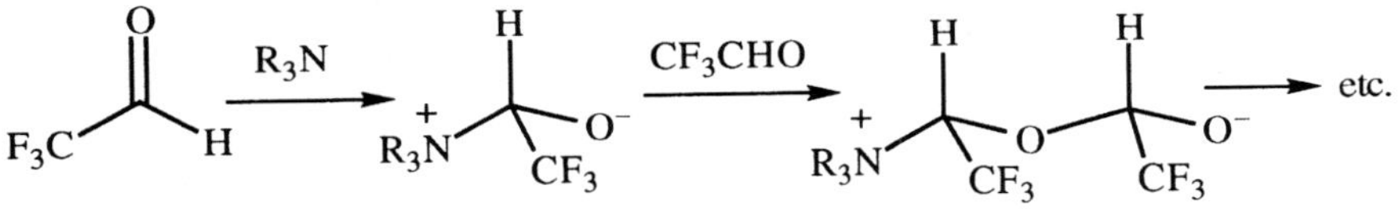

Scheme 4. Polymerization of fluoral

Ethyl acrylate provided ~20% yield of the product along with the polymer (Scheme 5). Polymerization is faster than the reaction with acrylonitrile (17). Two of the reactive olefins, acrolein and methyl vinyl ketone provided 50-70% yields of the product allyl alcohols (Scheme 6). Indeed, the reaction of these reactive olefins and aldehydes was complete in THF at –25 °C within 1 h (Reddy, M. V. R.; Rudd, M. T., unpublished results).

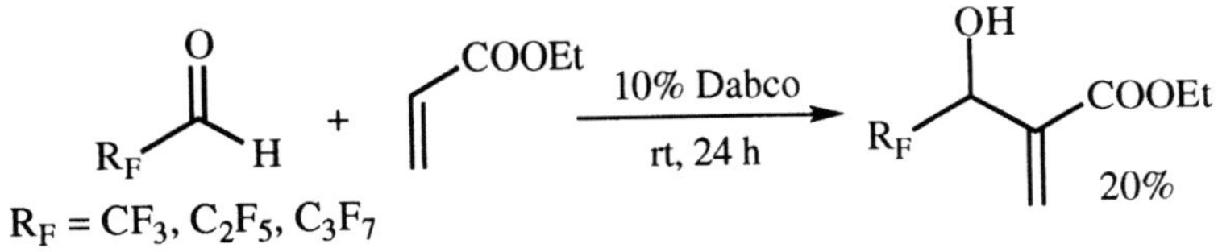

R$_F$ = CF$_3$, C$_2$F$_5$, C$_3$F$_7$

20%

Scheme 5. MBH reaction of perfluoro-aldehydes with ethyl acrylate

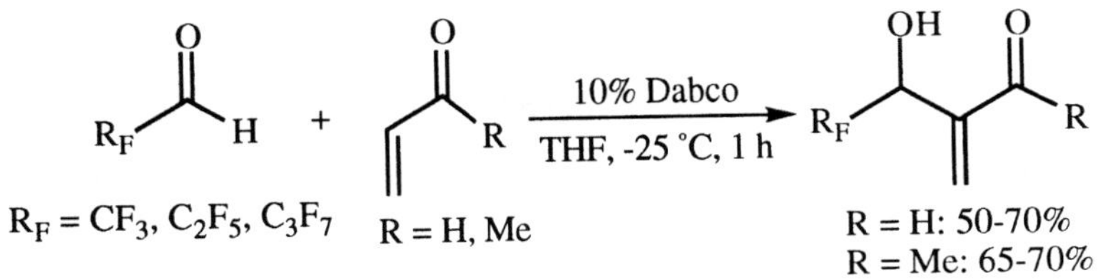

R$_F$ = CF$_3$, C$_2$F$_5$, C$_3$F$_7$ R = H, Me

R = H: 50-70%
R = Me: 65-70%

Scheme 6. MBH reaction of perfluoro-aldehydes

We observed similar results with two other perfluoroalkyl aldehydes, 2,2,3,3,3-pentafluoropropionaldehyde, and 2,2,3,3,4,4,4-heptafluorobutyraldehyde.

MBH Reaction of Aromatic Trifluoromethyl Ketones. The MBH reaction of hexafluoroacetone and the corresponding N-benzoyl imine is known (Scheme 7) (*10*). We were engaged in the reaction of perfluoroalkyl alkyl and aryl ketones due to our interest in asymmetric synthesis of fluoroalkyl allylic alcohols.

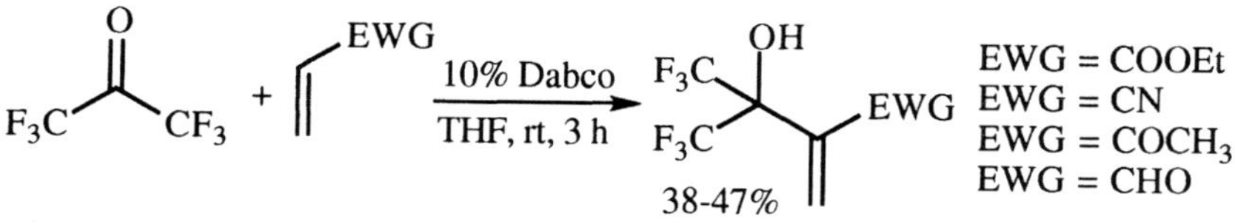

10% Dabco
THF, rt, 3 h
38-47%
EWG = COOEt
EWG = CN
EWG = COCH3
EWG = CHO

Scheme 7. MBH reaction of hexafluoroacetone

It has been reported that 1,1,1-trifluoroacetone trimerizes in the presence of amines (Scheme 8) (*18*). Although we observed some reaction in the case of perfluoroalkyl aldehydes, ethyl acrylate and methyl vinyl ketone failed to react with 1,1,1-trifluoroacetone. Acrolein and acrylonitrile provided ~10% yield of the product along with the trimer.

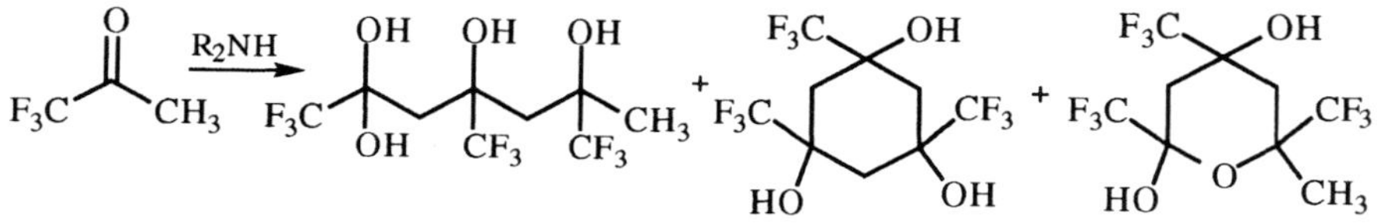

R2NH

Scheme 8. Trimerization of 1,1,1-trifluoroacetone

Aromatic trifluoromethyl ketones react readily with ethyl acrylate within 7 d and provide the products in 65-70% yields. Acrylonitrile reacts even faster, within 24 h, and the products were obtained in 82-94% yields (Scheme 9).

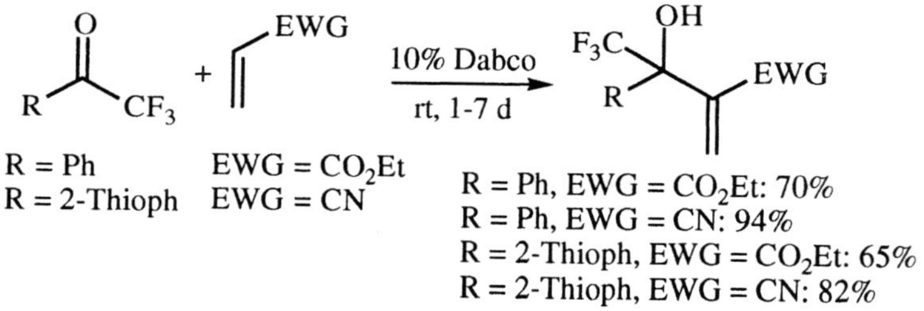

10% Dabco
rt, 1-7 d
R = Ph
R = 2-Thioph
EWG = CO2Et
EWG = CN
R = Ph, EWG = CO2Et: 70%
R = Ph, EWG = CN: 94%
R = 2-Thioph, EWG = CO2Et: 65%
R = 2-Thioph, EWG = CN: 82%

Scheme 9. MBH reaction of aryl trifluoromethyl ketones

However, acrolein underwent polymerization and methyl vinyl ketone dimerized under the reaction conditions (Scheme 10) (*19*).

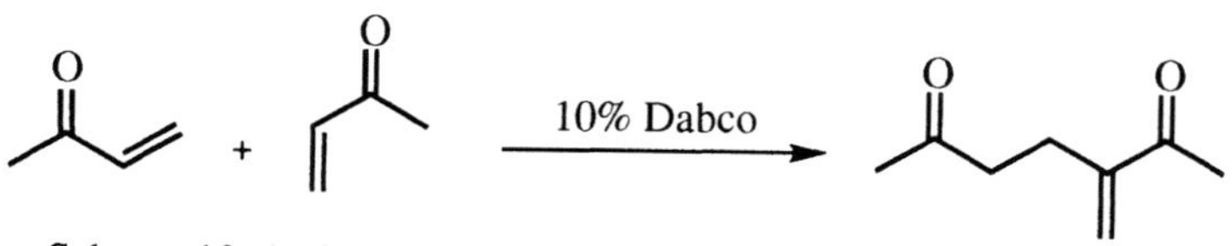

10% Dabco

Scheme 10. Dabco-catalyzed dimerization of methyl vinyl ketone

In contrast to aryl trifluoromethyl ketones, 2,2,2-trichloroacetophenone underwent an extremely slow reaction providing several products. 2-Chloro-2,2-difluoroacetophenone reacted readily with the activated olefins (Scheme 11).

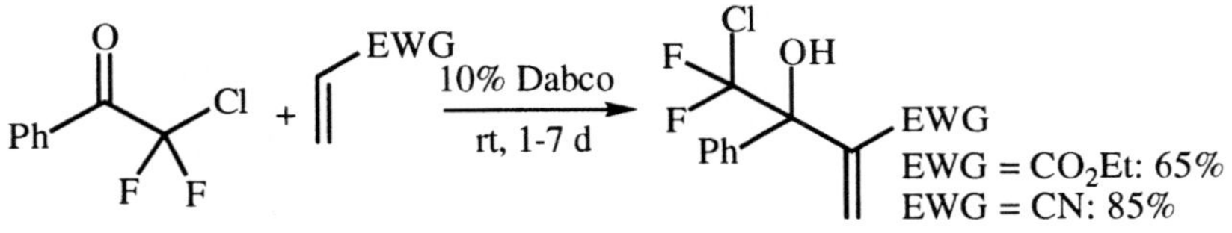

Scheme 11. MBH reaction of 2-chloro-2,2-difluoroacetophenone

Trifluoroacetylpyrrole and -indole underwent 1,4-addition to activated olefins rather than the MBH reaction (Scheme 12).

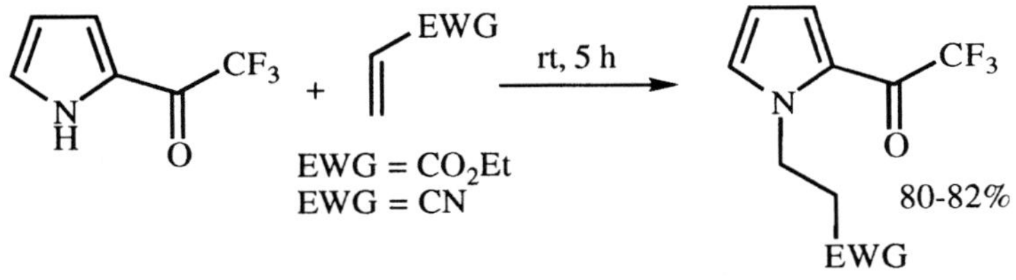

Scheme 12. 1,4-Addition of heteroaryl trifluoromethyl ketones

MBH Reaction of α-Acetylenic Trifluoromethyl Ketones. We exploited the reactivity of α-trifluoromethyl α'-acetylenic ketones for MBH reaction. In this case, all four activated olefins reacted comfortably, and the product enynols were obtained in 40-75% yield. In fact, acrolein and methyl vinyl ketone reacted almost instantaneously with the trifluoromethyl acetylenic ketones (Scheme 13).

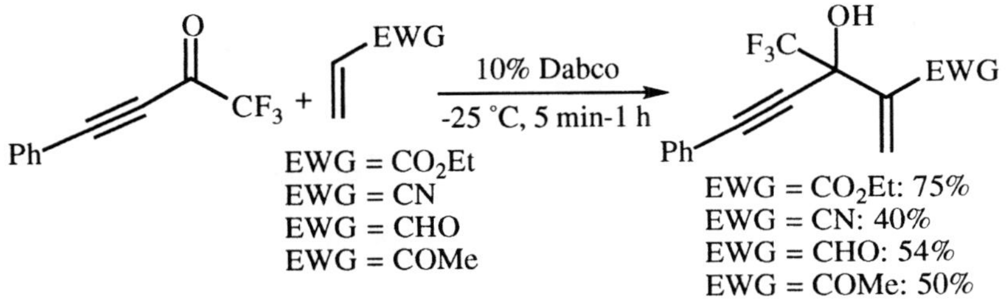

Scheme 13. MBH reaction of α-acetylenic α'-trifluoromethyl ketones

Vinylmetalation of Fluoro-carbonyls

Vinylalumination of Fluoro-aldehydes and Ketones. One of the problems of MBH reaction is the lack of reativity of β-substituted activated olefins. We applied the vinylalumination reaction reported by Tsuda and co-workers (13) for the synthesis of fluorinated MBH products (Scheme 14) (17). Due to the high reactivity of fluorinated aldehydes and ketones, we could achieve good yields of the products within 4 h. Contrary to ordinary ketones, fluorinated ketones did not require Lewis acid catalysis. This is much superior to the MBH procedure which requires several days for a reaction. Moreover, this procedure is amenable to combinatorial synthesis.

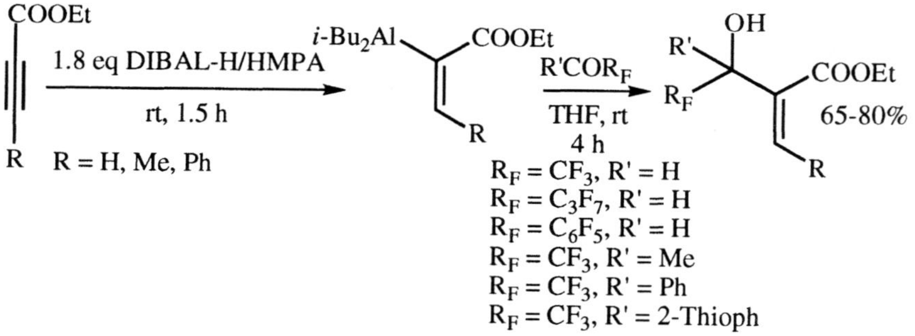

Scheme 14. Vinylalumination of fluoro-carbonyl compounds

Vinylcupration of Fluoro-aldehydes and Ketones. Marino and Lindermann (20) have reported the addition of Corey's vinylcopper reagents (21) to aldehydes and ketones. We carried out vinylcupration of fluorinated aldehydes and ketones to prepare functionalized fluorinated β-substituted allyl alcohols (Scheme 15). Unlike vinylalumination, vinylcupration of fluoro-carbonyls is not stereoselective (Pitre, S. V., unpublished results).

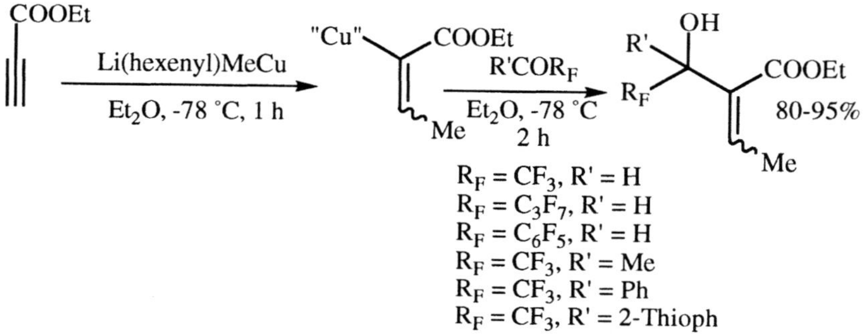

Scheme 15. Vinylcupration of fluoro-carbonyl compounds

Asymmetric MBH Reaction

Asymmetric MBH reaction has encountered only sporadic successes. While it is apparent that high rewards are to be expected from developing an asymmetric version of this reaction, only a few attempts have been made. Drews (22-23), Basavaiah (24), Isaacs (25), and Roos (26) utilized one or more of the following chiral auxiliaries (Fig 1) in the activated olefin to achieve asymmetry in the product. However, success has eluded them.

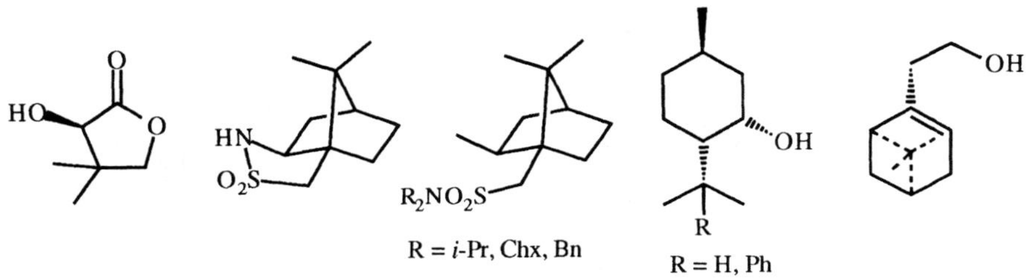

Figure 1. Chiral auxiliaries tested in the literature
for asymmetric MBH reaction

Kundig and co-workers achieved near quantitative enantioselectivities in the reaction of tricarbonylchromium complexes of *ortho*-substituted benzaldehydes (Scheme 16) (*27*).

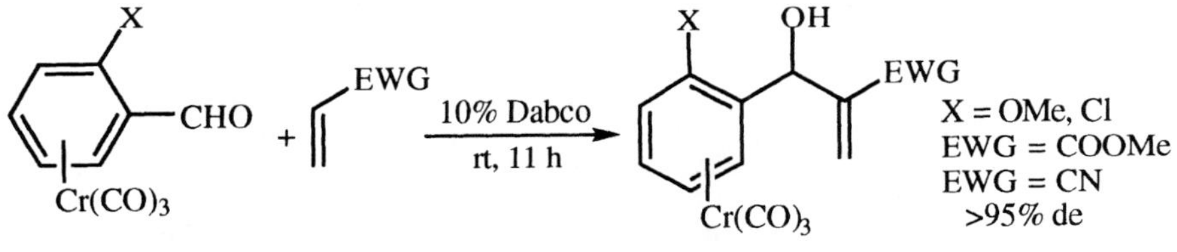

Scheme 16. Asymmetric MBH reaction of diastereomeric benzaldehydes

Leahy and co-workers recently reported that camphor sultam is an excellent chiral auxiliary to achieve almost quantitative enantioselectivity in the reaction of aliphatic aldehydes (Scheme 17) (*28*). However, aromatic aldehydes fail to undergo reaction under their conditions.

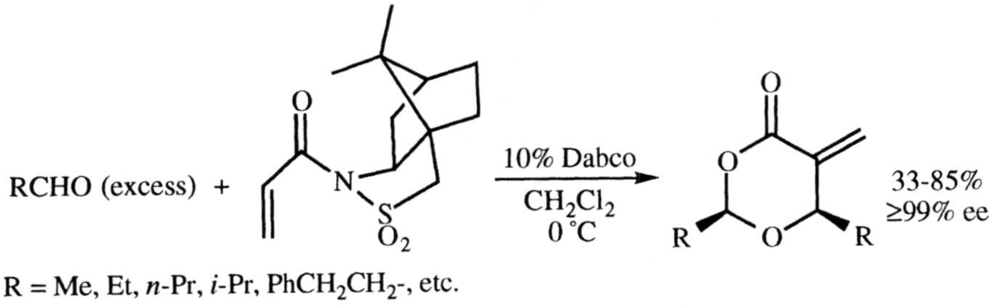

R = Me, Et, *n*-Pr, *i*-Pr, PhCH$_2$CH$_2$-, etc.

Scheme 17. Camphor sultam based asymmetric MBH reaction

Marko and co-workers studied the effect of pressure on the enantioselectivity using (-)-3-hydroxyquiniclidine, brucine, strychnine, cinchonidine, cinchonine, quinine, quinidine, *N*-methylprolinol, and *N*-methyl ephedrine as chiral catalysts (Figure 2) (*29*). Hirama et al. reported the employment of synthetic C$_3$-symmetric Dabco derivatives (Figure 2) as catalysts in the reaction (*30*). Nevertheless, none of these attempts achieved high chiral induction.

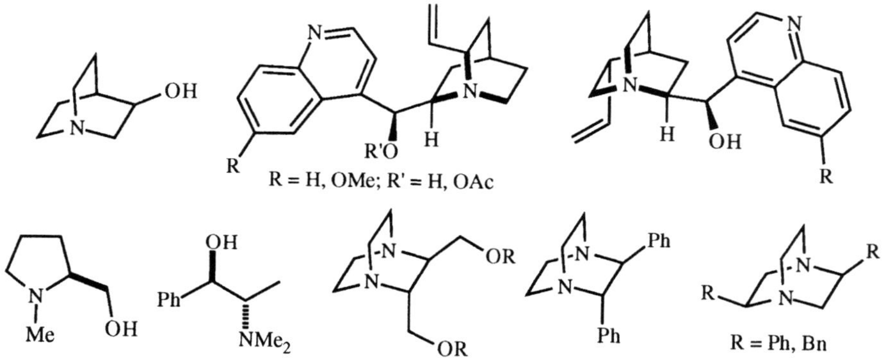

Figure 2. Chiral catalysts tested in the literature for asymmetric MBH reaction

Exploratory Study of Terpenyl Alcohols for MBH Reaction of Fluorobenzaldehydes

We have carried out several asymmetric transformations using terpenes, such as α-pinene and carenes as chiral auxiliaries in our organoborane program (*31-32*). We examined the asymmetric MBH reaction of isopinocampheyl- (Ipc) and 4-isocaranyl (4-Icr) acrylates with a series of fluorinated benzaldehydes and perfluoroalkyl aldehydes. The reactions were very slow, requiring 15-30 days for completion. Unfortunately, perfluoroaldehydes and 1,1,1-trifluoroacetone underwent polymerization, and fluorobenzaldehydes provided the products in very poor ee (5-13%) (Scheme 18) (Rudd, M. T.; Reddy, M. V. R., unpublished results). Considering that isopinocampheyl acrylate provides only 13% ee with benzaldehyde, no influence of fluorine substitution is observed in asymmetric MBH reaction. Currently, none of the chiral auxiliaries are successful in inducing high asymmetric induction for benzaldehyde in the MBH reaction. Our results suggest that development of this chemistry should be applicable to MBH reaction of fluorine containing substrates as well.

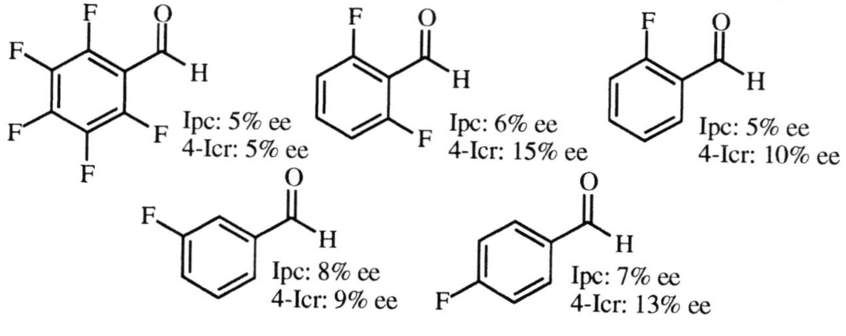

Scheme 18. Asymmetric MBH reaction of fluorobenzaldehydes

Exploratory Study of Terpenyl Alcohols for Vinylalumination Reactions

Greene and co-workers examined a series of chiral alcohols, such as menthol, 2-phenylcyclopentanol, 2-phenylcyclohexanol, etc. as chiral auxiliaries in the asymmetric vinylalumination reaction of benzaldehyde and activated aldimines (*15*). They reported that 2-phenyl-cyclohexanol is the best chiral auxiliary for aldimines. We examined isopinocampheol and 4-isocaranol as chiral auxiliaries for the asymmetric vinylalumination reaction so that ready comparison with asymmetric MBH reaction described above can be made.

Since it is metal mediated, the reactions could be carried out at low temperatures, and were complete within 15 h–2 d. Amine sensitive substrates, such as fluoral and 1,1,1-trifluoroacetone, also underwent reaction to provide the products in good yields. However, we obtained the product allyl alcohols in only 15-43% ee (Scheme 19) (Rudd, M. T.; Reddy, M. V. R., unpublished results). We did not attempt to determine the configurations of these allylic alcohols due to their low ee.

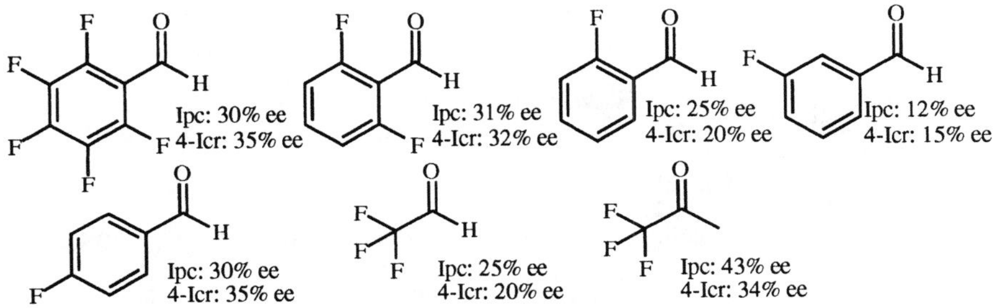

Scheme 19. Asymmetric vinylalumination of fluoro-carbonyl compounds

Recently, Li and co-workers reported superior enantioselectivity in vinylcupration of a series of aldehydes starting with menthyl propiolates (*33*). While MBH reaction of menthyl acrylates provides ≤ 20% ee, vinylcupration provides 50-87% ee. The enantioselectivities obtained in the MBH reaction of benzaldehyde with menthyl, isopinocampheyl and isocaranyl acrylates were in the same range (≤ 20%). Comparison of Greene's results (*15*) on asymmetric vinylalumination, and Li's asymmetric vinylcupration of benzaldehyde (*33*) starting with menthyl propiolate shows that vinylcupration provides better enantioselectivity. These results point to the fact that asymmetric vinylcupration of fluoro-carbonyl compounds with these auxiliaries should be a worthwhile project.

Conclusions

We have reviewed our study on Morita-Baylis-Hillman (MBH) and vinylmetalation reactions of fluoro-carbonyl compounds for the synthesis of achiral and chiral, functionalized, fluorinated allyl alcohols. Vinylmetalation of fluoro-carbonyls with aluminum and copper reagents is an efficient alternative to MBH reaction. We have discussed an exploratory study of terpenyl alcohols as chiral auxiliaries in asymmetric MBH and vinylmetalation reactions. Although we did not obtain high asymmetric induction, there is reason to believe that rich rewards should be obtained by a careful examination of several known and new chiral auxiliaries. Asymmetric vinylalumination provides higher ee compared to asymmetric MBH reaction. Vinylcupration of nonfluorinated carbonyls has shown that even higher ees were obtained as compared to vinylalumination reaction. All of these point to the fact that development of a good chiral auxiliary for MBH reaction could possibly lead to successful vinylmetalation reactions.

126

Acknowledgment

We gratefully acknowledge the financial support from Purdue Borane Research Fund.

Literature Cited

(1) Ciganek, E. in *Organic React.* **1997**, *51*, 201.
(2) Basavaiah, D.; Rao, P. D.; Hyma, R. S.*Tetrahedron* **1996**, *52*, 8001.
(3) Drewes, S. E.; Roos, G. H. P. *Tetrahedron* **1988**, *44*, 4653.
(4) Crouch, R. D.; Nelson, T. D. *J. Chem. Ed.* **1995**, *72 (1)*, A6.
(5) Amri, H.; Rambud, M.; Villieras, J. *Tetrahedron Lett.* **1989**, *30*, 7381.
(6) Masuyama, Y.; Nimura, Y.; Kurusu, Y. *Tetrahedron Lett.* **1991**, *32*, 225.
(7) Hoffman, H.M. R.; Rabe, J. *Angew. Chem. Int. Ed. Engl.* **1985**, *24*, 94.
(8) Hall, A. W.; Lacey, D.; Hill, J. S.; McDonnel, D. G. *Supramolecular Science* **1994**, *1*, 21.
(9) Grundke, C.; Hoffman, H. M. R. *Chem. Ber.* **1987**, *120*, 1461.
(10) Basavaiah, D.; Bharathi, T. K.; Gowriswari, V. V. L. *Tetrahedron Lett.* **1987**, *28*, 4351.
(11) Golubev, A. S.; Galakhov, M. V.; Kolomiets, A. F.; Fokin, A. V. *Izv. Akad. Nauk Ser. Khim.* **1992**, 2763.
(12) Strunz, G. M.; Bethell, R.; Sampson, G.; White, P. *Can. J. Chem.* **1995**, *73*, 1666.
(13) Tsuda, T.; Yoshida, T.; Saegusa, T. *J. Org. Chem.* **1988**, *53*, 1037.
(14) Zu, L. H.; Kundig, E. P. *Helv. Chim. Acta* **1994**, *77*, 1480.
(15) Génisson, Y.; Massardier, C.; Gautier-Luneau, I.; Greene, A. E. *J. Chem. Soc. Perkin Trans. I* **1996**, 2869.
(16) Busfield, W. K.; Whalley, E. *Polymer* **1966**, *7*, 541.
(17) Ramachandran, P. V.; Reddy, M. V. R.; Rudd, M. T.; de Alaniz, J. R. *Tetrahedron Lett.* **1998**, *39*, 8791.
(18) Dhingra, M. M.; Tatta, K. R. *Org. Mag. Res.* **1977**, *9*, 23.
(19) Basavaiah, D.; Gowriswari, V. V. L.; Bharathi, T. K. *Tetrahedron Lett.* **1987**, *28*, 4591.
(20) Marino, J. P.; Linderman, R. J.; *J. Org. Chem.* **1983**, *48*, 4621.
(21) Corey, E. J.; Katzenellenbogen, J. A. *J. Am. Chem. Soc.* **1969**, *91*, 1853.
(22) Khan, A. A.; Esmslie, N. D.; Drews, S. E.; Field, J. S.; Ramesar, N. *Chem. Ber.* **1993**, *126*, 1477.
(23) Drewes, S. E.; Emslie, N. D.; Khan, A. A. *Synth. Commun.* **1993**, *23*, 1215.
(24) Basavaiah, D.; Gowriswari, V. V. L.; Sarma, P. K. S.; Rao, P. D. *Tetrahedron Lett.* **1990**, *31*, 1621.
(25) Gilbert, A.; Heritage, T. W.; Isaacs, N. S. *Tetrahedron: Asym.* **1991**, *2*, 969.
(26) Jensen, K. N.; Roos, GHP. *S. Afr. J. Chem.* **1992**, *45*, 112.
(27) Kundig, E. P.; Xu, L. H.; Romanens, P.; Bernardinelli, G. *Tetrahedron Lett.* **1993**, *34*, 7049.
(28) Brzezinski, L. J.; Rafel, S. Leahy, J. W. *J. Am. Chem. Soc.* **1997**, *119*, 4517.
(29) Marko, I. E.; Giles, P. R.; Hindley, N. J. *Tetrahedron* **1997**, *53*, 1015.
(30) Oishi, T.; Oguri, H.; Hirama, M. *Tetrahedron: Asym.* **1995**, *6*, 1241.
(31) Brown, H. C.; Ramachandran, P. V. *J. Organometal. Chem.* **1995**, *500*, 1.
(32) Brown, H. C.; Ramachandran, P. V. in *Advances in Asymmetric Synthesis*, Vol. 1. Chapter 5, Hassner, A. Ed. JAI Press, Greenwich, CT, **1994**.
(33) Wei, H. X.; Willis, Li, G. *Tetrahedron Lett.* **1998**, *39*, 8203.

Chapter 10

Stereoselective Synthesis of β-Fluoroalkyl β-Amino Alcohol Units

Matteo Zanda[1], Pierfrancesco Bravo[2], and Alessandro Volonterio[1]

[1]Dipartimento di Chimica del Politecnico, Via Mancinelli 7, I–20131 Milano, Italy
[2]CNR-Centro di Studio sulle Sostanze Organiche Naturali, via Mancinelli 7, I–20131 Milano, Italy

Stereochemically defined β-fluoroalkyl β-amino alcohol units feature a great potential as key components of a wide range of biologically and biomedicinally interesting compounds, such as fluorinated peptide isosteres, *ephedra*-alkaloids and amino acids. Two synthetic strategies to build the title compounds are described, the stereochemical aspects of the key asymmetric reactions, as well as their mechanism, are discussed. One exploits a synergistic combination of two asymmetric processes: a) the chiral sulfoxide controlled additions of nucleophiles to fluorinated imines; b) the "non oxidative" *Pummerer* reaction, which allows for totally stereocontrolled substitution of the sulfinyl auxiliary with a hydroxyl. The other extends the *Evans* aldol reaction to strongly electrophilic ketimines derived from trifluoropyruvate, and allows for the highly stereoselective preparation of densely functionalized building blocks having a quaternary α-fluoroalkylamino stereogenic center.

Molecules incorporating a stereochemically defined β-fluoroalkyl β-amino alcohol unit hold a great potential for the preparation of new biologically and pharmaceutically important substrates, such as peptide mimetics, enzyme inhibitors, anti-cancer drugs (for example analogs of Taxol) and antibacterials (*1-3*). However, few methods are available for achieving their synthesis. Some selected examples are summarized in Scheme 1. An approach to racemic β-fluoroalkyl β-amino alcohols, protected as oxazolidin-2-ones or Δ-2-oxazolines, was reported in 1989, by *Laurent* and *Mison*, who used the sodium iodide promoted isomerization of *N*-activated aziridines (*4*). *Kozikowski* published the synthesis of fluorine-containing isosteres of sphingosine, as potential inhibitors of protein kinase C, starting from the aldehyde derived from (*S*)-3-fluoro alanine (*5*). *Seebach* described a new approach to the racemic title compounds based on the condensation of silyl nitronates of 2,2,2-trifluoronitroethane to aldehydes

(*6,7*). *Uno* and *Suzuki* reported the boron trifluoride assisted perfluoroalkylation of α-alkoxy imines (*8*), producing optically active β-fluoroalkyl β-amino alcohols with moderate to good stereoselectivity. The synthesis of β-trifluoromethyl isoserinate was reported by *Bégué* (1996) (*9,10*), who exploited the stereoselective [2 + 2] cycloaddition between *N-p*-methoxyphenyl (PMP) imine of fluoral and the appropriate ketene. Very recently (1998) *Uneyama* described a novel strategy to prepare racemic β-trifluoromethyl isoserinates (*11*), based on the stereoselective intramolecular rearrangement of fluoroalkyl imino ethers.

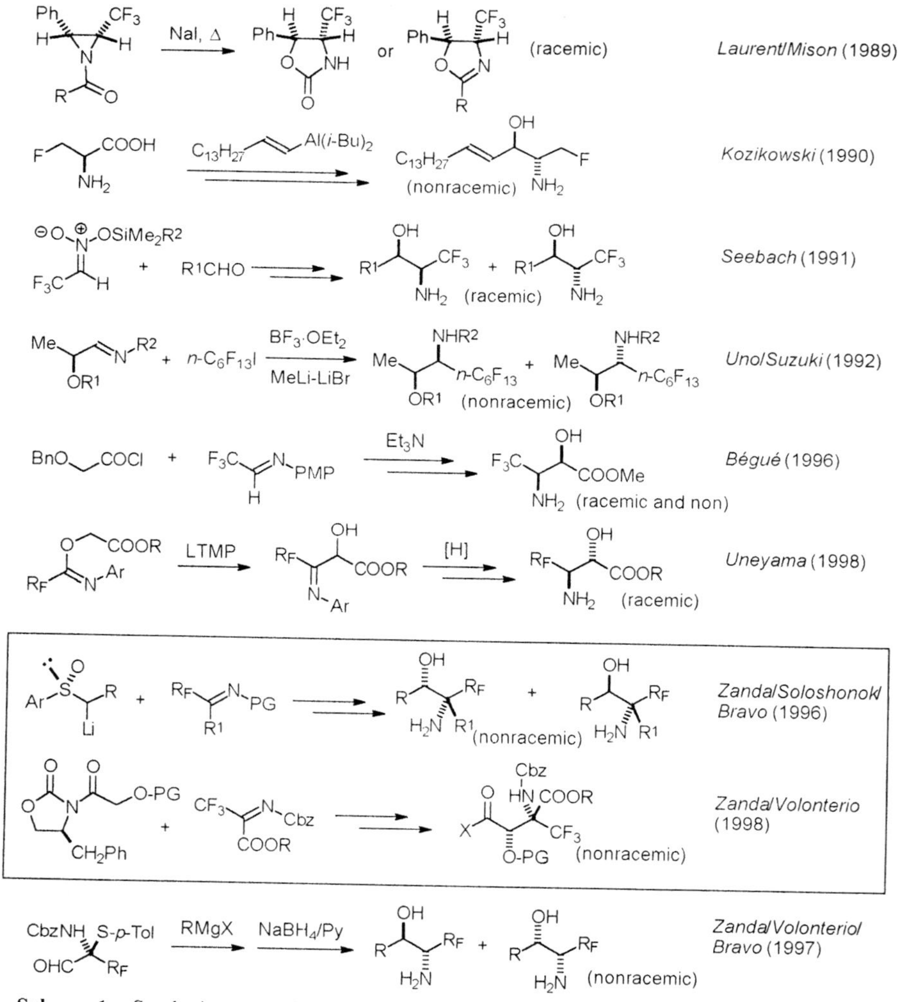

Scheme 1. Synthetic approaches to racemic and nonracemic β-fluoroalkyl β-amino alcohols

Considering the biological interest and the challenge connected with the stereocontrolled synthesis of β-fluoroalkyl β-amino alcohols, we became interested in developing new asymmetric methodologies to prepare these compounds (*12-14*). Since 1994 we have disclosed three general stereoselective approaches to the target: 1) a synergistic combination of two reactions: a) stereocontrolled C-C bond formation by additions of metalated sulfinyl nucleophiles to fluorinated imines (*15-16*); b) S_N2-type displacement of the sulfinyl auxiliary with a hydroxy group, by means of the "Non-Oxidative" *Pummerer* reaction (*17-19*). 2) *Evans* aldol reaction involving *N*-Cbz imines of trifluoropyruvic esters, leading to functionalized β-fluoroalkyl β-amino alcohol units having a quaternary amine center (*20*). 3) Using fluoropyruvaldehyde *N,S*-ketals, new electrophilic nonracemic molecules prepared in our laboratories (*21-23*), as stereochemically stable synthetic equivalents of α-amino α-fluoroalkyl aldehydes (*24,25*).

This chapter will discuss the first two methods (highlighted in Scheme 1), while the third one will be presented in the chapter entitled "Asymmetric Synthesis of Fluoro Amino Compounds *via* Chiral Sulfoxides".

Additions of Lithium Alkyl Aryl Sulfoxides to *N*-PMP Fluoroalkyl/aryl Imines: Synthesis of Stereodefined Fluoroalkyl/aryl Amines, β-Amino Alcohols and Amino Acids.

In this section we describe an original two-step approach to a wide range of α-fluoroalkyl/aryl-amino compounds, most of them incorporating the title unit.

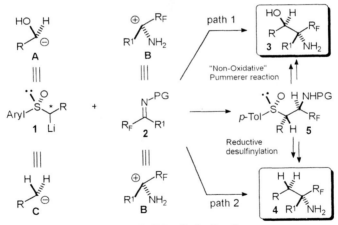

R_F = Fluoroalkyl, Fluoroaryl; PG = Protecting Group

Scheme 2. Use of lithiated sulfoxides in the stereoselective synthesis of β-fluoroalkyl β-amino alcohols and fluoroalkyl amines

According to this strategy, α-lithium alkyl sulfoxides **1** (Scheme 2) have been used as chiral α-hydroxyalkyl carbanion equivalents **A**, with fluorinated *N*-protected imines **2**, synthetic equivalents of α-amino fluoroalkyl/aryl carbocations **B** (path 1), to give the target β-fluoroalkyl β-amino alcohols **3**. This represents an unprecedented application of lithium alkyl sulfoxides in asymmetric synthesis, which have been previously used only as chiral alkyl anion equivalents **C** (*26-28*): for example, we have used the latter protocol to prepare chiral β-fluoro amines **4** (path 2, Scheme 2). The first asymmetric step (path 1) consists in the formation of a new C-C bond by assembling of the lithiated alkyl sulfoxide **1** with the fluoro imine **2**, to provide the intermediate *N*-protected β-sulfinyl amine **5** (*15*). *p*-Methoxyphenyl (PMP) is of choice as nitrogen protecting group, because it provides configurational stability to the imine and leaves alive the nitrogen lone-pair for coordination by the lithium, which are very important features for achieving stereocontrol in the condensation (*16*).

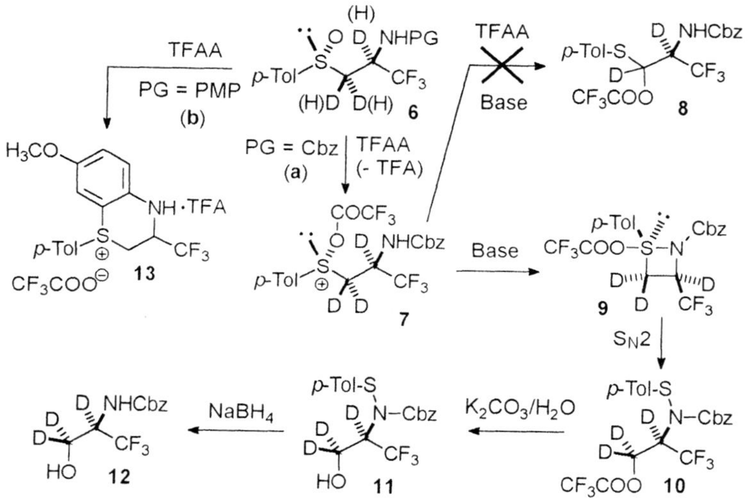

Scheme 3. Mechanism of the "Non-Oxidative" Pummerer Reaction

The second step is the one-pot S_N2-type displacement of the sulfinyl auxiliary with a hydroxyl, by means of the "Non-Oxidative" *Pummerer* reaction (NOP Reaction) (*17-19*), a key methodology recently disclosed in our laboratories that allows for a full exploitation of the stereochemical informations of the intermediate **5**, that can be directly converted into the target β-fluoroalkyl β-amino alcohol **3**. The mechanism of this reaction, that occurs when *N*-alkoxycarbonyl (in particular PG = Cbz) β-sulfinyl amines like **6a** (Scheme 3) are treated with trifluoroacetic anhydride (TFAA) in the presence of a base like *sym*-collidine, was assessed by means of deuterium labeling experiments. The classical *Pummerer* reaction promoted by TFAA/*sym*-collidine of **6a** would result in an internal redox process producing the masked α-amino aldehyde **8** *via* reduction of the sulfinyl group to sulfenyl and oxidation of the carbon attached to sulfur (*29-31*). However, this outcome was not observed at all. Treatment of **6** with TFAA produced the expected formation of the cation **7**, but *sym*-collidine triggered the formation of the chiral σ-sulfurane **9**, *via* deprotonation of the acidic aminic proton. Neither of the D_2C-S deuterium atoms was removed, as would have occurred in the classical rearrangement. Intermediate **9** underwent a fast intramolecular rearrangement, consisting in a highly stereospecific S_N2-type displacement of the *p*-tolylthio residue, which migrates to the nitrogen, by the trifluoroacetoxy group. The isolable trifluoroacetoxy sulfenamide **10** was cleanly produced in this rearrangement. Unlike the classical *Pummerer* reaction, the carbon attached to sulfur is not oxidized, therefore we have proposed the name "Non-Oxidative" *Pummerer* reaction.

The S_N2-like stereochemical course of this process was confirmed performing the reaction on *N*-Cbz β-sulfinyl amines having a stereogenic center in β-position (*vide infra*) (*16,18*). Finally, compound **10** could be transformed readily and *in situ* into the target deuterated trifluoroalaninol (*R*)-**12**, by treatment with aqueous K_2CO_3, that produces the intermediate sulfenamido alcohol **11**, and then with $NaBH_4$, that cleaves the N-S bond. Key factor of this fascinating reaction is the neighbouring participation by the amino group monoprotected as NH-Cbz. We believe that also other carbamate and even amidic protections of nitrogen should allow the process to take place, but with NH-aryl β-sulfinyl amines a completely different outcome was observed. For example, the *N*-PMP derivative **6b** (Scheme 3) upon treatment with TFAA provided the isolable sulfonium trifluoroacetate **13** as the major product (*32*). Consequently, NH-PMP sulfinyl amines **5** (Scheme 2) must be transformed into the corresponding NH-Cbz derivatives before undertaking the NOP reaction. This is readily achieved by oxidative cleavage of the PMP with ceric ammonium nitrate (CAN) (*33*), followed by reprotection with benzyl chloroformate.

First application of our methodology is the preparation of a series of α-fluoroalkyl and α-(fluoro)aryl glycinols **20** (Scheme 4), carried out in collaboration with *Soloshonok* (*34,35*). Condensation of lithium derivative of (*R*)-methyl *p*-tolyl sulfoxide **14** with several *N*-PMP imines derived from fluoroalkyl aldehydes **15a,b** or (fluoro)benzaldehydes **15c-e** represents the first step. These C-C bond forming reactions occurred with good yields and stereoselection providing the diastereomeric β-sulfinyl *N*-PMP-amines **16** as major products. Transformation into the corresponding NH-Cbz derivatives **17** was accomplished in high overall yields by chemoselective oxidative cleavage of the PMP group (the sulfinyl was not affected), followed by *N*-reprotection with benzyl chloroformate. Removal of the sulfinyl and the Cbz group from the trifluoromethyl derivative, provided trifluoro *iso*-propylamine (*R*)-**18**. More interestingly, the NOP reaction applied on all the NH-Cbz β-sulfinyl amines **17** provided a convenient and high yielding entry to the corresponding (*R*)-alaninols and (*S*)-aryl glycinols **20**, *via* the intermediate trifluoroacetoxy sulfenamides **19**, which were generally not isolated. It is worth noting that the NOP reaction can be equally applied to substrates having a fluoroalkyl, a fluoroaryl and even a fluorine-free aryl group in α-position.

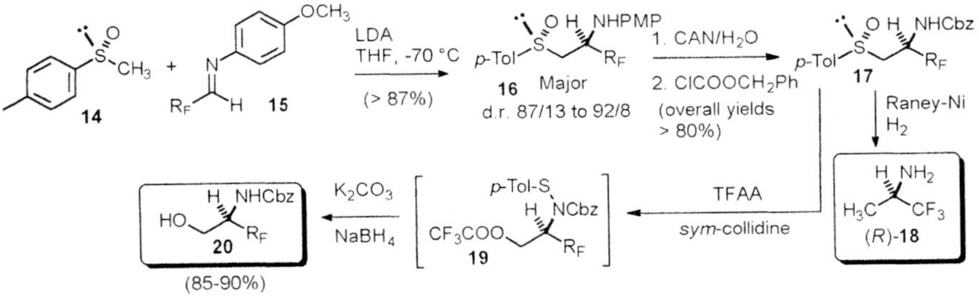

R_F = (**a**) CF_3, (**b**) CF_2CF_3, (**c**) 2-F-C_6H_4, (**d**) 4-F-C_6H_4, (**e**) C_6H_5

Scheme 4. Stereoselective synthesis of (*R*)-β-fluoroalaninols and (*S*)-fluoroaryl glycinols

The method has been recently improved, in collaboration with *Fustero* (Scheme 5) (*36*). In fact, lithium derivative of (*S*)-naphthyl methyl sulfoxide **21** was found to provide much better stereocontrol in the condensation with *N*-PMP fluoroalkyl imines **15a,b,f**. The resulting *N*-PMP naphthyl derivatives **22** were transformed into the enantiomeric (*S*)-β-fluoroalaninols **20** following an identical sequence of reactions, in high overall yields and stereoselectivity. Oxidation of the hydroxymethylene group, according to the method of *Sharpless* (*37*), extended the field of application of this strategy to the synthesis of enantiopure β-fluoro alanines **23**. The wide scope of the NOP reaction is demonstrated by the fact that also naphthyl sulfoxides, like **22**, can be successfully submitted to the reaction.

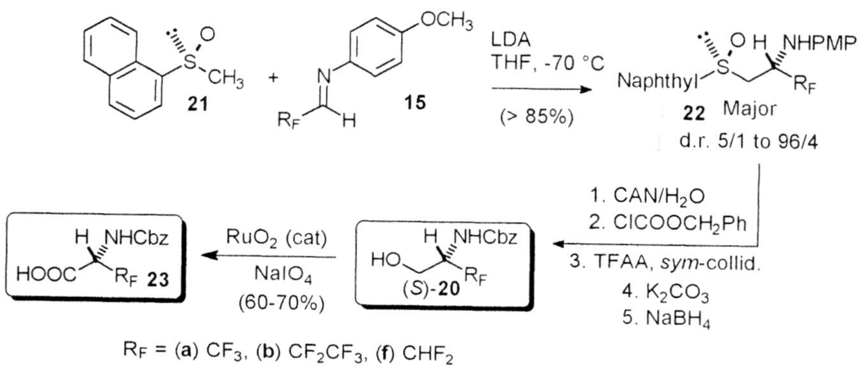

Scheme 5. Stereoselective synthesis of (*S*)-β-fluoroalaninols and alanines

Since *N*-PMP imines **15** should be (*E*)-geometrically homogeneous (*38*), the reactions above are likely to take place through the dominant *Zimmerman-Traxler* (aldol) chair-like transition state (TS) **D** (*39*), in which the fluoroalkyl/aryl group R_F points down away from the rest of the substituents, minimizing any non-bonding interaction (*40-42*). Accordingly, the better stereocontrol observed with naphthyl sulfoxide **21** could arise from the increased steric bulk of the naphthyl with respect to the *p*-tolyl group, that further stabilizes the TS **D**, having the aryl ligand attached to sulfur in *pseudo*-equatorial position.

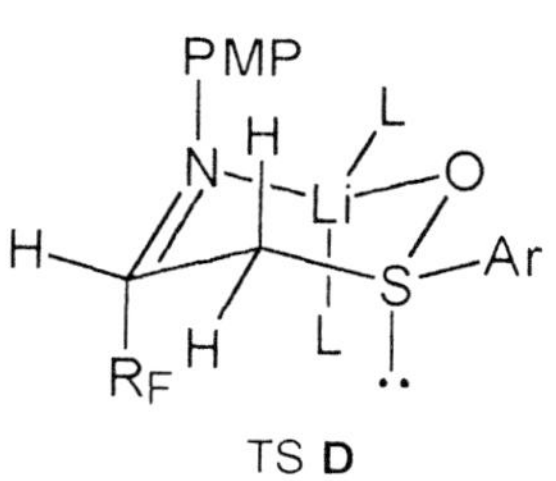

TS **D**

The first stereoselective application of the NOP reaction is described in Scheme 6 (*16,34*). (1*R*,2*R*)-NH-Cbz trifluoronorephedrine **26** was synthesized with high stereocontrol starting from lithiated (*R*)-benzyl *p*-tolyl sulfoxide **24**, used as chiral α-hydroxy benzyl carbanion equivalent **E**, and the *N*-PMP imine of fluoral **15a**. Condensation between lithiated **24** and **15a** provided efficiently one of four possible diastereoisomeric β-sulfinyl amines **25**. Transformation into the corresponding NH-Cbz derivative and stereoselective NOP reaction, carried out under the usual conditions with TFAA/*sym*-collidine, provided with good yields, total stereocontrol and inversion of configuration at the carbon center originally bound to sulfur, the NH-Cbz trifluoronorephedrine **26**, from which the free compound can be quantitatively obtained by standard hydrogenolysis of the Cbz. Alternatively, using more conventional chemistry, the Tfm-benzyl amine (*R*)-**27** was obtained in good yields by reductive desulfinylation of **25**. The stereochemical outcome of the condensation between lithium benzyl *p*-tolyl sulfoxide **24** and imine **15a** might be rationalized by supposing that **24** reacts predominantly in its less thermodynamically favorable *syn* stereoisomeric form (Scheme 6), having the *p*-tolyl and the phenyl groups *cis* with respect to the plane defined by the O-S-C-Li bonds, *via* an aldol-type TS, analogous to the TS **D** described above (*26*).

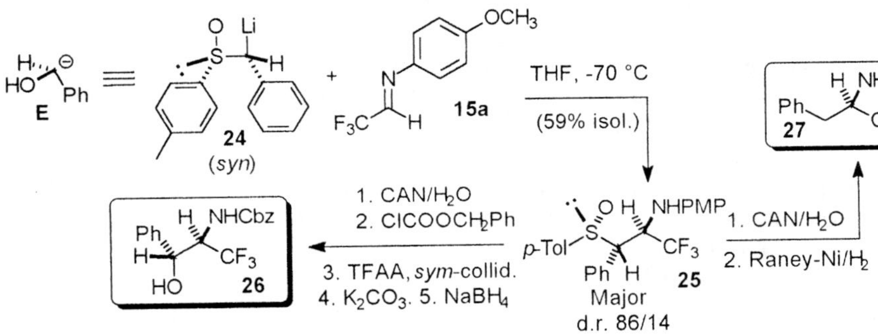

Scheme 6. Stereoselective synthesis of trifluoronorephedrine

The orthogonally protected γ-Tfm GABOB **32** (Scheme 7), a new hydroxymethylene (statine) dipeptide isostere with potential applications for the preparation of protease inhibitors, was obtained from lithiated (*R*)-3-butenyl *p*-tolyl sulfoxide **28**, used as chiral equivalent of the β-carbanion of β-hydroxy propionic acid **F**, and the imine **15a** (*43*). The condensation provided two diastereisomers **29** and **30** out of four possible, in 1.0/2.8 d.r. and quantitative yields. Conversion into the corresponding diastereomeric NH-Cbz derivatives, separation and stereoselective NOP reaction on the major diastereomer afforded in good overall yields and total stereocontrol the β-Tfm β-amino alcohol **31**. Protection of the hydroxyl and oxidative cleavage of the C=C double bond with KMnO₄ delivered the desired enantiopure NH-Cbz *O*-benzoyl γ-Tfm GABOB **32**. The preferential formation of the diastereomer **30** in the condensation step could be explained by the fact that the lithiated butenyl sulfoxide **28** reacts mainly in the thermodynamically more stable *anti*-geometry (see Scheme 7).

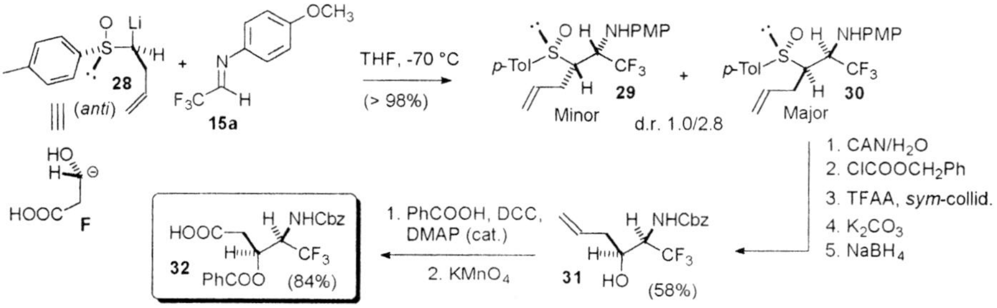

134

Scheme 7. Stereoselective synthesis of γ-Tfm-GABOB

α-Trifluoromethyl α-amino acids (α-Tfm-AAs) are synthetic analogues of naturally occurring amino acids, which attracted much attention due to the fact that some of them exhibit anticancer, antibacterial or antihypertensive properties, ability to behave as potent suicide inhibitors of pyridoxalphosphate-dependent enzymes (transaminases, decarboxylases), and that some peptides containing α-Tfm-AAs have increased metabolic stability and biological activity (44-46). Several enantiopure α-Tfm-AAs have been synthesized from lithium alkyl aryl sulfoxides **34,38,41** and N-Cbz imines of trifluoropyruvate **33** (Schemes 8 and 9), prepared by *Staudinger* (aza-Wittig) reaction (47-49) between the iminophosphorane Ph₃P=NCbz and trifluoropyruvate.

Both enantiomers of α-Tfm-alanine **36** (Scheme 8) were obtained by reaction of lithium (R)-methyl p-tolyl sulfoxide **34** with the imine **33**. This paper published in 1994 represented our first work in the field of the asymmetric synthesis of fluoro-amino compounds (15). Unfortunately, the reaction was not stereoselective providing both the diastereomeric β-sulfinyl amines **35**, which were separated and submitted to reductive desulfinylation, and then to hydrolysis providing both (R)- and (S)-**36**. The same diastereomeric intermediate **35** was submitted to the NOP reaction, followed by hydrolysis, to afford both enantiomers of the previously unknown α-Tfm-serine **37** (17). Both enantiomers of α-Tfm-phenylalanine **40** (Scheme 8) were obtained starting from lithium (R)-benzyl p-tolyl sulfoxide **38** and the same imine **33** (50). No stereocontrol was observed in the formation of the quaternary amine center of the adduct **39**, although a moderate stereoselectivity was observed in the formation of the center attached to the sulfinyl group. Deoxygenation with TFAA/NaI (*Oae/Drabowicz* protocol) (51), reductive desulfenylation and hydrolysis of mixtures of diastereomers having homogenous stereochemistry at the aminic center, afforded both enantiomers of the target α-Tfm-AA **40**.

Lithiation of (R)-ethyl p-tolyl sulfoxide **41** and condensation with the imine **33** afforded the β-sulfinyl amines **42,43** in almost equimolar ratio (50). These compounds are key intermediates for the preparation of both enantiomers of α-Tfm-butyrine **44** (Scheme 9), as well as of densely functionalized α-Tfm-allo-threoninate **45** and threoninate **46**. In the synthesis of compounds **45,46** the lithium derivative of sulfoxide

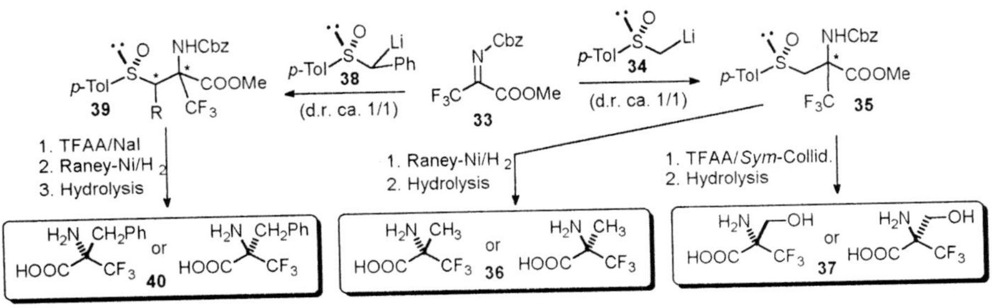

Scheme 8. Synthesis of both enantiomers of α-Tfm-phenylalanine, -alanine and -serine

41 is used as chiral hydroxyethyl anion equivalent. Noteworthy, the stereochemical outcome of the NOP reaction proved to be totally independent of the stereochemistry of the aminic stereocenter, in agreement with the mechanism proposed in Scheme 3, involving a S_N2-like process.

The lack of stereocontrol in the formation of the quaternary amine center, featured by all the condensations involving lithiated sulfoxides and N-Cbz imines of trifluoropyruvate, is likely to be due to the unavailability of the lone-pair of the strongly electronpoor iminic nitrogen toward coordination by the lithium atom. These reactions should not take place through a chelated Zimmerman-Traxler TS (*39*). An efficient solution to this problem of stereocontrol is reported in the next section.

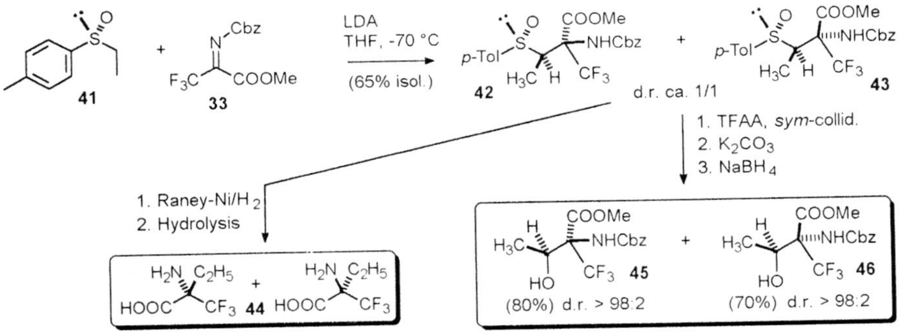

Scheme 9. Synthesis of enantiopure α-Tfm-butyrine, -threoninate and -allo-threoninate

In summary, lithium alkyl aryl sulfoxides can be efficiently used as equivalents of chiral α-hydroxy alkyl carbanion in the synthesis of β-fluoroalkyl β-amino alcohols. On the other hand, fluorinated imines are versatile and readily available electrophilic sources of fluoroalkyl/fluoroaryl amino groups. Excellent stereocontrol and reactivity can be achieved using aldimines N-protected with a PMP group, which also makes fluorinated aldimines electronrich enough to be easy to handle. Furthermore, the PMP group can be cleaved readily and under mild conditions, thus representing the protecting group of choice. Alternatively, for more sterically congested fluoroalkyl ketimines like those derived from trifluoropyruvate, an activating Cbz group can be conveniently employed as N-protector. As an additional advantage of using Cbz, the resulting β-sulfinyl amine intermediates can be directly submitted to the NOP reaction, whereas the NH-PMP intermediates must be preliminarly converted into the corresponding NH-Cbz compounds.

Stereoselective Synthesis of β-Fluoroalkyl β-Amino Alcohols Having a Quaternary Amine Center, via *Evans* Aldol Reaction with Imines of Trifluoropyruvate.

The stereocontrolled preparation of functionalized β-fluoroalkyl β-amino alcohols, particularly those having a quaternary stereogenic amine center, is one of the main goals of our research, due to the potentially high biological interest of such molecules, and to the stimulating challenge that their synthesis presents. Recently we undertook a study toward the synthesis of enantiomerically pure 2-fluoroalkyl analogs of sphingolipids, in order to evaluate the effect of the fluoroalkyl group on the biological activity of this class of compounds (*52,53*). According to the retrosynthetic analysis shown in Scheme 10, the template **G** was envisaged as key intermediate (*20*).

We decided to investigate the *Evans* aldol reaction (*54,55*) of the chiral α-hydroxy acetic anion equivalent **47** with the N-Cbz imine of trifluoropyruvate **33**. Screening of several enolates and reaction conditions revealed that titanium enolate of the oxazolidinone **47**, prepared with 1 equiv. of TiCl$_4$ and 1.1 equiv. of i-Pr$_2$NEt, reacts with excellent stereocontrol affording the "*Evans*" *syn* diastereomer **48**, in 88% isolated yield, having the correct stereochemistry to be used as intermediate for the synthesis of the targeted 2-Tfm-sphingolipids.

Much worse results, both in terms of yields and stereoselectivities, were obtained using boron, tin (II) and (IV) and lithium enolates of **47**, or different reaction conditions. To our knowledge, this represents the first example of highly stereoselective reaction involving imines of trifluoropyruvate (*56,57*), as well as a powerful methodology for the preparation of templates **G**. Although the exact reaction mechanism is not yet well understood, the reaction is likely to involve a TS in which the imine **33** is tightly chelated by the titanium atom. This chelation might involve one or both the carboxylic groups of the imine **33**, namely the COOEt and the COOCH$_2$Ph, rather than the almost inert nitrogen lone pair. This hypothesis is supported by the reports of *Iseki* and *Kobayashi* on the *Evans* aldol reactions involving analogous titanium enolates and strongly electrophilic and poorly coordinable fluorinated carbonyl electrophiles, such as trifluoroacetone and fluoral, which were found to occur with reverse selectivity, providing the corresponding non-*Evans*-products, *via* nonchelated open transition states (*58,59*).

The versatility of the adduct **48** is shown in Scheme 12, that summarizes the preparation of a number of functionally diverse building blocks by removal, and high yielding recovery, of the oxazolidinone auxiliary.

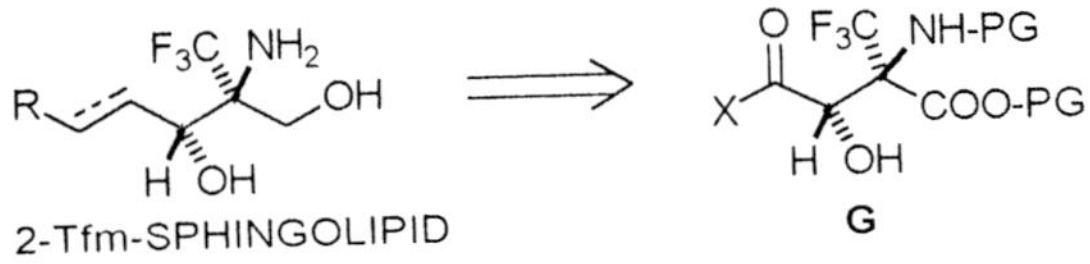

2-Tfm-SPHINGOLIPID

G

R = long aliphatic chain
PG Protecting group

Scheme 10. Retrosynthetic approach to
2-Tfm-sphingolipids

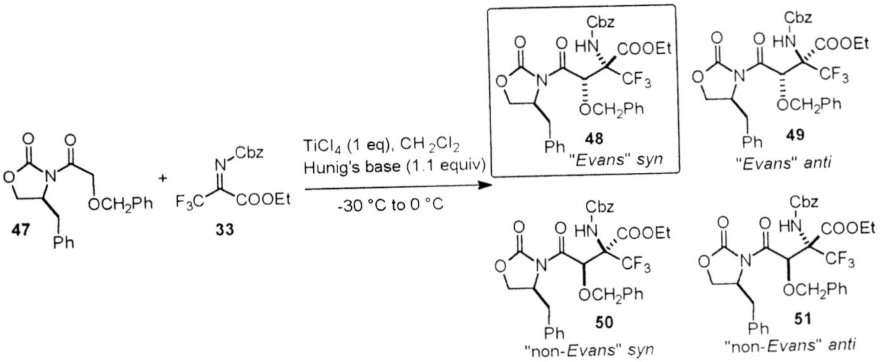

Scheme 11. The stereoselective tandem imino-aldol reaction

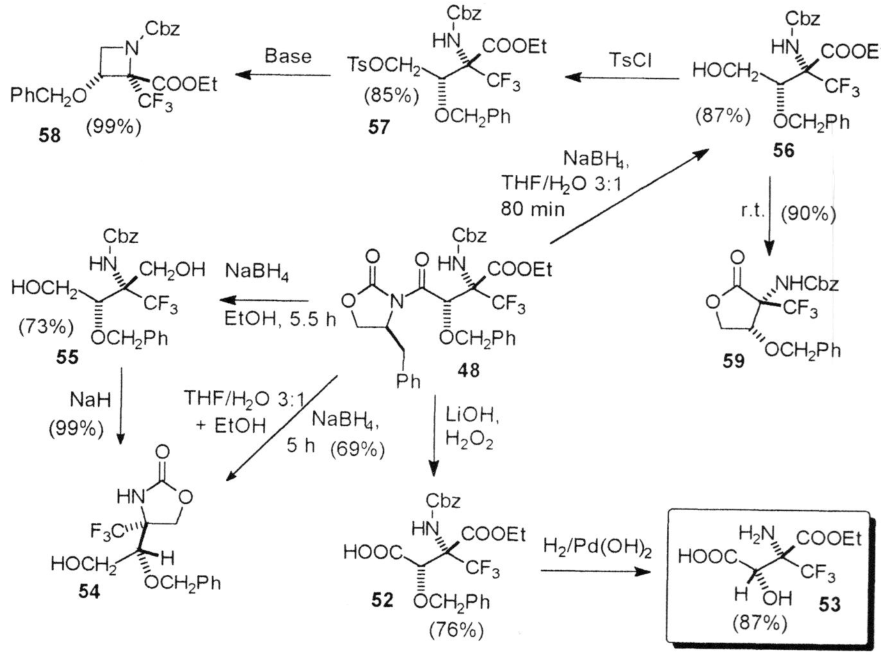

138

Scheme 12. Synthesis of enantiopure α-Tfm-β-hydroxy aspartic units and derivatives

The new amino acid α-Tfm β-hydroxy aspartic acid **53** was obtained by removal of the auxiliary with LiOH/H₂O₂ and hydrogenolysis of the resulting carboxylic acid **52**. Cleavage of the oxazolidinone by means of NaBH₄ in THF/water or ethanol, which represents a novel reductive removal of this auxiliary, produced selectively several different products, depending on the equivalents of NaBH₄ employed and the reaction conditions. For example, reduction of both amidic and carboxylic functions, followed by intramolecular cyclization and elimination of benzyl alcohol to afford the oxazolidinone **54** was achieved using 5 equiv of NaBH₄ in a THF/H₂O mixture, in the presence of ethanol. With 5 equiv. in pure ethanol only the first two events occurred, leading to the diol **55**. A large excess of NaBH₄ (16 equiv) in THF/H₂O after 80 min. resulted in selective reduction to give the γ-hydroxy ester **56**. This compound provided the interesting azetidine **58** in excellent overall yields by base promoted intramolecular cyclization of the intermediate tosylate **57**. The γ-lactone **59** could be obtained in high yields by spontaneous lactonization of the γ-hydroxy ester **56**.

Ongoing work is directed toward the total synthesis of the targeted 2-Tfm sphingolipids.

Acknowledgements. This study has been made possible by the passionate and creative work of many scientists and students. Above all we would like to thank Dr. Vadim A. Soloshonok, Dr. Marcello Crucianelli, Dr. Fiorenza Viani, Prof. Stefano V. Meille and Prof. Santos Fustero-Lardies. Thanks are due also to all the authors of our papers cited in the references below.

Literature Cited.

1. Bégué, J.-P.; Bonnet-Delpon, D. in *Biomedical Frontiers of Fluorine Chemistry*; Ojima, I.; McCarthy, J.R.; Welch, J.T., Eds.; ACS Books, American Chemical Society: Washington, D.C., 1996, pp 59-72.
2. Ojima, I.; Kuduk, S. D.; Slater, J. C.; Gimi, R. H.; Sun, C. M.; Chakravarty, S.; Ourevitch, M.; Abouabdellah, A.; Bonnet-Delpon, D.; Bégué, J.-P.; Veith, J. M.; Pera, P.; Bernacki, R. J *ibid.*, Chapter 17, pp 228-243.
3. Ojima, I.; Slater, J. C.; Pera, P.; Veith, J. M.; Abouabdellah, A.; Bégué, J.-P.; Bernacki, R. J. *Bioorg. Med. Chem. Lett.* **1997**, *7*, 133.
4. Quinze, K.; Laurent, A.; Mison, P. *J. Fluorine Chem.* **1989**, *44*, 233.
5. Kozikowski, A. P.; Wu, J.-P. *Tetrahedron Lett.* **1990**, *31*, 4309.
6. Beck, A. K.; Seebach, D. *Chem. Ber.* **1991**, *124*, 2897.
7. Marti, R. E.; Heinzer, J.; Seebach, D. *Liebigs Ann.* **1995**, 1193.
8. Uno, H.; Okada, S.; Ono, T.; Shiraishi, Y.; Suzuki, H. *J. Org. Chem.* **1992**, *57*, 1504.
9. Abouabdellah, A.; Bégué, J.-P.; Bonnet-Delpon, D. *Synlett* **1996**, 399.
10. Abouabdellah, A.; Bégué, J.-P.; Bonnet-Delpon, D.; Thanh Nga, T. T. *J. Org. Chem.* **1997**, *62*, 8826.
11. Uneyama, K.; Hao, J.; Amii, H. *Tetrahedron Lett.* **1998**, *39*, 4079.
12. For other recent reports in which β-fluoroalkyl β-amino alcohol units are synthesized (see also ref. 13,14): Behr, J.-B.; Evina, C. M.; Phung, N.; Guillerm, G. *J. Chem. Soc., Perkin Trans 1* **1997**, 1597.
13. Andersen, S. M.; Ebner, M.; Ekhart, C. W.; Gradnig, G.; Legler, G.; Lundt, I.; Stütz, A. E.; Withers, S. G.; Wrodnigg, T. *Carbohydr. Res.* **1997**, *301*, 155.
14. Bravo, P.; Cavicchio, G.; Crucianelli, M.; Poggiali, A.; Zanda, M. *Tetrahedron: Asymmetry* **1997**, *8*, 2811.
15. Bravo, P.; Capelli, S.; Kukhar, V.P.; Meille, S.V.; Soloshonok, V.A.; Viani, F.; Zanda, M. *Tetrahedron: Asymmetry* **1994**, *5*, 2009.
16. Bravo, P.; Farina, A.; Kukhar, V. P.; Markovsky, A. L.; Meille, S. V.; Soloshonok, V. A.; Sorochinsky, A. E.; Viani, F.; Zanda, M.; Zappalà, C. *J. Org. Chem.* **1997**, *62*, 3424.
17. Arnone, A.; Bravo, P.; Bruché, L.; Crucianelli, M.; Vichi, L.; Zanda, M. *Tetrahedron Lett.* **1995**, *36*, 7301.
18. Bravo, P.; Zanda, M.; Zappalà, C. *Tetrahedron Lett.* **1996**, *37*, 6005.
19. Arnone, A.; Bravo, P.; Capelli, S.; Fronza, G.; Meille, S. V.; Zanda, M.; Cavicchio, G.; Crucianelli, M. *J. Org. Chem.* **1996**, *61*, 3375; corrigenda: *J. Org. Chem.* **1996**, *61*, 9635.
20. Bravo, P.; Guidetti, M.; Volonterio, A.; Zanda, M. *12th International Conference on Organic Synthesis*, Venezia, **1998**, 352.
21. Bravo, P.; Crucianelli, M.; Fronza, G.; Zanda, M. *Synlett* **1996**, 249.
22. Volonterio, A.; Zanda, M.; Bravo, P.; Fronza, G.; Cavicchio, G.; Crucianelli, M. *J. Org. Chem.* **1997**, *62*, 8031.
23. Bjørsvik, H.-R.; Bravo, P.; Crucianelli, M.; Volonterio, A.; Zanda, M. *Tetrahedron: Asymmetry* **1997**, *8*, 2817.
24. Volonterio, A.; Bravo, P.; Capelli, S.; Meille, S. V.; Zanda, M. *Tetrahedron Lett.* **1997**, *38*, 1847.

25. Volonterio, A.; Vergani, B.; Crucianelli, M.; Zanda, M.; Bravo, P. *J. Org. Chem.* **1998**, in press.

26. Walker, A. J. *Tetrahedron: Asymmetry* **1992**, *3*, 961.

27. Carreño, M. C. *Chem. Rev. (Washington, D. C.)* **1995**, *95*, 1717.

28. Bravo, P.; Zanda, M. In *Enantiocontrolled Synthesis of Fluoro-Organic Compounds: Stereochemical Challenges and Biomedicinal Targets*; Chapter #4, Soloshonok, V. A., Ed.; Wiley: Chichester, scheduled to appear in 1998.

29. De Lucchi, O.; Miotti, U.; Modena, G. *Organic Reactions*; Paquette, L. A. Ed.; John Wiley and Sons: New York, 1991, Vol. 40.

30. Oae, S. *Organic Sulfur Chemistry: Structure and Mechanism*, CRC Press: Boca Raton, 1991, pp 380-400.

31. Pummerer, R. *Ber.* **1909**, *42*, 2282.

32. Arnone, A.; Bravo, P.; Bruché, L.; Crucianelli, M.; Zanda, M.; Zappalà, C. *J. Chem. Res.* **1997**, 416.

33. Bravo, P.; Cavicchio, G.; Crucianelli, M.; Markovsky, L. A.; Volonterio, A.; Zanda, M. *Synlett* **1996**, 887.

34. Bravo, P.; Guidetti, M.; Viani, F.; Zanda, M.; Markovsky, A. L.; Sorochinsky, A. E.; Soloshonok, I. V.; Soloshonok, V. A. *Tetrahedron* **1998**, *54*, 12789.

35. Guidetti, M.; Sorochinsky, A. E.; Bravo, P.; Soloshonok, V. A.; Zanda, M. *et al. Tetrahedron* **1998**, in press.

36. Fustero, S.; Bravo, P.; Volonterio, A.; Zanda, M. *et al. 12th European Symposium on Fluorine Chemestry*, Berlin, **1998**, PI-5.

37. Carlsen, P. H. J.; Katsuki, T.; Martin, V. S.; Sharpless, K. B. *J. Org. Chem.* **1981**, *46*, 3936.

38. Yamamoto, Y.; Nishii, S.; Maruyama, K.; Komatsu, T.; Ito, W. *J. Am. Chem. Soc.* **1986**, *108*, 7778 and references therein.

39. Zimmerman, H. E.; Traxler, M. D. *J. Am. Chem. Soc.* **1957**, *79*, 1920.

40. Pyne, S. G.; Boche, G. *J. Org. Chem.* **1989**, *54*, 2663.

41. Pyne, S. G.; Dikic, B. *J. Chem. Soc., Chem. Commun.* **1989**, 826.

42. Pyne, S. G.; Dikic, B. *J. Org. Chem.* **1990**, *55*, 1932.

43. Bravo, P.; Corradi, E.; Pesenti, C.; Vergani, B.; Viani, F.; Volonterio, A.; Zanda, M. submitted for publication.

44. *Fluorine-Containing Amino Acids: Synthesis and Properties*; Kukhar, V. P.; Soloshonok, V. A. Eds.; Wiley: Chichester, 1994.

45. Koksch, B.; Sewald, N.; Jakubke, H.-D.; Burger, K. "Synthesis and Incorporation of α-Trifluoromethyl-Substituted Amino Acids into Peptides". In the book of Ref. 1.

46. Dal Pozzo, A.; Muzi, L.; Moroni, M.; Rondanin, R.; de Castiglione, R.; Bravo, P.; Zanda, M. *Tetrahedron* **1998**, *54*, 6019.

47. Staudinger, H.; Meyer, J. *Helv. Chim. Acta* **1919**, *2*, 635-646.

48. Molina, P.; Vilaplana, M. J. *Synthesis* **1994**, 1197.

49. Johnson, A. W.; Kasha, W. C.; Starzewsky, K. A. O.; Dixon, D. A. "Iminophosphoranes and Related Compounds". In: *Ylides and Imines of Phosphorus*; New York: Wiley, 1993, pp 403-483.

50. Bravo, P.; Viani, F.; Zanda, M.; Kukhar, V.P; Soloshonok, V.A.; Fokina, N.; Shishkin, O.V.; Struchkov, Yu.T. *Gazz. Chim. Ital.* **1996**, *126*, 645.

51. Drabowicz, J.; Oae, S. *Synthesis* **1977**, 404.

52. Kolter, T.; Sandhoff, K. *Chem. Soc. Rev.* **1996**, 371.

53. Koskinen, P. M.; Koskinen, A. M. P. *Synthesis* **1998**, 1075.
54. Evans, D. A.; Urpi, F.; Somers, T. C.; Clark, J. S.; Bilodeau, M. T. *J. Am. Chem. Soc.* **1990**, *112*, 8215 and references therein.
55. Crimmins, M. T.; King, B. W.; Tabet, E. A. *J. Am. Chem. Soc.* **1997**, *119*, 7883.
56. For the addition of titanium enolates to other imines (see also ref. 57): Annunziata, R.; Benaglia, M.; Cinquini, M.; Cozzi, F.; Raimondi, L. *Tetrahedron* **1994**, *50*, 9471.
57. Abrahams, I.; Motevalli, M.; Robinson, A. J.; Wyatt, P. B. *Tetrahedron* **1994**, *50*, 12755.
58. Makino, Y.; Iseki, K.; Fujii, K.; Oishi, S.; Hirano, T.; Kobayashi, Y. *Tetrahedron Lett.* **1995**, *36*, 6527.
59. Iseki, K.; Oishi, S.; Taguchi, T.; Kobayashi, Y. *Tetrahedron Lett.* **1993**, *34*, 8147.

Chapter 11

Diastereoselective Construction of Novel Sugars Containing Variously Fluorinated Methyl Groups as Intermediates for Fluorinated Aldols

Takashi Yamazaki, Shuichi Hiraoka, and Tomoya Kitazume

Department of Bioengineering, Tokyo Institute of Technology, 4259 Nagatsuta-cho, Midori-ku, Yokohama 226–8501, Japan

Highly diastereoselective introduction of an appropriately fluorinated methyl group into D-glucose has been realized by heterogeneous catalytic hydrogenation with the aid of relatively rigid cyclic conformations, and an *exo*-difluoromethylene group played an pivotal role in the preparation of mono-, di-, and trifluoromethylated materials *via* the divergent pathways.

Substitution of an appropriate hydrogen atom or a hydroxy moiety for fluorine has recently been one of the major strategies for realizing the effective modification or enhancement of the inherent biological activities of organic compounds (*1-3*). From such a point of view, aldol structures, widely found in the important naturally occurring biologically active substances, would be one of the most intriguing targets for the fluorine modification. However, as long as the authors concern, only a few examples (*4-7*) have been reported in the literature on the formation of this specific structure with introducing fluorine(s) to a methyl group at an appropriate position (*8-12*), and especially the corresponding nonracemic materials seem difficult to be obtained (*13,14*). This might be mainly because of the less ready availability (*15-20*) as well as the inherent instability of the requisite 3-fluorinated propionyl derivatives (Figure 1). Thus, in the case of β-fluorinated carbonyl compounds, hydrogen α to a carbonyl group becomes sufficiently acidic by the double activation by this moiety and fluorine(s), but once anionic species is formed during the course of deprotonation, then a fluorine atom is ready to depart from the original molecule to furnish α,β-unsaturated materials usually in an irreversible manner (*21*).

On the other hand, application of the traditional fluorination techniques (*22,23*)

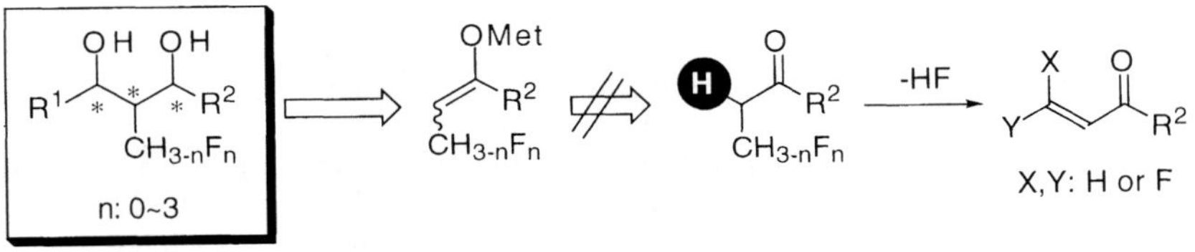

Figure 1 A Retrosynthetic Scheme to the Target Aldol Structures

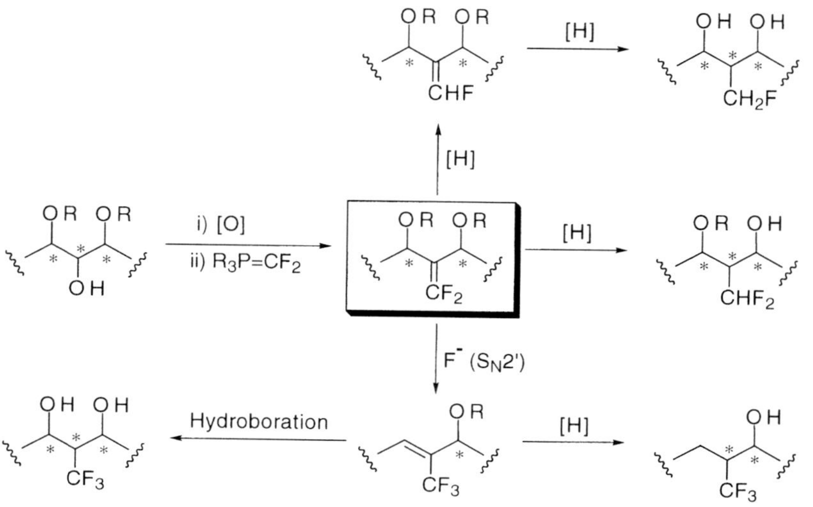

Figure 2 Synthetic Plans to Access the Targets

or the direct introduction methods of fluorine-containing groups (24-30) would not be recognized as the alternative solution for this purpose in view of the still unsolved problems on the regio-, stereo-, and/or chemoselectivities.

In our continuing work on the preparation of various types of optically active fluorinated building blocks (31-33) we have recently started the new synthetic study of fluorine-possessing aldol structures for opening a new route to access such interesting substances by way of the synthetic plans depicted in Figure 2. The key intermediate is the *exo*-difluoromethylene compound which, starting from optically active 1,2,3-triols with appropriate protections both at 1 and 3 positions, seems to be conveniently synthesized by oxidation of the free OH moiety at the central carbon atom, followed by difluoromethylenation (34,35). Moreover, this intermediate, on treatment with hydride (36,37) or fluoride (38) according to the Tellier and Sauvêtre's reports, would be further transformed into the corresponding *exo*-fluoromethylene compound or trifluoromethylated *endo*-olefin which would be eventually converted to mono- or trifluoromethylated aldols, respectively. For this purpose, D-glucose was selected as the precursor of chiral 1,2,3-triols because, in addition to its ready availability as an optically pure form, i) regioselective protection of OH moieties at various positions has been extensively studied, and ii) usage of the different site would lead to the possible formation of diastereoisomeric structures.

Preparation of Difluoromethyl Compounds

The requisite key intermediates **4**, **6a**, and **8** were prepared as described in Figure 3 from easily accessible methyl 4,6-*O*-benzylidene-α-D-glucopyranoside **1** (39) only by two steps from D-glucose. Reaction of **1** with *tert*-butyldimethylsilyl (TBS) chloride, already performed by the Fraser-Reid's group (40), produced an easily separable regioisomeric mixture of **2a** and **2b** almost in a 1:1 ratio and an OH

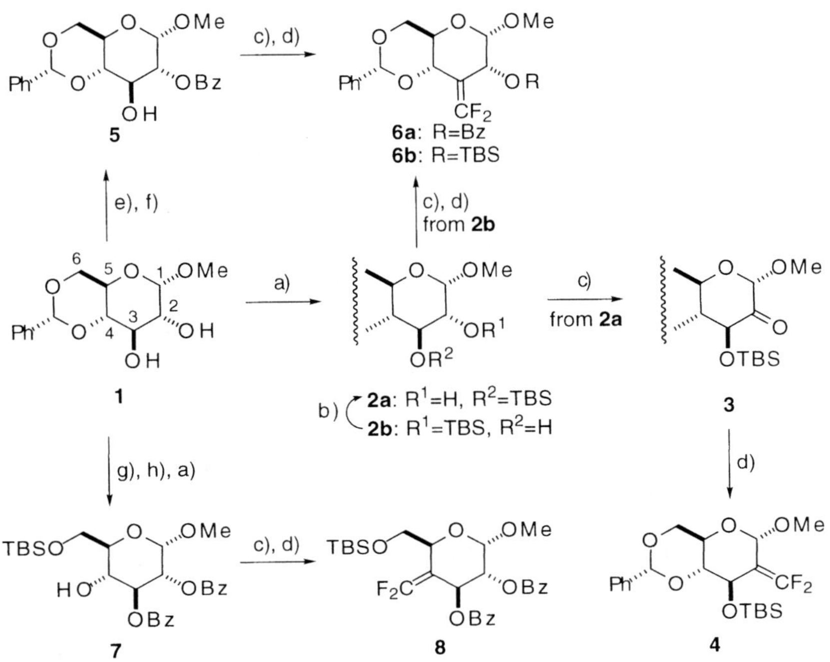

a) TBSCl, imidazole, b) K_2CO_3/MeOH, c) PDC, Ac_2O,
d) CBr_2F_2, $(Me_2N)_3P$, MS 4Å, e) n-Bu_2SnO/MeOH,
f) BzCl, g) BzCl, pyr., h) cat. p-TsOH/MeOH

Figure 3 Preparation of Sugars Containing
a Difluoromethylene Group at 2, 3, and 4 Positions

group at 2 position of the former was oxidized with PDC in acetic anhydride. Conversion of **3** to **4** was first tried with the well-known Ishikawa-Burton protocol (*34,41*) and then its modified method (*35*) but **4** was obtained only in moderate yields (32 and 45%, respectively). However, our investigation of the reaction conditions in detail demonstrated that employment of MS 4Å guaranteed the much higher yield (91%) as well as reproducibility. The regioisomer **2b** was utilized for the formation of **6b** or partial isomerization to **2a** under base-catalyzed conditions (**2a**:**2b**=65:35).

The 3-difluoromethylenated counterpart **6a**, on the other hand, was accessible by way of tin acetal-mediated regioselective benzoylation at 2 position (*42,43*), followed by the above-depicted PDC oxidation-difluoromethylene Wittig reaction processes. For introduction of a CF_2=C group at 4 position, after benzylidene acetal was cleaved and the resultant diol was regioselectively protected at 6 position by TBSCl, the routine oxidation-difluoromethylenation procedure yielded the desired compound **8**.

Three types of the key *exo*-difluoromethylenated materials in hand, they were at the next stage subjected to hydrogenation conditions in the presence of 10% Pd/C

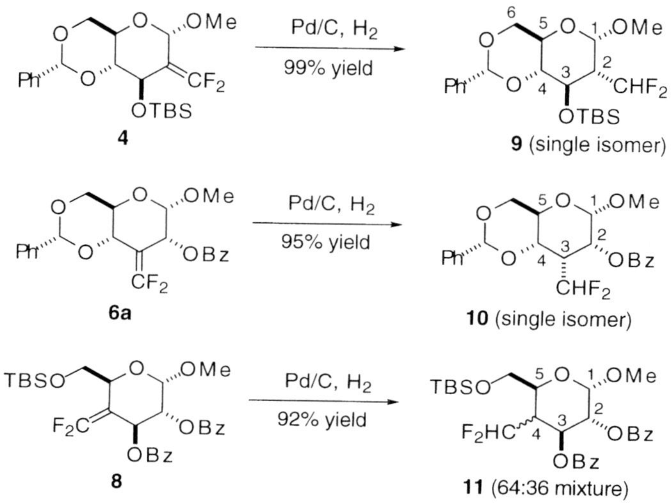

Figure 4 Hydrogenation of Difluoromethylenated Materials

under atmospheric pressure of hydrogen at room temperature (Figure 4). Substrates **4** and **6a** with an *exo*-difluoromethylene group at 2 and 3 positions, respectively, were smoothly transformed into the corresponding saturated forms **9** and **10** almost in quantitative yields, both as single stereoisomers with the benzylidene acetal part intact. The stereochemistries at the newly created chiral sp^3 carbon atoms were deduced from their ^{1}H NMR. Thus, the coupling constant of 10.0 Hz between H^2 and H^3 in **9** allowed us to assume that these protons have an *antiperiplanar* relationship and the CHF$_2$ moiety at 2 position should occupy the equatorial position. On the other hand, NMR analysis of **10** revealed the coupling constants of H^3-H^4 and H^4-H^5 as 4.9 and 10.5 Hz, respectively. Since the latter value unambiguously indicated that the H^4 proton was axially disposed, the 4.9 Hz ^3J coupling between H^3 and H^4 should support the equatorial-axial relationship of these protons, and thus the CHF$_2$ moiety of **10** was suggested to be located at the axial position. Stereochemistry could not be determined for the compound **11** due to overlap and complexity of NMR peaks.

For clarification of the origin of the excellent diastereofacial selectivity on hydrogenation obtained as above, conformational analysis for the compound **6a** was performed by use of MOPAC AM1 (*44*) with some modification (BzO→AcO and Ph→Me) for the convenience of computation. The most stable structure of the model compound **12** calculated is depicted in Figure 5 in two ways, one by the ball and stick model on the left and the other by the CPK model so as to place the F-C-F atoms in the X-Z plane (horizontally perpendicular to the paper) on the right hand. One would easily understand that attack at the CF$_2$=C moiety preferentially occurs from the direction of the more vacant upper *si* face. Although an acetoxy group at 2 position and the ethylidene acetal part are not likely to exert the significant steric hindrance towards incoming reagents (*C=C-C-O*Ac: 13.3°, *C=C-C-O-C*(O)Me: 161.4°; *C=C-C-O*CHMe: 5.8°, *C=C-C-O-C*HMe: 178.0°), a palladium catalyst would suffer from the

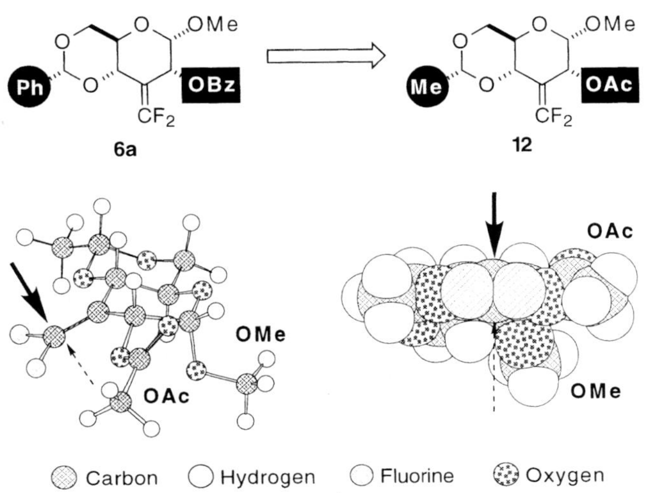

Figure 5 The Most Stable Conformer **12** Obtained from AM1 Calculations

sterically repulsive interaction with the anomeric methoxy moiety (the angle between $F_2C=C$ and C-OMe of 69.4°) when it approaches from the bottom face. The CPK model also explicitly illustrates the exclusive *si* face selection trend at the olefinic reaction site.

Preparation of Mono- and Trifluoromethylated Materials

The corresponding monofluoromethylated materials were also prepared in a similar manner (Figure 6). In this case, the partial reduction of an *exo*-difluoromethylene group was required before palladium-mediated hydrogenation. We first used Red-Al (*34*) for this purpose, but the reaction actually occurred was the unexpected S_N2' type hydride attack to furnish **13**. In quite sharp contrast, DIBAH was proved to be effective only for the reductive cleavage of the benzoyl group at 2 position, and the Red-Al reduction of the resultant intermediary alcohol **15** (a 61:39 mixture, stereochemistry was not clarified yet), followed by hydrogenation smoothly produced the desired monofluoromethylated substance **16** again as a single isomer. From the ^{1}H NMR analysis, the stereochemistry at 3 position was manifested to be the same as the case of the corresponding CHF_2 form **10**. On the other hand, conversion of a difluoromethylene group in the substrate **4** to the corresponding *exo*-fluoromethylene moiety was facilitated by Red-Al, and a single stereoisomer was obtained by the subsequent hydrogenation. The same explanation towards CHF_2 materials would be also operative here for the realization of the high level of facial selectivity because, on the basis of the most stable conformation of **12** displayed in Figure 5, substitution of one of two fluorine atoms for hydrogen seems not to significantly alter the conformational preference. It is interesting to mention that the substrate **6c** was successfully transformed into **15** by Red-Al reduction in a direct fashion and such reactivity difference between **6a** and **6c** caused by a slight change

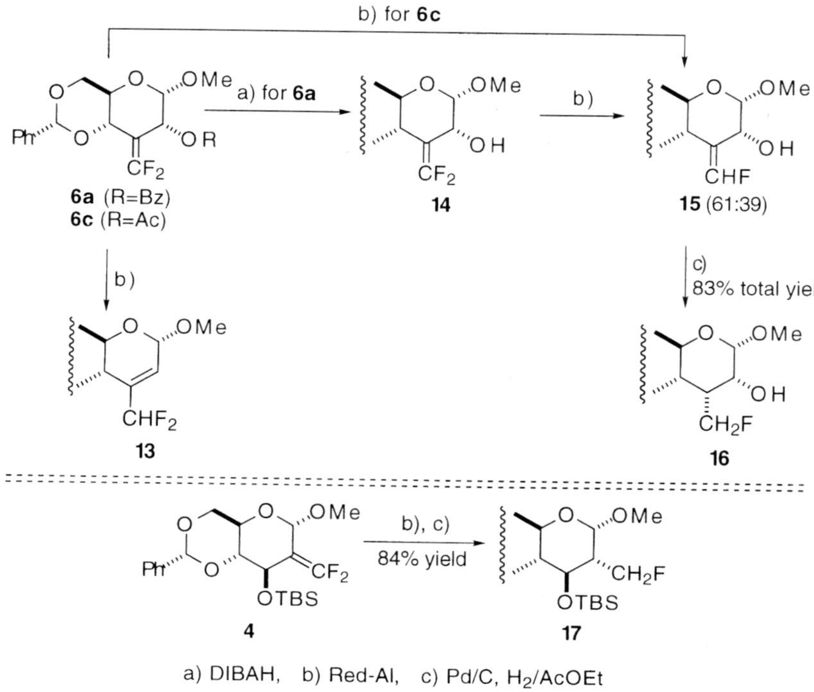

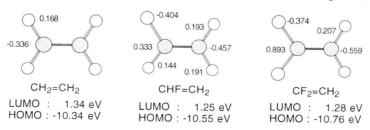

a) DIBAH, b) Red-Al, c) Pd/C, H₂/AcOEt

Figure 6 Hydrogenation of *exo*-Monofluoromethylenated Materials

would be, at least qualitatively, elucidated in terms of the leaving group ability (pKa values of benzoic and acetic acids in H_2O at 25 °C of 4.20 and 4.76 *(45)*, respectively).

At this stage, *ab initio* calculation of 1,1-difluoroethylene as well as its mono- and non-fluorinated counterparts for comparison was carried out by Gaussian 94 *(46)* at the HF/6-311++G** level of theory for better understanding of the above reactivity difference highly dependent on the nature of hydride reagents used (Figure 7).   It is

CH₂=CH₂	CHF=CH₂	CF₂=CH₂
LUMO : 1.34 eV	LUMO : 1.25 eV	LUMO : 1.28 eV
HOMO : -10.34 eV	HOMO : -10.55 eV	HOMO : -10.76 eV

Figure 7 NBO Charges and Frontier Orbital Energy Levels of
Ethylenes Obtained by *Ab initio* (6-311++G**) Calculation

quite intriguing to note that the carbon-carbon double bond of this material was highly polarized due to the electronic repulsion between lone pairs of two fluorine atoms and the olefin π electron, and the fluorine-possessing carbon atom was charged almost like a usual cation (the NBO charge of 0.89 at this site). On the other hand, its LUMO was calculated to be basically at the same level as the one of nonfluorinated ethylene. So, the S_N2' type attack of hydride to **6a** by Red-Al was occurred not because the $F_2C=C$ part was activated by decrease of the LUMO energy level but merely because the "strongly cationic" carbon increased the ability to accept the attack by a nucleophilic ate complex Red-Al. Mechanism of Lewis acidic DIBAH reduction would be interpreted as the exclusive complexation with the benzoyl carbonyl group at the first stage due to difficult coordination at the electron-deficient olefinic site, and then delivery of hydride would be occurred to affect the reductive deprotection of the BzO moiety.

In the following Figures 8 and 9 was described the preparation of the CF_3-containing substances. As was already discussed, DAST plays a pivotal role in converting an *exo*-difluoromethylene group into the corresponding *endo*-trifluoro-methylated olefin *via* S_N2' type attack of fluoride (*38*). Because **6a** includes two latent terminally difluorinated allylic alcohol structures two different types of saturated

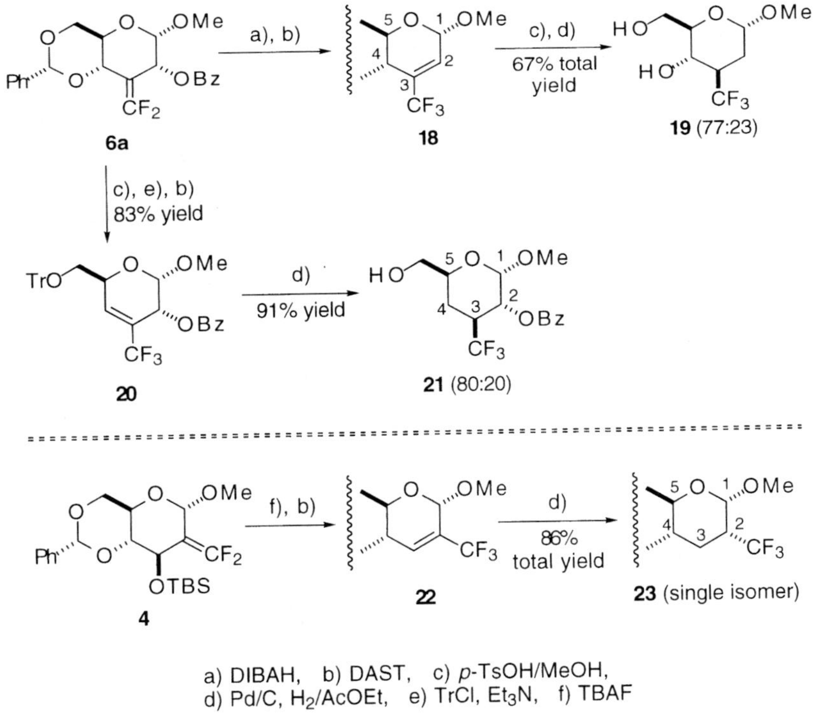

a) DIBAH, b) DAST, c) *p*-TsOH/MeOH,
d) Pd/C, H$_2$/AcOEt, e) TrCl, Et$_3$N, f) TBAF

Figure 8 Hydrogenation of 2- and 3-CF$_3$ Olefins

CF_3 compounds would be obtained. For example, the ^{2,3} type olefin **18** was isolated by reductive cleavage of a benzoyl moiety with DIBAH, followed by DAST fluorination, and was further hydrogenated to give **19** as a 77:23 diastereomer mixture. On the other hand, the regioisomeric ^{3,4} olefin **20** was constructed by the successive conversion of **6a** by deprotection of the benzylidene acetal moiety, the regioselective protection of the primary alcohol at 6 position, and S_N2' type DAST fluorination, and was then subjected to hydrogenation conditions to produce the corresponding saturated material **21** as an 80:20 diastereomeric mixture. A trifluoromethyl moiety of both major products, **19** and **21**, were determined to have the equatorial location by NMR coupling constants of H^{2ax}-H^3=13.4 Hz and H^2-H^3= 11.2 Hz, respectively.

The compound **23** with a CF_3 moiety at 2 position was efficiently prepared by the three-step sequence as shown in Figure 8, and the newly created sp^3 center was proved to possess R stereochemistry (an equatorial CF_3 group) on the basis of the strong NOE observed between H^2 and H^4.

An interesting reaction was encountered during the same transformation starting from **8**. Thus, benzoyl deprotection with DIBAH and benzoylation allowed us to obtain a requisite terminally difluorinated allylic alcohol structure **24**, which, on

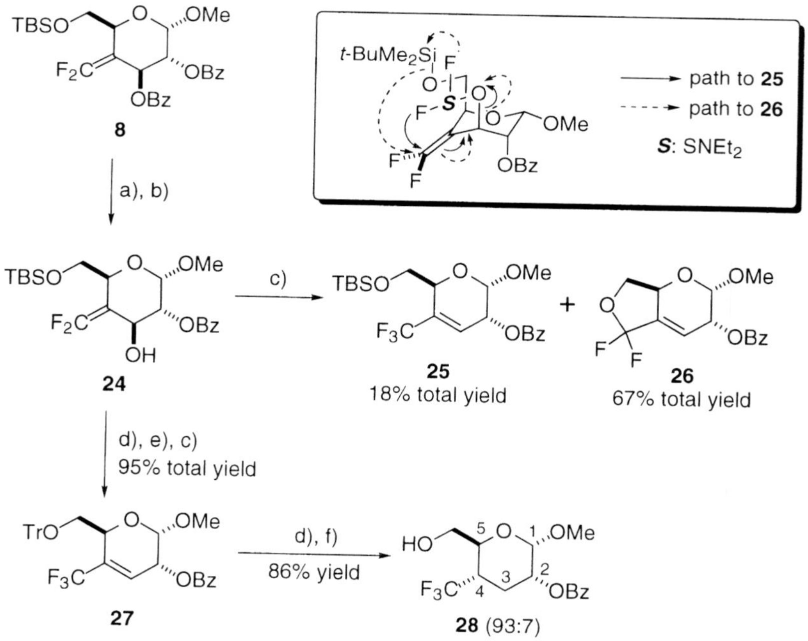

a) DIBAH, b) BzCl, pyr., c) DAST, d) p-TsOH/MeOH,
e) TrCl, Et₃N, f) Pd/C, H₂/AcOEt

Figure 9 Hydrogenation of 4-CF_3 Olefin

treatment with DAST, afforded the desired trifluorinated *endo*-olefin **25** only in 18% yield.   On the other hand, the main product **26** was manifested to possess the bicyclic structure as depicted in Figure 9 and its formation would be considered as the result of the activation of ether oxygen at 6 position by intramolecular attack of fluoride to silicon, followed by cyclization in a *5-endo-trig* manner known as the disfavored mode in terms of the Baldwin's rule (*47-49*).

Due to such a reason, the *endo*-olefin **27** was synthesized by changing the protective group at 6 position of **24** to a trityl group (*50*), and the following hydrogenation after removal of a trityl group led to the formation of **28** as a 93:7 diastereomer mixture.   The stereostructure of the major isomer at 4 position was deduced as *S* (CF$_3$ at an equatorial position) from the H^4-H^5 coupling constant of 10.3 Hz, showing these protons in an *antiperiplanar* relationship.

Different from the cases of **22** and **27** attaining highly diastereoselective hydrogenation, an acceptable level of olefinic facial discrimination was not observed for materials **18** and **20** with a CF$_3$ group at 3 position, and thus the substituent effect at both 2 and 6 positions were briefly studied for the purpose of improving diastereo-selectivity towards **20** (Table 1).   Comparison of data in Entries 1-3 suggested larger R^1 protective groups at 6 position decreased the preference of the CF$_3$-eq isomer **21-eq** and the same tendency was noticed for the substituent at 2 position (Entry 2 vs 4).   Because the non-protected hydroxy group afforded the best result at the both sites, we changed the solvent to a protic EtOH aiming at attainment of the higher proportion of **21b-eq** even when the totally protected form **20b** was employed, and the selectivity was conveniently increased to an acceptable level as our expectation (Entry 2 vs 5).

In general, sterically demanding trityl (Tr, **20a**) or *tert*-butyldimethylsilyl (TBS, **20b**) moieties at 6 position are likely to restrict free movement to fix molecules at specific conformations.   If this is the case, effective discrimination between two olefinic faces would be anticipated and possibly lead to better diastereoselectivity on hydrogenation.   We again performed AM1 computation of the model material **22** (modified at a benzoyl group at 2 position of **20c** with an acetoxy moiety) to obtain information on its spatial shape preference.   The most stable conformer was, as described in Figure 10, revealed to possess the half-chair conformation with pseudo-

Table 1   Substituent Effect on the Hydrogenation Stereoselectivity

Entry	Substrate	R^1	R^2	Solvent	Diastereomeric Ratio	
					21-eq	**21-ax**
1	**20a**[a)]	Tr	Bz	AcOEt	80	20
2	**20b**	TBS	Bz	AcOEt	82	18
3	**20c**	H	Bz	AcOEt	100	0
4	**20d**	TBS	H	AcOEt	94	6
5	**20b**	TBS	Bz	EtOH	97	3

a) The trityl group was deprotected during the course of hydrogenation.

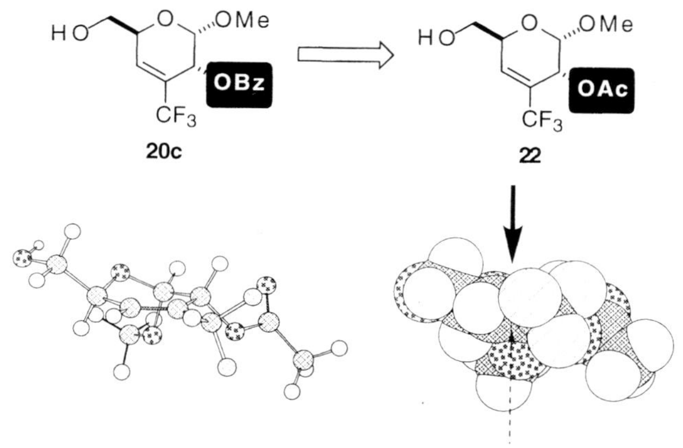

Figure 10 The Most Stable Conformer **22**
Obtained by AM1 Calculations

equatorially disposed acetoxy and hydroxymethyl groups at 2 and 6 positions, respectively ($C=C(CF_3)-C-O$Ac: 133.2°, $HOH_2C-C-C=C(CF_3)$: 138.0°), and an anomeric methoxy group in an axial position.   In consequence, the factor controlling the direction of incoming reagents was expected to be the 1-OMe group ($C=C(CF_3)-C-C-O$Me: 81.7°, $C=C(CF_3)-C-C-O$Me: 101.0°) as clearly shown by the CPK model, and approach from the upper *si* face would be preferable to the one from the other side. However, this discussion anticipated the formation of the wrong diastereomer **21-ax**.

Hydrogenation Mechanism

This type of heterogeneous catalytic hydrogenation is in general considered to be a multistep equilibration sequence (*51*), and taking the previous report on the rate-determining step (*52*) into account, it is most likely that the observed stereoselectivity is the reflection of the thermodynamic stability difference of the intermediates just prior to proceed reductive elimination of the Pd catalyst.

In Figure 11 was shown the possible scheme.   As was pointed out, the upper face of the substrate **20** seems to have more vacancy than the opposite bottom side, and the palladium catalyst will preferentially approach from the "U" direction to furnish the π complex **Int-U** at the first stage.   By way of the oxidative addition of Pd-related species (abbreviated as Pd in the Figure for clarity), this intermediate would be further converted to either regioisomeric pair of **Int-Ua** or **Int-Ub**, but both of them should suffer from the significantly destabilizing 1,3-diaxial type repulsive interactions between methoxy and trifluoromethyl moieties.   So, it is anticipated that the energy barriers between **Int-U** and **Int-Ua** or **Int-Ub** are rather high.

On the other hand, because of the axially disposed anomeric methoxy group, the "B" side of **20** is more congested than the "U" side and reluctant to accept reagents. However, the energetically less favorable **Int-B** is transformed to the next inter-mediates, **Int-Ba** and **Int-Bb**, at the oxidative addition step, and although **Int-Ba** is annoyed by the 1,3-diaxial repulsion similar to the case of **Int-Ua** or **Ub**, all the

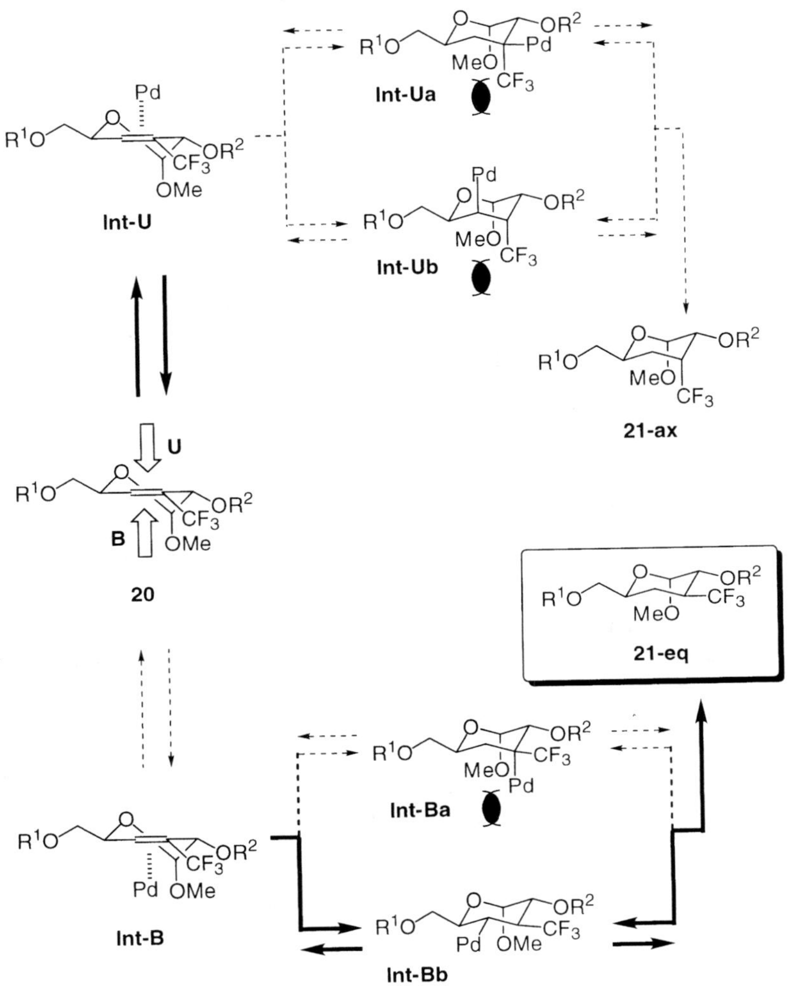

Figure 11 Possible Mechanism on Hydrogenation of **20**

substituents in **Int-Bb** locate the favorable equatorial positions except for the methoxy moiety at the axial position, which, in turn, stabilized by the well-known anomeric effect (53-55).   We at present believe that the present result was obtained on the basis of the higher energy barrier between **Int-U** and **Int-Ua** (or **Ub**) relative to the one between **20** and **Int-B**, leading to the exclusive formation of **21-eq** *via* **Int-Bb**.

Ring Opening of the Cyclic Compounds Obtained

Sugar derivatives with variously fluorinated methyl groups at specific positions synthesized as above were further transformed into the corresponding acyclic forms and the representative example was described in Figure 12.   The glycosidic bond and

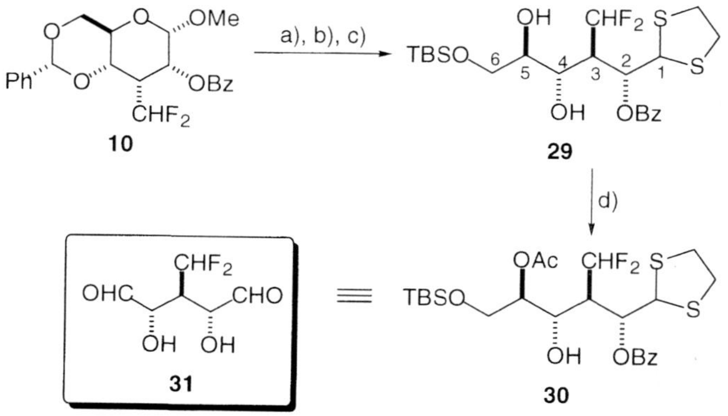

a) BBr_3, b) $(HSCH_2)_2$, $BF_3 \cdot OEt_2$, c) TBSCl, imidazole, d) AcCl, pyr.

Figure 12 Ring Opening of the CHF_2 Compound 10

benzylidene acetal of 10 were simultaneously cleaved by BBr_3, followed by the successive protection of the resultant hemiacetal as 1,3-dithiolane and the primary alcohol at 6 position as a TBS ether to give diol 29. Interestingly, its acetylation exclusively occurred at the 5-OH moiety, which would be elucidated on the basis of the less nucleophilic ability of the hydroxy group at 4 position due to the proximity of the electron-withdrawing CHF_2 group (56). Compounds 29 or 30 thus obtained would be recognized as the useful *anti,anti*-aldol unit 31 and because of its latent *meso* structure, they would be easily employed for the formation of the corresponding enantiomers.

As described above, a novel preparation of homochiral mono-, di-, as well as trifluoromethylated aldol structures starting from the readily available chiral pool compound, D-glucose, has been demonstrated. Further synthetic utilization of optically active cyclic materials as well as investigation of the similar route with different sugars is under way in our laboratory.

Acknowledgment

This work was financially supported by the Ministry of Education, Science and Culture of Japan [Grant-in-Aid No. 08651002 and 085269]. One of the authors, S. H., is grateful to JSPS Fellowship for Japanese Junior Scientists.

Literature Cited

1) Kukhar, V. P.; Soloshonok, V. A. *Fluorine-containing Amino Acids Synthesis and Properties*; John Wiley & Sons: New York, 1995.
2) Resnati, G. *Tetrahedron* 1993, 42, 9385.

3) Welch, J. T.; Eswarakrishnan, S. *Fluorine in Bioorganic Chemistry*, John Wiley & Sons: New York, 1990.
4) Hanamoto, T.; Fuchikami, T. *J. Org. Chem.* **1990**, *55*, 4969.
5) Yokozawa, T.; Ishikawa, N.; Nakai, T. *Chem. Lett.* **1987**, 1971.
6) Tsushima, T.; Kawada, K.; Ishihara, S.; Uchida, N.; Shiratori, O.; Higaki, J.; Hirata, M. *Tetrahedron* **1988**, *44*, 5375.
7) Bey, P.; Vevert, J.-P.; van Dorsselaer, V.; Kolb, M. *J. Org. Chem.* **1979**, *44*, 2732.
8) For related materials, see, Yamazaki, T.; Yamamoto, T.; Kitazume, T. *J. Org. Chem.* **1989**, *54*, 83.
9) Ihara, M.; Taniguchi, N.; Kai, T.; Satoh, K.; Fukumoto, K. *J. Chem. Soc. Perkin Trans. 1* **1992**, 221.
10) Ishihara, T.; Ichihara, K.; Yamanaka, H. *Tetrahedron Lett.* **1995**, *36*, 8267.
11) Kuroboshi, M.; Ishihara, T. *Bull. Chem. Soc. Jpn.* **1990**, *63*, 1191.
12) Ishihara, T.; Yamaguchi, K.; Kuroboshi, M.; Utimoto, K. *Tetrahedron Lett.* **1994**, *35*, 5263.
13) Konno, T.; Yamazaki, T.; Kitazume, T. *Tetrahedron* **1996**, *52*, 199.
14) Lavaire, S.; Plantier-Royon, R.; Portella, C. *J. Carbohydr. Chem.* **1996**, *15*, 361.
15) Yamanaka, H.; Takekawa, T.; Morita, K.; Ishihara, T. *Tetrahedron Lett.* **1996**, *37*, 1829.
16) Shi, G.-Q.; Huang, X.-H.; Hong, F. *J. Chem. Soc., Perkin Trans 1* **1996**, 763.
17) Iseki, K.; Nagai, T.; Kobayashi, Y. *Tetrahedron Lett.* **1993**, *34*, 2169.
18) Bouillon, J.-P.; Maliverney, C.; Merényi, R.; Viehe, H. G. *J. Chem. Soc. Perkin Trans. 1* **1991**, 2147.
19) Chen, Q.-Y.; Wu, S.-W. *J. Chem. Soc., Chem. Commun.* **1989**, 705.
20) Rico, I.; Cantacuzène, D.; Wakselman, C. *Tetrahedron Lett.* **1981**, *22*, 3405.
21) When an amine is used as a base, eliminated HF possibly adds to the resultant olefin in some specific case by way of formation of an amine-HF complex known as a potential fluorinating reagent. Martin, V.; Molines, H.; Wakselman, C. *J. Org. Chem.* **1992**, *57*, 5530.
22) Hudlicky, M. *Org. React.* **1988**, *35*, 513.
23) Wang, C.-L. *J. Org. React.* **1985**, *34*, 31.
24) Barhdadi, R.; Troupel, M.; Périchon, J. *Chem. Commun.* **1998**, 1251.
25) Yokoyama, Y.; Mochida, K. *Synlett* **1997**, 907.
26) Hagiwara, T.; Fuchikami, T. *Synlett* **1995**, 717.
27) Krishnamurti, R.; Bellew, D. R.; Surya Prakash, G. K. *J. Org. Chem.* **1991**, *56*, 984.
28) Umemoto, T.; Ishihara, S. *Tetrahedron Lett.* **1990**, *31*, 3579.
29) Gassman, P. G.; O'Reilly, N. J. *J. Org. Chem.* **1987**, *52*, 2481.
30) Kitazume, T.; Ishikawa, N. *J. Am. Chem. Soc.* **1985**, *109*, 5186.
31) Shinohara, N.; Haga, J.; Yamazaki, T.; Kitazume, T.; Nakamura, S. *J. Org. Chem.* **1995**, *60*, 4363.
32) Mizutani, K.; Yamazaki, T.; Kitazume, T. *J. Chem. Soc., Chem. Commun.* **1995**, 51.
33) Yamazaki, T.; Iwatsubo, H.; Kitazume, T. *Tetrahedron : Asym.* **1994**, *5*, 1823.
34) Hayashi, S.; Nakai, T.; Ishikawa, N.; Burton, D. J.; Naae, D. G.; Kesling, H. S. *Chem. Lett.* **1979**, 983.
35) Vinson, W. A.; Prickett, K. S.; Spahic, B.; Ortiz de Montellano, P. R. *J. Org. Chem.* **1983**, *48*, 4661.
36) Tellier, F.; Sauvêtre, R. *Tetrahedron Lett.* **1995**, *36*, 4223.

156

37) Tellier, F.; Sauvêtre, R. *J. Fluorine Chem.* **1996**, *76*, 181.
38) Tellier, F.; Sauvêtre, R. *J. Fluorine Chem.* **1993**, *62*, 183.
39) Hall, D. M. *Carbohydr. Res.* **1980**, *86*, 158.
40) Tulshian, D. B.; Tsang, R.; Fraser-Reid, B. *J. Org. Chem.* **1984**, *49*, 2347.
41) See also, Houlton, J. S.; Motherwell, W. B.; Ross, B. C.; Tozer, M. J.; Williams, D. J.; Slawin, A. M. Z. *Tetrahedron* **1993**, *49*, 8087.
42) Lipshutz, B. H.; Nguyen, S. L.; Elworthy, T. R. *Tetrahedron* **1988**, *44*, 3355.
43) Maki, T.; Iwasaki, F.; Matsumura, Y. *Tetrahedron Lett.* **1998**, *39*, 5601.
44) Calculations were performed by CACheMechanics v. 3.8 (MM), CAChe-Dynamics V. 3.8 (MD), and CACheMOPAC v. 94.10 (MOPAC) implemented in CAChe Worksystem (SONY/Tektronix Corporation). Conformers, from the MD-MM sequence for finding out the possible ring conformations, followed by the rigid search method with rotating the free single bond in 15° increment for all the obtained cyclic structures, were optimized with MM and further with MOPAC (AM1) by the eigenvector following minimization (EF) method with the extra keyword "gnorm = 0.01", final gradient norm being less than 0.01 kcal/Å. The obtained conformers were eventually employed for the vibrational frequency calculation for their confirmation as the stationary points.
45) Dean, J. A. In *Lange's Handbook of Chemistry (14th edition)*: McGraw-Hill, Inc.: New York, 1992, p. 8.19 and 8.27.
46) Gaussian 94, Revision C.3, Frisch, M. J.; Trucks, G. W.; Schlegel, H. B.; Gill, P. M. W.; Johnson, B. G.; Robb, M. A.; Cheeseman, J. R.; Keith, T.; Petersson, G. A.; Montgomery, J. A.; Raghavachari, K.; Al-Laham, M. A.; Zakrzewski, V. G.; Ortiz, J. V.; Foresman, J. B.; Cioslowski, J.; Stefanov, B. B.; Nanayakkara, A.; Challacombe, M.; Peng, C. Y.; Ayala, P. Y.; Chen, W.; Wong, M. W.; Andres, J. L.; Replogle, E. S.; Gomperts, R.; Martin, R. L.; Fox, D. J.; Binkley, J. S.; Defrees, D. J.; Baker, J.; Stewart, J. P.; Head-Gordon, M.; Gonzalez, C.; Pople, J. A. Gaussian, Inc., Pittsburgh PA, 1995.
47) Baldwin, J. E.; Cutting, J.; Dupont, W.; Kruse, L.; Silberman, L.; Thomas, R. C. *J. Chem. Soc., Chem. Commun.* **1976**, 736.
48) Baldwin, J. E. *J. Chem. Soc., Chem. Commun.* **1976**, 734.
49) The similar type of cyclization was independently found out. See, Ichikawa, J.; Wada, Y.; Okauchi, T.; Minami, T. *Chem. Commun.* **1997**, 1537.
50) Treatment of **1** with TrCl and Et$_3$N instead of the combination of TBSCl and imidazole would be the alternative and direct route to the requisite compound **27**.
51) Siegel, S. In *Comprehensive Organic Synthesis*: Trost, B. M., Ed.; Pergamon Press, Inc.: New York, 1991, vol. 8, p. 417.
52) Gonzo, E. E.; Boudart, M. *J. Catal.* **1978**, *52*, 462.
53) Graczyk, P. P.; Mikolajczyk, M. *Top. Stereochem.* **1994**, *21*, 159.
54) Juaristi, E.; Cuevas, G. *Tetrahedron* **1992**, *48*, 5019.
55) Box, V. G. S. *Heterocycles* **1990**, *31*, 1157.
56) Yamazaki, T.; Oniki, T.; Kitazume, T. *Tetrahedron* **1996**, *52*, 11753.

Synthesis of Fluoroorganic Targets

Chapter 12

Synthesis of Enatiopure Fluorine-Containing Taxoids and Their Use as Anticancer Agents as well as Probes for Biomedical Problems

Iwao Ojima[1], Tadashi Inoue[1], John C. Slater[1], Songnian Lin[1], Scott D. Kuduk[1], Subrata Chakravarty[1], John J. Walsh[1], Thierry Cresteil[2], Bernard Monsarrat[2], Paula Pera[3], and Ralph J. Bernacki[3]

[1]Department of Chemistry, State University of New York at Stony Brook, Stony Brook, NY 11794–3400
[2]Laboratoire de Pharmacologie et de Toxicologie Fondamentales, 205 route de Nabonne, F–31400 Toulouse, France
[3]Department of Experimental Therapeutics, Grace Cancer Drug Center, Roswell Park Cancer Institute, Elm and Carlton Streets, Buffalo, NY 14263

A series of fluorine-containing analogs of paclitaxel and docetaxel are synthesized through the coupling of (3R,4S)-1-acyl-β-lactams of high enantiomeric purity with various baccatin derivatives as the key step. Some taxoids bearing CF_3 or CF_2H at the C-3' position are up to three orders of magnitude more porent than paclitaxel or doxorubin against human breast cancer cell lines, including those that are drug resistant. Metaboism studies show that fluorinated taxoid block the action of cytochrome P-450s. A combination of ^{19}F and ^1H NMR analyses with molecular modeling has disclosed a previously unrecognized conformer that may be the first structure to be recognized by microtubles and the one tightly bound to the protein.

Taxol® (paclitaxel, **1**), a highly functionalized naturally occurring diterpenoid, is currently considered one of the most important drugs in cancer chemotherapy (*1-4*). A tremendous amount of research focusing on the science and applications of paclitaxel has been performed since its initial isolation in 1966 from the bark of the Pacific yew tree (*Taxus brevifolia*) and subsequent structural determination in 1971 (*5*). Paclitaxel was approved by the FDA for the treatment of advanced ovarian cancer (December 1992) and metastatic breast cancer (April 1994). A semisynthetic analog of paclitaxel, Taxotère® (docetaxel, **2**) (*6*), was also approved by the FDA in 1996 for the treatment of advanced breast cancer. These two taxane anticancer drugs are currently undergoing Phase II and III clinical trials worldwide for a variety of other cancers as well as for combination therapy with other agents.

Paclitaxel (1): R^1 = Ph, R^2 = Ac
Docetaxel (2): R^1 = t-BuO, R^2 = H

Paclitaxel and docetaxel have been shown to act as spindle poisons, causing cell division cycle arrest, based on a unique mechanism of action (7-10). These drugs bind to the β-subunit of the tubulin heterodimer, the key constituent protein of cellular microtubules (spindles). The binding of these drugs accelerates the tubulin polymerization, but at the same time stabilize the resultant microtubules, thereby inhibiting their depolymerization. The inhibition of microtubule depolymerization between the prophase and anaphase of mitosis results in the arrest of the cell division cycle, which eventually leads to the apoptosis of cancer cells.

In the course of our study on the rational design, synthesis and structure-activity relationships of new antitumor taxoids, we became interested in incorporating fluorine(s) into paclitaxel and taxoids in order to investigate the effects of fluorine on cytotoxicity, blocking of known metabolic pathways, and its use as a probe for identifying bioactive conformation(s) of taxane antitumor agents.

Studies on the metabolism of docetaxel have shown that the *para*-position of the C-3' phenyl, the *meta*-position of the C-2 benzoate, the C-6 methylene, and the C-19 methyl groups are the primary sites of hydroxylation by the P450 family of enzymes (11,12). The predominant one among these is the hydroxylation of the C-3' phenyl at the *para*-position, presumably by the cytochrome 3A family (11). Substitution of a C-H bond with a C-F bond has been shown to substantially slow down the enzymatic oxidation in general (13). Accordingly, we synthesized a series of fluorine-containing taxoids with the expectation of obtaining analogs with improved metabolic stability and cytotoxicity (14,15). Also discussed in this section is the successful use of fluorine-containing taxoids as probes for the study of solution and solid-state (microtubule-bound) conformations of paclitaxel. The conformational analyses of fluorine-containing taxoids based on VT-^{1}H-NMR, VT-^{19}F-NMR, and molecular modeling have provided extremely valuable information towards the identification of the recognition and binding conformations of paclitaxel at its binding site on the microtubules (16,17).

Fluorine-Containing Taxoids: Synthesis and Biological Activity

Fluorine-containing baccatins were prepared from 10-deacetylbaccatin III (DAB) through 2-debenzoylbaccatin 3 (Scheme 1). Thus, the protection of the 7,10, and 13-hydroxyl groups of DAB using TES-Cl and imidazole was followed by the removal of the 2-benzoyl group using Red-Al to yield baccatin 3 quantitatively, which was coupled with 4-fluoro-3-(trifluoromethyl)benzoic acid or 3-fluorobenzoic acid in the presence of DIC/DMAP or DCC/DMAP to afford baccatins 4 in 80% yield. Conversion of 4 to C-2 modified baccatins 5a-d was accomplished by the following

Scheme 1

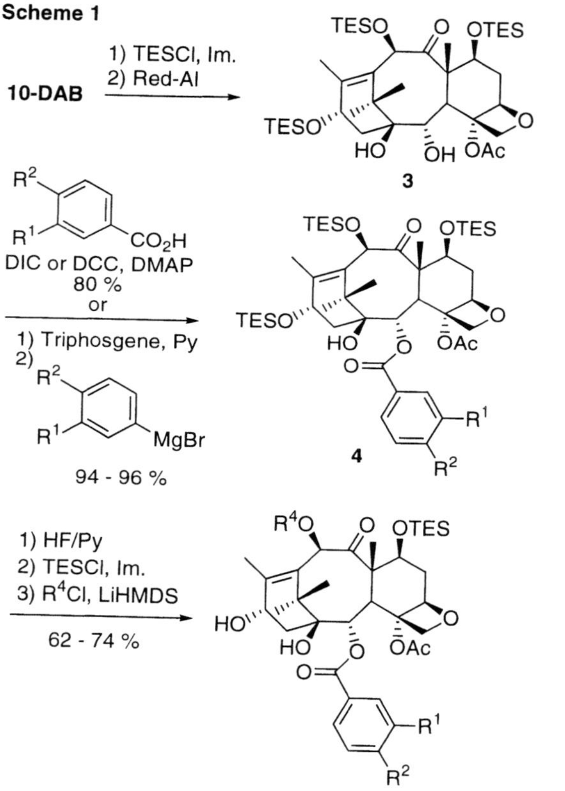

10-DAB

1) TESCl, Im.
2) Red-Al

3

R^2, R^1, CO$_2$H
DIC or DCC, DMAP
80 %
or
1) Triphosgene, Py
2) R^2, R^1, MgBr
94 - 96 %

4

1) HF/Py
2) TESCl, Im.
3) R^4Cl, LiHMDS
62 - 74 %

5a: R^1 = CF$_3$, R^2 = F, R^4 = Ac
5b: R^1 = F, R^2 = F, R^4 = TES
5c: R^1 = H, R^2 = F, R^4 = Ac
5d: R^1 = Me, R^2 = F, R^4 = Ac

sequence: (1) removal of the TES protection with HF/pyridine, (2) selective reprotection of the C-7 hydroxyl group with TES and, (3) protection of the C-10 hydroxyl group with an acetyl or silyl moiety (Scheme 1). Alternatively, the ring-opening reaction of the cyclic 1,2-carbonate, which was prepared from 7,10,13-triTES-2-debenzoylbaccatin **3** and triphosgene, is also useful to synthesize C-2 modified baccatins. The reaction of this cyclic carbonate with 4 equiv. of fluorinated phenyl magnesium bromide provided the TES-protected DABs in good yields (Scheme 1). Removal of the TES protection with HF/pyridine, followed by selective reprotection of the C-7 hydroxyl group with TES and acetylation at C-10 provided baccatins **5c** and **5d** in good yields. Fluorine-containing baccatins can also be prepared using a modified version of Chen's method (*18*), *i.e.*, DCC coupling of 13-oxo-2-debenzoyl-7-TES-baccatin **6** using 3 equiv. of mono- or difluorobenzoic acid followed by NaBH$_4$ reduction of the 13-keto moiety afforded **5e** and **5f** in good yield (Scheme 2).

The syntheses of fluorine-containing taxoids **8** were carried out through the coupling of 1-acyl-β-lactams **7** and C-2 modified baccatins **5** with proper protecting groups at C-7 and/or C-10 by means of the protocol developed in our laboratories (*19-22*) (Scheme 3).

Bioassay for new fluorine-containing taxoids. The *in vitro* cytotoxicity of the new fluorine-containing taxoids **8c**, **8e**, and **8f** were evaluated against a standard set of five human tumor cell lines, A121 (ovarian), A549 (non-small cell lung), HT-29 (colon), MCF-7 (breast), and a drug-resistant cell line MCF7-R (breast). Results are shown in Table I with the values for paclitaxel, docetaxel and some fluorine-containing taxoids previously reported from these laboratories for comparison.

As Table I shows, several of the new fluorine-containing taxoids exhibit similar or slightly weaker activity as compared to paclitaxel and docetaxel. 3'-(4-Fluorophenyl)paclitaxel (**SB-T-3101**) shows slightly decreased activity compared to the parent compound, while 3'-(4-fluorophenyl)docetaxel (**SB-T-3001**) and 10-acetyl-3'-(4-fluorophenyl)docetaxel (**SB-T-3002**) show comparable activity to docetaxel. 3'-(*N*-Fluorobenzoyl)paclitaxel (**SB-T-13103**) is similar to **SB-T-3101** in activity, while 3'-*N*-fluorobenzoyl-3'-(4-fluorophenyl)paclitaxel (**SB-T-31031**) is about 10-fold weaker as compared to paclitaxel, showing that the incorporation of two fluorines at these positions is unfavorable to the activity. Thus, it appears that fluorine incorporation at the *para*-position on the side chain aromatic rings of paclitaxel decreases the activity, while exhibiting little effect on docetaxel analogs.

Modification at the C-2 position also exhibits interesting effects on the cytotoxicity. Compound **8c**, which is an **SB-T-3002** analog with the addition of a fluorine at the *para*-position of the C-2 benzoate, has weaker activity than that of docetaxel but similar to that of paclitaxel. Difluoro-paclitaxel **8e** showed drastically weaker cytotoxicity. 2-(3,5-Difluorobenzoyl)taxoid **8f** was one of the most active analogs, showing that substitution at the *meta*-position is tolerated very well. These results are consistent with other reports that substitution at the *para*-position of the C-2 benzoate (cyano, chloro, methoxy, and azido) significantly decreases the activity but *meta*-substitution enhances activity (*23*).

Scheme 2

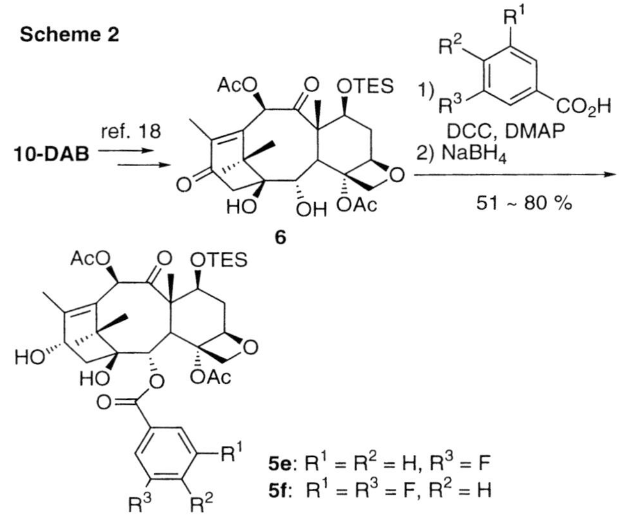

5e: $R^1 = R^2 = H$, $R^3 = F$
5f: $R^1 = R^3 = F$, $R^2 = H$

Scheme 3

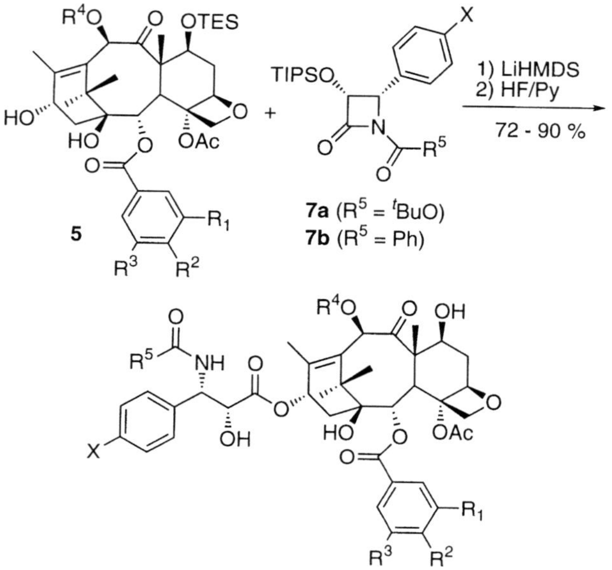

7a ($R^5 = {}^tBuO$)
7b ($R^5 = Ph$)

8a: $R^1 = CF_3$, $R^2 = F$, $R^3 = H$, $R^4 = Ac$, $R^5 = {}^tBuO$, $X = F$
8b: $R^1 = F$, $R^2 = R^3 = R^4 = H$, $R^5 = {}^tBuO$, $X = H$
8c: $R^1 = R^3 = H$, $R^2 = F$, $R^4 = Ac$, $R^5 = {}^tBuO$, $X = F$
8d: $R^1 = Me$, $R^2 = F$, $R^3 = H$, $R^4 = Ac$, $R^5 = {}^tBuO$, $X = H$
8e: $R^1 = R^2 = H$, $R^3 = F$, $R^4 = Ac$, $R^5 = Ph$, $X = F$
8f: $R^1 = F$, $R^2 = H$, $R^3 = F$, $R^4 = Ac$, $R^5 = {}^tBuO$, $X = F$

Table I. Cytotoxicities (IC_{50} nM)[a] of Fluorine Containing Taxoids

F-Taxoid	A121	A549	HT-29	MCF7	MCF7R
Paclitaxel	6.3	3.6	3.6	1.7	300
Docetaxel	1.2	1.0	1.2	1.0	235
SB-T-3101	6.3	4.2	14.5	5.1	>1000
SB-T-3001	1.3	0.5	3.9	0.5	477
SB-T-3002	1.2	0.5	3.5	0.4	315
SB-T-31031	76	35	51	45	>1000
SB-T-13103	6.3	9.7	9.8	3.6	>1000
8c	6.8	5.0	5.3	3.7	518
8e	130	42	24	15	>1000
8f	1.3	0.4	1.2	0.7	80

[a] The concentration of compound which inhibits 50% (IC_{50}, nM) of the growth of human tumor cell line, A121 (ovarian carcinoma), A549 (non-small cell lung carcinoma), HT-29 (colon carcinoma), MCF7 (mammary carcinoma), and MCF7-R (mammary carcinoma cells 180 fold resistant to doxorubicin) after 72 h drug exposure according to the method developed by Skehan et al. (24) The data represent the mean values of at least three separate experiments.

Metabolism study of new fluorine-containing taxoids. Enzyme inhibition studies have shown that the cytochrome P-450 (CYP) family of enzymes is playing a key role in the biotransformation of paclitaxel and docetaxel. It has been found that the CYP 2C enzyme subfamily is responsible for the 6α-hydroxylation of paclitaxel, while the CYP 3A subfamily plays a key role in the hydroxylation of the *t*Boc group of docetaxel as well as that of the 3'-phenyl and 2-benzoyl groups of paclitaxel (Figure 1) (11,25)

Novel fluorine-containing paclitaxel and docetaxel analogs, **8b**, **8e**, and **SB-T-3101** have been used as substrates to probe the metabolic pathways of taxane anticancer drugs (Figure 2). Human liver microsomes were used for the enzymatic degradation, and in most cases, additional experiments involving recombinant CYP enzymes were also performed. In the case of 3'-(4-fluorophenyl)paclitaxel (**SB-T-3101**), there was no sign of hydroxylation at the 3'-phenyl ring, which clearly indicates that the fluorine at the *para*-position of C-3' phenyl group completely blocks the enzymatic hydroxylation by CYP as expected. A striking result was observed for 2-(3-fluorobenzoyl)paclitaxel **8e**, *i.e.*, the hydroxylation occurred only at the 6-position, while no hydroxylation was observed at the 3'-phenyl ring, suggesting a strong allosteric effects of the fluorine-containing 2-benzoate moiety on the action of the enzyme. The fluorine in the C-2 benzoate moiety appears to block the hydroxylation on the 3'-phenyl ring, which is consistent with the "hydrophobic cluster" conformation that is strongly suggested for both paclitaxel and docetaxel in aqueous media. 2-(3-Fluorobenzoyl)docetaxel **8b** underwent the usual side-chain hydroxylation, which may well be due to the greater reactivity of the *tert*-butyl group as compared to the 3'-phenyl of paclitaxel towards enzymatic hydroxylation by CYP. Further studies are in progress.

Synthesis and biological activity of 3'-trifluoromethyl-docetaxel analogs (15,26). During the course of the structure-activity relationship study of fluorine-containing taxoids, we designed a series of taxoids bearing a trifluoromethyl moiety at the C-3' position in place of the phenyl group. For the synthesis of a series of 3'-CF$_3$-taxoids

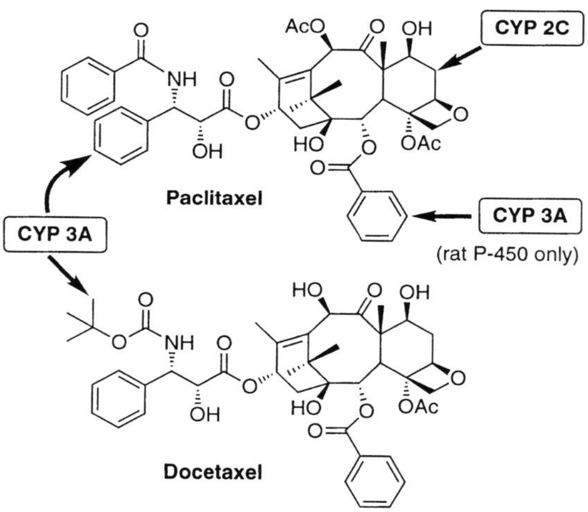

Figure 1. Primary sites of hydroxylation on paclitaxel and docetaxel by the P450 family of enzymes.

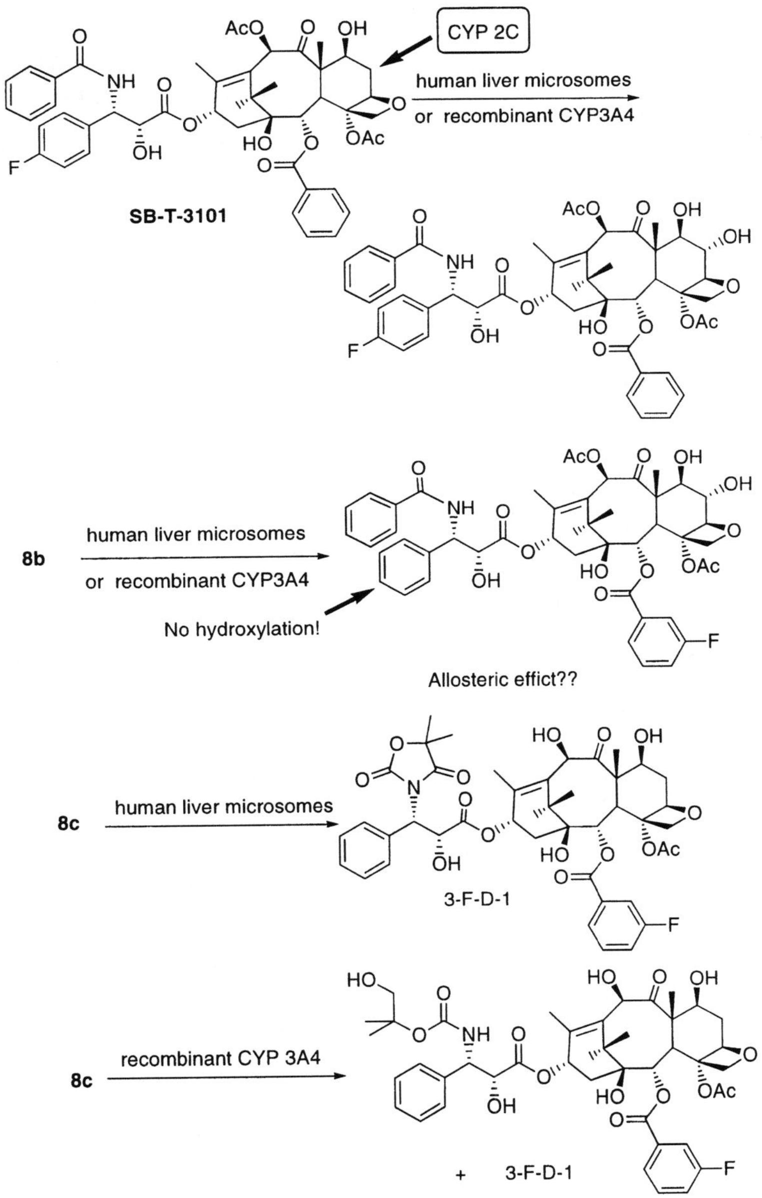

Figure 2. Metabolism of fluorine-containing taxoids by P450 enzymes.

11, the ring-opening coupling of racemic N-^{t}Boc-β-lactam 10 (excess) with 7-TES-baccatins 9 was examined with the expectation of the occurrence of high level kinetic resolution at temperatures in the range of -40 °C to -20 °C (Scheme 4, Table II). The coupling reactions proceeded smoothly to give the desired 3'-CF$_3$-taxoids 11 with diastereomeric ratios of 9:1 to >30:1 (by ^{19}F NMR analysis) in fairly good overall yields after deprotection. In particular, (2'R,3'R) taxoids 11a, 11b, and 11c were obtained exclusively through the reactions of 10 with 9a, 9b, and 9c (R = Me$_2$N-CO), respectively. In addition, the reaction of 9g (R = Et-CO) with 10 afforded 11g with >30:1 diastereomeric ratio.

Table II. Syntheses of 3'-CF$_3$-taxoids 11 Through Kinetic Resolution

Taxoid	R	Conversion of Coupling (%)[a]	Yield (%)[b]	Isomer Ratio[c] (2'R,3'S): (2'S,3'R)
11a	H	80	54 (67)	single isomer
11b	Ac	72	41 (57)	single isomer
11c	Me$_2$N-CO	100	63	single isomer
11d	cyclopropane-CO	100	64	10:1
11e	MeO-CO	93	54 (58)	24:1
11f	morpholine-4-CO	100	60	23:1
11g	Et-CO	100	74	>30:1
11h	CH$_3$(CH$_2$)$_3$-CO	100	56	9:1
11i	(CH$_3$)$_3$CCH$_2$-CO	91	59 (65)	22:1

[a]Based on consumed baccatin. [b]Two-step yield. Values in parentheses are conversion yields. [c]Determined by ^{19}F NMR analysis.

Mechanistically, the nucleophilic ring-opening of the β-lactam by the lithium alkoxide of the baccatin requires the carbonyl of the β-lactam be aligned with the lithium-oxygen bond in the transition state. It is shown that when the lithium is located outside of the concave face of the baccatin, due to its solvation by THF, the β-lactam is situated underneath the baccatin. This alignment gives rise to the four possible orientations depicted in Figure 3(a)-(d).

When the (3R,4R)-CF$_3$-β-lactam, that is the precursor to the correct stereochemistry in the C-13 side chain, is oriented in the manner shown in Figure 3(a), steric interactions are minimized relative to other possible orientations. The C-3 and C-4 substituents are facing away from the baccatin core and the bulky ^{t}Boc group is located outside of the baccatin framework.

The positioning of the enantiomer with its substituents facing the baccatin [Figure 3(b)], creates significant steric crowding between the C-3 and C-4 substituents of the β-lactam and the baccatin. In addition, there are serious steric interactions involving the ^{t}Boc group and the baccatin.

In Figure 3(c), the ^{t}Boc group of the (3S,4S)-CF$_3$-β-lactam is situated underneath the baccatin core, thereby creating steric crowding in the same fashion as shown in Figure 3(b). In addition, when the orientation of the β-lactam is inverted to

Scheme 4

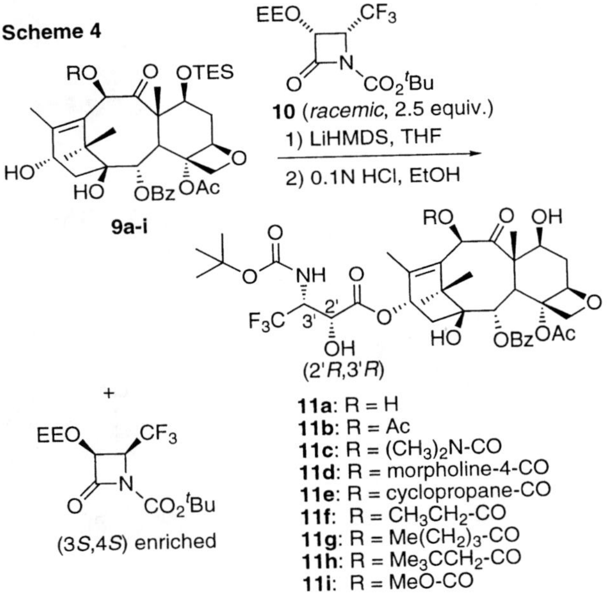

9a-i

10 (*racemic*, 2.5 equiv.)

1) LiHMDS, THF

2) 0.1N HCl, EtOH

(2'*R*,3'*R*)

+

(3*S*,4*S*) enriched

11a: R = H
11b: R = Ac
11c: R = (CH₃)₂N-CO
11d: R = morpholine-4-CO
11e: R = cyclopropane-CO
11f: R = CH₃CH₂-CO
11g: R = Me(CH₂)₃-CO
11h: R = Me₃CCH₂-CO
11i: R = MeO-CO

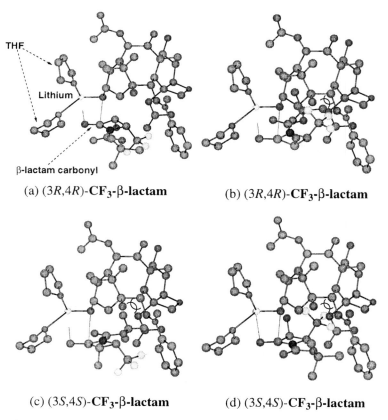

(a) (3*R*,4*R*)-**CF₃-β-lactam**

(b) (3*R*,4*R*)-**CF₃-β-lactam**

(c) (3*S*,4*S*)-**CF₃-β-lactam**

(d) (3*S*,4*S*)-**CF₃-β-lactam**

Figure 3. Possible transition states for the kinetic resolution of 4-CF₃-*β*-lactam. (Reproduced with permission from reference 26. Copyright 1997 Wiley.)

allow the ^{t}Boc group to reside outside of the baccatin core, the C-3 and C-4 substituents are facing towards the baccatin and result in obvious steric crowding.

The C-3 and C-4 substituents are believed to create a large enough steric hindrance to render orientations 3(b) and 3(d) insignificant in terms of their contributions to the coupling reaction. Therefore, the relative orientations of the ^{t}Boc group in the approach of the β-lactam in Figures 3(a) and 3(c) give rise to the kinetic resolution observed during these coupling reactions.

Cytotoxicities of the 3'-CF$_3$-taxoids **11** thus obtained were evaluated against human cancer cell lines and the results are summarized in Table III. As Table III shows, all 3'-CF$_3$-taxoids possess excellent activities, and are substantially more potent than either paclitaxel or docetaxel in virtually every case. The most remarkable results are, however, the one order of magnitude better activities of the 10-acyl-taxoids **11a-i** as compared to paclitaxel and docetaxel against the drug-resistant breast cancer cell line, MCF7-R. The comparison of 10-OH-taxoid **11a** with 10-acyl-taxoids **11b-i** clearly indicates that the observed remarkable activity enhancement against MCF7-R can be attributed to the modifications at the C-10 position, while this modification has little effect on the activity against the normal cancer cell lines, *i.e.*, there is no apparent relationship between the activity against the normal cancer cell lines and that against the drug-resistant cell line.

Table III. Cytotoxicity of 3'-CF$_3$-taxoids (IC$_{50}$ nM)[a]

Taxoid	R	A121	A549	HT-29	MCF-7	MCF7-R
Paclitaxel	–	6.3	3.6	3.6	1.7	299
Docetaxel	–	1.2	1.0	1.2	1.0	235
11a	H	1.15	0.44	0.65	0.44	156
11b	Ac	0.3	0.2	0.4	0.6	17
11c	Me$_2$N-CO	0.3	0.2	0.4	0.3	21
11d	cyclopropane-CO	0.4	0.4	0.5	0.5	16
11e	MeO-CO	0.3	0.2	0.4	0.4	21
11f	morpholine-4-CO	0.5	0.4	0.4	0.4	48
11g	Et-CO	0.3	0.3	0.4	0.3	14
11h	CH$_3$(CH$_2$)$_3$-CO	0.7	0.8	1.4	0.7	26
11i	(CH$_3$)$_3$CCH$_2$-CO	0.5	0.5	0.6	0.5	12

[a]See footnote of Table I.

Synthesis and biological activity of 3'-difluoromethyl-docetaxel analogs. We have also synthesized a series of 3'-difluoromethyl-taxoids. (3R,4R)-3-TISPO-4-(2-methyl-1-propenyl)-β-lactam **13** (*27*) was used as the key compound toward the synthesis of difluoro-β-lactam **16** (Scheme 5). Ozonolysis of β-lactam **13** afforded 4-formyl-β-lactam **14** in quantitative yield. Treatment of β-lactam **14** with 3.0 equiv. of diethylaminosulfur trifluoride (DAST) afforded N-PMP-3-TIPSO-4-difluoro-methyl-β-lactam **15** in good yield. There was no detectable epimerization of the β-lactam during fluorination. Subsequent CAN oxidation of the N-PMP group followed by *tert*-Boc introduction afforded 1-^{t}Boc-3-TIPSO-4-CF$_2$H-β-lactam **16** in good yield.

Treatment of a mixture of baccatins **9a-c,f,g** and β-lactam **16** in THF with NaHMDS at -40~-20 °C afforded the coupling products in moderate to good yield.

Scheme 5

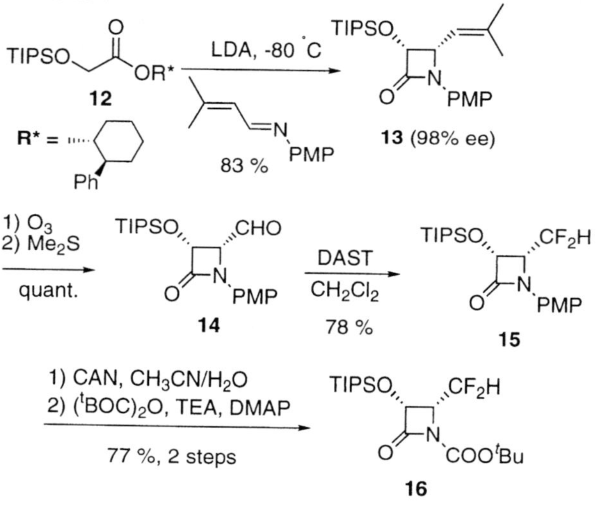

TIPSO–CH₂–C(=O)–OR*

12

R* = cyclohexyl with Ph

LDA, -80 °C

83 %

CH₃–C(CH₃)=CH–CH=N–PMP

TIPSO / CH=C(CH₃)₂ azetidinone, N-PMP

13 (98% ee)

1) O₃
2) Me₂S

quant.

TIPSO, CHO azetidinone N-PMP

14

DAST
CH₂Cl₂

78 %

TIPSO, CF₂H azetidinone N-PMP

15

1) CAN, CH₃CN/H₂O
2) (ᵗBOC)₂O, TEA, DMAP

77 %, 2 steps

TIPSO, CF₂H azetidinone N-COOᵗBu

16

Desilylation with HF/pyridine gave 3'-CF$_2$H-docetaxel analogs **17a-c,f,h** in good yields (Scheme 6).

As Table IV shows, virtually all 3'-CF$_2$H analogs exhibit activities greater than either paclitaxel or docetaxel, **17b** and **17c** being the most cytotoxic agents against the drug-sensitive breast cancer call line LCC6-WT. These second generation fluorine-containing taxoids exhibit single digit nM level IC$_{50}$ values against the drug-resistant human breast cancer cell line LCC6-MDR, which are *two orders of magnitude* more potent than paclitaxel (Taxol®) and three orders of magnitude more potent than doxorubicin.

Table IV. Cytotoxicity of 3'-CF$_2$H-taxoids (IC$_{50}$ nM)[a]

Taxoid	R	LCC6-WT	LCC6-MDR
Paclitaxel	–	3.1	346
Docetaxel	–	1.0	120
Doxorubicin	–	180	2900
17a	H	0.96	43.5
17b	Ac	0.50	8.79
17c	Me$_2$N-CO	0.48	8.20
17g	Et-CO	1.0	5.69
17h	(CH$_3$)$_3$CCH$_2$-CO	1.34	5.04

[a] The concentration of compound which inhibits 50% of the growth tumor cell line, LCC6-WT (breast carcinoma) and LCC6-MDR (drug-resistant breast carcinoma) after 72h drug exposure according to the method developed by Skehan et al. (*24*)

The Fluorine Probe Approach: Solution-phase Structure and Dynamics of Taxoids. The rational design of new generation taxoid anticancer agents would be greatly facilitated by the development of reasonable models for the biologically relevant conformations of paclitaxel. In this regard, we recognized that the design and synthesis of fluorine-containing taxoids would have an extremely useful offshoot of providing us with the capability of studying bioactive conformations of taxoids using a combination of ^{19}F/^{1}H-NMR techniques and molecular modeling (*16*).

Previous studies in the conformational analysis of paclitaxel and docetaxel have largely identified two major conformations, with minor variations between studies (*28*). Structure **A**, characterized by a gauche conformation with a H2'-C2'-C3'-H3' dihedral angle of ca. 60°, is based on the X-ray crystal structure of docetaxel (*29*), and is believed to be commonly observed in aprotic solvents (Figure 4) (*28*). Structure **B**, characterized by the *anti* conformation with a H2'-C2'-C3'-H3' dihedral angle of ca. 180°, has been observed in theoretical conformational analysis (*30,31*) as well as 2D NMR analyses (*32*), and has also been found in the X-ray structure of the crystal obtained from a dioxane/H$_2$O/xylene solution (Figure 4) (*33*). Despite extensive structural studies, no systematic study on the dynamics of these two and other possible bioactive conformations of paclitaxel had been reported when we started our study on this problem. The relevance of the "fluorine probe" approach to study dynamic properties prompted us to conduct a detailed investigation into the solution dynamics of fluorine-containing paclitaxel and docetaxel analogs.

The use of ^{19}F-NMR for a variable temperature (VT) NMR study of fluorine-containing taxoids is obviously advantageous over the use of ^{1}H-NMR because of

172

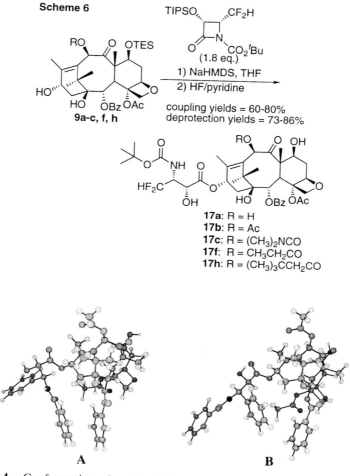

Scheme 6

9a-c, f, h

(1.8 eq.)
1) NaHMDS, THF
2) HF/pyridine

coupling yields = 60-80%
deprotection yields = 73-86%

17a: R = H
17b: R = Ac
17c: R = (CH₃)₂NCO
17f: R = CH₃CH₂CO
17h: R = (CH₃)₃CCH₂CO

A   B

Figure 4. Conformation of paclitaxel based on the X-ray structure of docetaxel and proposed for non-polar aprotic organic solvents (structure **A**). Conformation based on X-ray structure of paclitaxel and proposed for aqueous solvents (structure **B**).

the wide dispersion of the ^{19}F chemical shifts that allows fast dynamic processes to be frozen out. Accordingly, F$_2$-paclitaxel **SB-T-31031** and F-docetaxel **SB-T-3001** were selected as probes for the study of the solution structures and dynamic behavior of paclitaxel and docetaxel, respectively, in protic and aprotic solvent systems (*16*).

SB-T-31031   SB-T-3001

Analysis of the low temperature VT-NMR (^{19}F and ^{1}H) and ^{19}F–^{1}H heteronuclear NOE spectra of **SB-T-31031** and **SB-T-3001** in conjunction with molecular modeling has revealed the presence of an equilibrium between two conformers in protic solvent systems. Interpretation of the temperature dependence of the coupling constants between H2' and H3' for **SB-T-31031** shows that one of these conformers (conformer C, Figure 5) has an unusual near-eclipsed arrangement around the H2'-C2'-C3'-H3' dihedral angle ($J_{H2'-H3'}$ = 5.2 Hz, corresponding to the H2'-C2'-C3'-H3' torsion angle of 124° based on the MM2 calculation), and is found to be more prevalent at ambient temperatures. The other one corresponds to the *anti* conformer (conformer B, $J_{H2'-H3'}$ = 10.1 Hz, corresponding to the H2'-C2'-C3'-H3' torsion angle of 178° based on the MM2 calculation) and is quite closely related to the structure **B** in Figure 2. These conformers are different from the one observed in aprotic solvents (conformer A, H2'-C2'-C3'-H3' torsion angle of 54°) that is related to the X-ray crystal structure of docetaxel represented by the structure **A** in Figure 2 (*29*). Figure 3 shows the Newman projections for these three conformers.

Restrained molecular dynamics (RMD) studies presented evidence for the hydrophobic clustering of the 3'-phenyl and 2-benzoate (Ph) for both conformers B and C. Although the conformer C possesses the rather unusual semi-eclipsed arrangement around the C2'-C3' bond, the unfavorable interaction associated with such a conformation is apparently offset by significant solvation stabilization, observed in the comparative RMD study in a simulated aqueous environment for the three conformers. The solvation stabilization term for the conformer C was estimated to be about 10 kcal/mol greater than those for the conformers A and B. Accordingly, the "fluorine probe" approach has succeeded in finding a new conformer that has never been predicted by the previous NMR and molecular modeling studies (*28*).

Strong support for the conformer C can be found in its close resemblance to a proposed solution structure of a water-soluble paclitaxel analog, paclitaxel-7-MPA (MPA = *N*-methylpyridinium acetate) (*34*) in which the H2'-C2'-C3'-H3' torsion angle of the *N*-phenylisoserine moiety is 127°, which is only a few degrees different from the value for the conformer C.

Thus, the "fluorine probe" approach has proved useful for the conformational analysis of paclitaxel and taxoids in connection with the determination of possible bioactive conformations (*16*). The previously unrecognized conformer C might be the molecular structure first recognized by the β-tubulin binding site on microtubules.

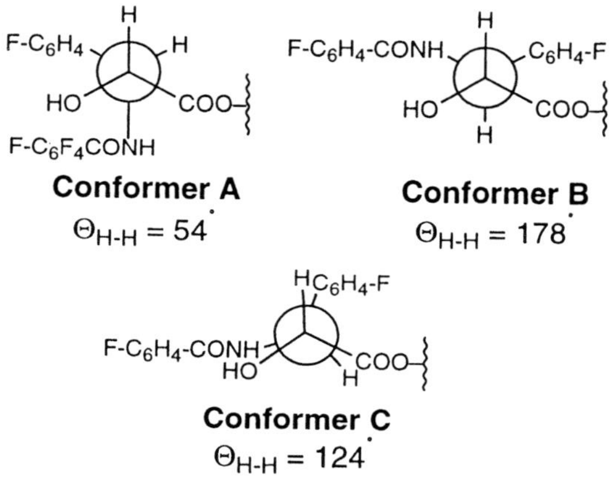

Figure 5. Newman projections of the three conformers for F2-paclitaxel **SB-T-31031**.

The fluorine probe approach (*16*) confirms hydrophobic clustering to be one of the main driving forces for the conformational stabilization of fluorine-containing paclitaxel and docetaxel analogs. It is obvious that the size of the hydrophobic moieties must play a very important role in determining the extent of clustering, and hence the conformational equilibrium. In this regard, we became interested in determining the conformational preference for the highly potent 3'-CF$_3$-taxoid **11b**. The ^{1}H and ^{19}F VT NMR study has revealed that **11b** does not exhibit different conformations in DMSO-D$_2$O, MeOD, and CD$_2$Cl$_2$ (*15*). In fact, the $J_{H2'-H3'}$ value remains ca. 2 Hz in all the protic and aprotic solvents examined. This value corresponds to a H2'-C2'-C3'-H3' torsion angle of ca. 60° with no detectable temperature dependence. These results clearly indicate that this molecule does not undergo drastic conformational change even the polarity of the solvent is dramatically varied. Systematic conformational search (Sybyl 6.04) for **11b** identifies the gauche conformers with H2'-C2'-C3'-H3' torsion angle of 60° and -60° to be more stable than the *anti* conformer (180°) by 2-5 kcal/mol, while the exact opposite is true for paclitaxel for which the *anti* conformer is more stable than the gauche conformers. Chem 3D representation of the most likely solution conformation of **11b** based on the H2'-C2'-C3'-H3' torsion angle of 60° is shown in Figure 6. This conformation is similar to the X-ray structure of docetaxel (*29*), although the 3'-phenyl is replaced by a sterically and electronically very different CF$_3$ group. The fact that **11b** and related CF$_3$-taxoids are extremely active warrants further investigation on the interaction of this molecule with microtubules using the CF$_3$ group as a probe.

Determination of the Binding Conformation of Taxoids to Microtubules Using Fluorine Probes. The knowledge of the solution structures and dynamics of paclitaxel and its analogs is necessary for a good understanding of the recognition and binding processes between paclitaxel and its binding site on the microtubules, which also provides crucial information for the design of future generation anticancer agents (*35*). However, the elucidation of the microtubule-bound conformation of paclitaxel is critical for the rational design of efficient inhibitors of microtubule disassembly. The lack of information about the three-dimensional tubulin binding site has prompted us to apply our fluorine probe approach to the determination of the F–F distances in the microtubule-bound F$_2$-taxoids, which should provide the relevant distance map for the identification of the bioactive (binding) conformation of paclitaxel.

We have successfully applied the fluorine probe approach to the estimation of the F–F distance in the microtubule-bound F$_2$-10-Ac-docetaxel **8c** using the solid-state magic angle spinning (SS MAS) ^{19}F NMR coupled with the radio frequency driven dipolar recoupling (RFDR) technique in our preliminary study in collaboration with L. Gilchrist, A. E. McDermott, K. Nakanishi (Columbia Univ.), S. B. Horwitz and M. Orr (A. Einstein College of Medicine) (*17*).

F$_2$-10-Ac-docetaxel **8c** was first studied in a polycrystalline form by the RFDR technique. Based on the standard simulation curves derived from molecules with known F–F distances (distance markers), the F–F distance of two fluorine

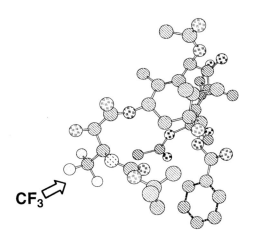

CF₃

Figure 6. Chem3D representation of 3′-CF₃-taxoid **11b**.

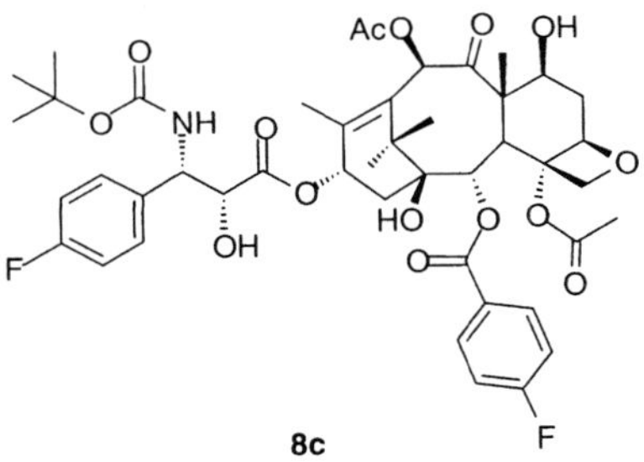

8c

atoms in **8c** was estimated to be 5.0 ± 0.5 Å (Figure 6). This value corresponds quite closely to the estimated F-F distances for the conformers B and C (F–F distance is ca. 4.5 Å for both conformers) based on our RMD studies for F_2-paclitaxel (**SB-T-31031**) (*vide supra*). This means that the microcrystalline structure of **8c** is consistent with the hydrophobic clustering conformer B or C, but not with the conformer A in which the F–F distance is ca. 9.0 Å.

The microtubule-bound complex of **8c** revealed the F–F distance to be 6.5 ± 0.5 Å (Figure 7), which is slightly larger than that observed in the polycrystalline form by ca. 1 Å. It is very likely that the microtubule-bound conformation of **8c** is achieved by a small distortion of the solution conformation (the *recognition conformation*), the latter being described by either conformer B or C.

Restrained high temperature molecular dynamics in vacuum were conducted for **8c** while maintaining a distance restraint of 6.5 Å between the two fluorine atoms in the minimization step for each sampled conformer. This study revealed that the distance of 6.5 Å between the two fluorine atoms could be maintained by energetically similar conformers with H2'–C2'–C3'–H3' torsions of 180°, 60° and -60° (Figure 8). Our investigation into the identification of the common pharmacophore of paclitaxel, epothilones, and eleutherobin has recently revealed a highly plausible common pharmacophore structure (35,36). When we screened the four low energy conformers shown in Figure 8 against the common pharmacophore, the conformer I was singled out as the most likely microtubule-bound conformation of paclitaxel.

We are currently evaluating several taxoid analogs containing fluorines at different positions by SS MAS ^{19}F-NMR. These studies are geared towards generating a detailed distance map that will help to finally pinpoint the microtubule-bound conformation of paclitaxel. The above account has demonstrated the power of the fluorine probe approach, that is clearly evident from its ability to supply extremely valuable and precise information about both bound and dynamic conformations of biologically active molecules, especially useful in the absence of knowledge about the three-dimensional structure of their binding site.

Acknowledgments. This research was supported by grants from the National Institutes of Health (GM417980 to I.O., and CA13038 to R.J.B.), Rhône-Poulenc Rorer (to I.O.). A generous support from Indena, SpA (to I.O.) is also gratefully acknowledged. The authors would like to thank Dr. Ezio Bombardelli, Indena, SpA

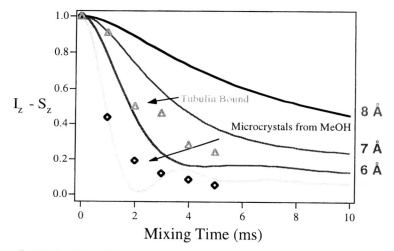

Figure 7. Estimation of F-F distances for microcrystalline and tubulin-bound forms of F2-10-Ac-docetaxel **8c** by the RFDR technique. Solid lines represent simulation curves for compounds with known F-F distances (distance markers).

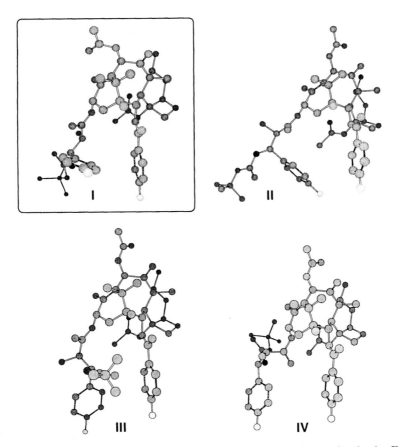

Figure 8. Energetically similar conformers of F2-taxoid **8c** that maintain the F-F distance of 6.5 Å based on the RMD study.

180

for providing them with 10-DAB. Two of the authors (S.D.K. and J.C.S.) would like to thank the U.S. Department of Education for GAANN fellowships.

Literature Cited

(1)   Georg, G. I.; Chen, T. T.; Ojima, I.; Vyas, D. M. *Taxane Anticancer Agents: Basic Science and Current Status*; American Chemical Society: Washington D.C., 1995.

(2)   Suffness, M. *Taxol: Science and Applications*; CRC Press: New York, 1995.

(3)   Rowinsky, E. K.; Onetto, N.; Canetta, R. M.; Arbuck, S. G., "Taxol: The First of the Taxanes, and Important New Class of Antitumor Agents", *Seminars in Oncology* **1992**, *19*, 646-662.

(4)   Rowinsky, E. K., *Ann. Rev. Med.* **1997**, *48*, 353-374.

(5)   Wani, M. C.; Taylor, H. L.; Wall, M. E.; Coggon, P.; McPhail, A. T., *J. Am. Chem. Soc.* **1971**, *93*, 2325-2327.

(6)   Guénard, D.; Guéritte-Vogelein, F.; Potier, P., *Acc. Chem. Res.* **1993**, *26*, 160-167.

(7)   Shiff, P. B.; Fant, J.; Horwitz, S. B., *Nature* **1979**, *277*, 665-667.

(8)   Shiff, P. B.; Horwitz, S. B *Proc. Natl. Acad. Sci., U. S. A.* **1980**, *77*, 1561-1565.

(9)   Vallee, R. B., *Taxol®: Science and Applications*; M. Suffness, Ed.; CRC Press: New York, 1995; pp 259-274.

(10)   Ringel, I.; Horwitz, S. B., *J. Natl. Cancer Inst.* **1991**, *83*, 288-291.

(11)   Vuilhorgne, M.; Gaillard, C.; Sanderlink, G. J.; Royer, I.; Monsarrat, B.; Dubois, J.; Wright, M., *Taxane Anticancer Agents: Basic Science and Current Status; ACS Symp. Series 583*; G. I. Georg; T. T. Chen; I. Ojima and D. M. Vyas, Ed.; American Chemical Society: Washington D. C., 1995; pp 98-110.

(12)   Monsarrat, B.; Mariel, E.; Cros, S.; Garès, M.; Guénard, D.; Guéritte-Voegelein, F.; Wright, M., *Drug Metab. Dispos.* **1990**, *18*, 895-901.

(13)   Filler, R.; Kobayashi, Y.; Yagupolskii, L. M. *Organofluorine Compounds in Medicinal Chemistry and Biomedical Applications*; Elsevier: Amsterdam, 1993; Vol. 48.

(14)   Ojima, I.; Kuduk, S. D.; Slater, J. C.; Gimi, R. H.; Sun, C. M., *Tetrahedron* **1996**, *52*, 209-224.

(15)   Ojima, I.; Slater, J. C.; Pera, P.; Veith, J. M.; Abouabdellah, A.; Bégué, J.-P.; Bernacki, R. J., *Bioorg. Med. Chem. Lett.* **1997**, *7*, 209-214.

(16)   Ojima, I.; Kuduk, S. D.; Chakravarty, S.; Ourevitch, M.; Bégué, J.-P., *J. Am. Chem. Soc.* **1997**, *119*, 5519-5527.

(17)   Gilchrist, L.; McDermott, A. E.; Ojima, I.; Kuduk, S. D.; Inoue. T.; Lin, S.; Walsh, J. J.; Chakravarty, S.; Nakanishi, K.; Orr, G.; Horwitz, S. B., 1997, unpublished results.

(18)   Chen, S.-H.; Farina, V.; Wei, J.-M.; Long, B.; Fairchild, C.; Mamber, S. W.; Kadow, J. F.; Vyas, D.; Doyle, T. W., *Bioorg. Med. Chem. Lett.* **1994**, *4*, 479-482.

(19)   Holton, R. A.; Somoza, C.; Kim, H.-B.; Liang, F.; Biediger, R. J.; Boatman, P. D.; Shindo, M.; Smith, C. C.; Kim, S.; Nadizadeh, H.; Suzuki, Y.; Tao, C.; Vu, P.; Tang, S.; Zhang, P.; Murthi, K. K.; Gentile, L. N.; Liu, J. H., *J. Am. Chem. Soc.* **1994**, *116*, 1597-1598.

(20)   Ojima, I.; Park, Y. H.; Sun, C.-M.; Fenoglio, I.; Appendino, G.; Pera, P.; Bernacki, R. J., *J. Med. Chem.* **1994**, *37*, 1408-1410.

(21)   Ojima, I.; Fenoglio, I.; Park, Y. H.; Pera, P.; Bernacki, R. J., *Bioorg. Med. Chem. Lett.* **1994**, *4*, 1571-1576.

(22)   Ojima, I.; Fenoglio, I.; Park, Y. H.; Sun, C.-M.; Appendino, G.; Pera, P.; Bernacki, R. J., *J. Org. Chem.* **1994**, *59*, 515-517.

(23)   Chaudhary, A. G., Gharpure, M. M.; Rimoldi, J. M.; Chordia, M. D.; Gunatilaka, A. A. L.; Kingston, D. G. I., *J. Am. Chem. Soc.* **1994**, *116*, 4097-4098.

(24)   Skehan, P.; Streng, R.; Scudierok, D.; Monks, A.; McMahon, J.; Vistica, D.; Warren, J. T.; Bokesch, H.; Kenney, S.; Boyd, M. R., *J. Nat. Cancer Inst.* **1990**, *82*, 1107-1112.

(25)   Monsarrat, B.; Royer, I.; Wright, M.; Cresteil, T., *Bull Cancer* **1997**, *84*, 125-133.

(26)   Ojima, I.; Slater, J. C., *Chirality* **1997**, *9*, 487-494.

(27)   Ojima, I.; Slater, J. C.; Kuduk, S. D.; Takeuchi, C. S.; Gimi, R. H.; Sun, C.-M.; Park, Y. H.; Pera, P.; Veith, J. M.; Bernacki, R. J. *J. Med. Chem.* **1997**, *40*, 267-278.

(28)   Georg, G. I.; Harriman, G. C. B.; Vander Velde, D. G.; Boge, T. C.; Cheruvallath, Z. S.; Datta, A.; Hepperle, M.; Park, H.; Himes, R. H.; Jayasinghe, L., *Taxane Anticancer Agents: Basic Science and Current Status*; G. I. Georg; T. T. Chen; I. Ojima and D. M. Vyas, Ed.; *ACS Symp. Series 583*;American Chemical Society: Washington D.C., **1995**; pp 217-232.

(29)   Guéritte-Voegelein, F.; Mangatal, L.; Guénard, D.; Potier, P.; Guilhem, J.; Cesario, M.; Pascard, C., *Acta Crstallogr.* **1990**, *C46*, 781-784.

(30)   Williams, H. J.; Scott, A. I.; Dieden, R. A.; Swindell, C. S.; Chirlian, L. E.; Francl, M. M.; Heerding, J. M.; Krauss, N. E., *Can. J. Chem.* **1994**, 252-260.

(31)   Williams, H. J.; Scott, A. I.; Dieden, R. A.; Swindell, C. S.; Chirlian, L. E.; Francl, M. M.; Heerding, J. M.; Krauss, N. E., *Tetrahedron* **1993**, *49*, 6545-6560.

(32)   Vander Velde, D. G.; Georg, G. I.; Grunewald, G. L.; Gunn, C. W.; Mitscher, L. A., *J. Am. Chem. Soc.* **1993**, *115*, 11650-11651.

(33)   Mastropaolo, D.; Camerman, A.; Luo, Y.; Brayer, G. D.; Camerman, N., "X-ray Structure of Paclitaxel", *Proc. Natl. Acad. Sci. USA* **1995**, *92*, 6920-6924.

(34)   Paloma, L. G.; Guy, R. K.; Wrasidlo, W.; Nicolaou, K. C., *Chem. Biol.* **1994**, *2*, 107-112.

(35)   Chakravarty, S.; Ojima, I., *The 214th American Chemical Society National Meeting, Las Vegas, September 8-12,* **1997**, *Abstracts*, MEDI 075.

(36)   Ojima, I., *7th International Kyoto Conference on Organic Chemistry, Kyoto, Japan, November 10-14,* **1997**, *Abstracts*, IL-19.

Chapter 13

The Synthesis of a Series of Fluorinated Tribactams

Ferenc Gyenes, Andrei Kornilov[1], and John T. Welch[2]

Department of Chemistry, University at Albany, Albany, NY 12222

A fluorinated building block was converted in a concise series of steps to a new series of fluorinated tribactams. During the course of the synthesis an interesting stereoselective addition, presumably influenced by the presence of fluorine was discovered. The products were carefully characterized by 2D NMR spectroscopic methods.

The search for novel antimicrobial agents containing the β-lactam motif has produced a plethora of novel chemical entities many of which are presently marketed as antibiotic agents. The report of the discovery of thienamycin opened a new frontier in β-lactam research. Since then, other advances in the chemistry and biology of the carbapenems have followed. Recently, scientists at Glaxo Wellcome have discovered the "trinem" family(1-5) of synthetic β-lactams (See below). Functionalization at the C-4 position of a cyclohexane ring fused to the carbapenem nucleus has been very effective at modifying the activity of the trinems. The main structural features of these tricyclic β-lactam derivatives (also known as tribactams) are the presence of a reactive β-lactam ring (A) with hydroxyethyl substituent, an unsaturated five-membered ring (B), and a six-membered carbocyclic ring (C).

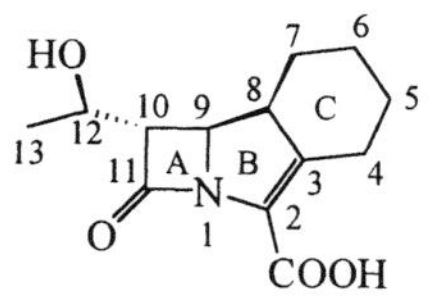

[1]Current address: Institute of Bioorganic Chemistry and Petrochemistry, National Academy of Science of the Ukraine, 1 Murmanskaya, Kiev 94, Ukraine 253660.
[2]Corresponding author.

Typically trinems have been prepared with optically active 4-acetoxy-2-azetidinone **1** as the crucial building block.

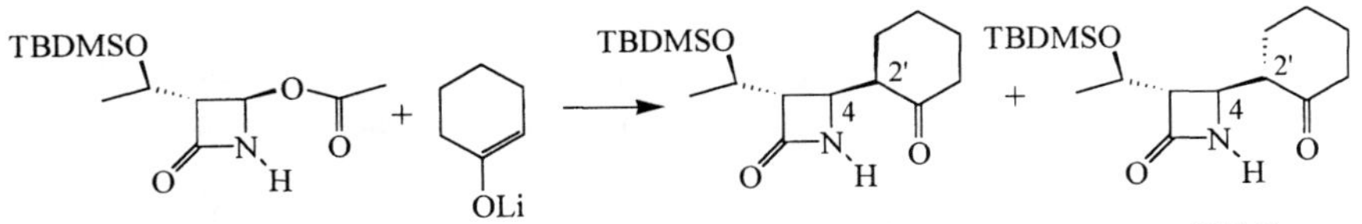

The more active β isomer (8S)-**4** was only slightly less than active than the clinical agent imipenem(6) **5** while the α isomer (8R)-**4** was less active.

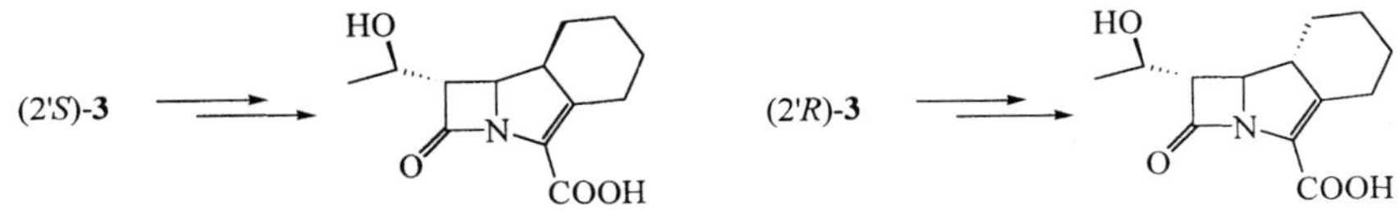

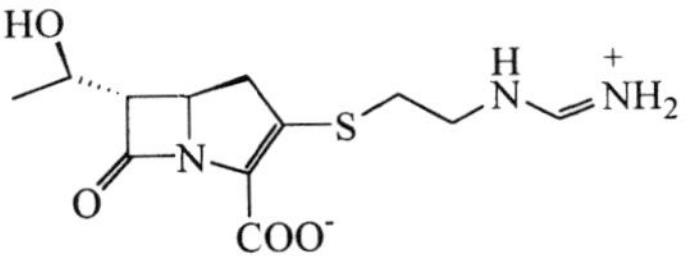

5 Imipenem

After establishing the optimum stereochemistry at C-8, modification of the C ring at position 4 led to the synthesis of 4-substituted derivatives. Of the four possible isomers the 4-α (4S), 8-β (8S) series of derivatives(7-9) has a comparable level of activity to imipenem **6**.

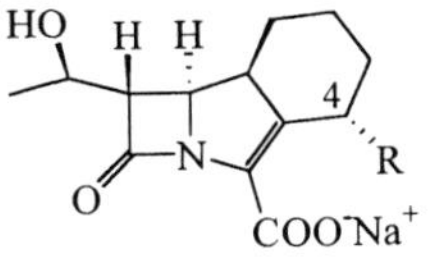

6a R = H
6b R = SCH₃
6c R = OCH₃
6d R = NHCH

Compound **6c**, also known as Sanfetrinem combines a particularly broad spectrum (including gram-negative and gram-positive aerobes and anaerobes) with high potency, resistance to β-lactamases, the bacterial enzymes that hydrolyze β-lactam antibiotics, and stability to dehydropeptidases.

Antimicrobial Activity

Sanfetrinem has in vitro activity equivalent to that of imipenem, equivalent to or better than that of a combination of Amoxicillin and clavulanic acid, and superior to that of ciprofloxacin, and erythromycin. Against gram-negative bacteria, this tribactam possessed activity similar to that of imipenem and cefpiran, but its activity was superior to that of a combination of Amoxicillin and clavulanic acid. Sanfetrinem is stable to all clinically relevant β-lactamases and is rapidly lethal to susceptible bacteria. Nonetheless, expression of β-lactamases is the likely means by which bacteria might develop resistant to Sanfetrinem. These enzymes are widely produced by gram-positive and gram-negative organisms and may be encoded either chromosomally or extrachromosomally, that is by means of plasmids or transposons.(7-9) Chromosomal β-lactamases are inducible enzymes but can be constituitively expressed (i.e., stable activated and expressed at all times) and are found in almost all gram-negative bacteria. Inducible chromosomal β-lactamases are especially problematic in the clinical setting. Since the induction of a β-lactamase is a chromosomal trait, all the organisms in a population produce the enzyme. Chromosomal β-lactamases are often resistant to inhibition by mechanism-based inactivators and unfortunately, the addition of a second antibiotic to the treatment regimen seems to have has little or no effect on the rate of emergence of resistance.

β-Lactamase Sensitivity. Although carbapenems have remained relatively insensitive to the hydrolytic action of many clinically relevant β-lactamases, a number of carbapenem-hydrolyzing enzymes have been reported in recent years. Carbapenem-hydrolyzing β-lactamases include the more recently described penicillin-interactive proteins.(10) The number of these enzymes relative to the variety of other β-lactamases remains low, only 17 carbapenem-hydrolyzing enzymes were included in the recent compilation of 190 functionally or molecularly distinct β-lactamases.(11) Most carbapenem-hydrolyzing β-lactamases confer resistance not only to carbapenems, but also to other β-lactams. Since the majority of the carbapenem-hydrolyzing β-lactamase genes are chromosomally encoded this has certainly mediated the spread of these enzymes, with a concomitant slow increase in β-lactamase-mediated resistance to carbapenems in clinical applications.

Two strategies have been employed to counter the problem of β-lactamase-mediated resistance. One approach to overcome the destructive acylation of the β-lactam is alteration of the β-lactam structure, rendering it insensitive to hydrolysis by the β-lactamase while maintaining potency as an antibiotic. Alternatively, an auxiliary agent that inactivates the β-lactamase can be employed, in synergy with a β-lactam antibiotic that is otherwise susceptible to the enzyme. The strategy proposed in this work is to combine the two effects by preparing a series of fluorine-containing trinems

7. Attempted modifications of penicillin and cephalosporin often led to molecules that were more resistant to the β-lactamase but were unfortunately less effective antibiotics. This is not surprising since at least some of the enzymes of cell-wall biosynthesis which are acylated by penems show a convincing homology(12) with residues involved in

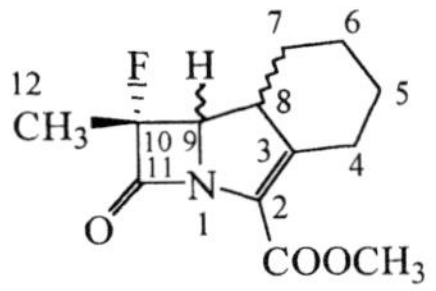

7

acyl-enzyme formation by the β-lactamase.(13) Carbapenems generally satisfy the requirements necessary for an inhibitor, ready acylation of the active site of β-lactamases, but sluggish deacylation, combined with antibacterial activity. The trinems also possess these desirable features.(1-5)

Synthesis of A Fluorinated Tribactam

Our synthesis of the β-lactam nucleus was based upon the highly efficient [2+2] cycloaddition approach.(14) Thus fluoroacetyl chloride is allowed to react with an optically active imine **8**, which was obtained by condensation of p-anisidine with (D)-glyceraldehyde acetonide. The product of this cycloaddition reaction, a single diastereomer of 3-fluoro-2-azetidinone **9**, was formed in 68% yield in 99% ee.(14) The absolute stereochemistry of this product was confirmed by single-crystal X-ray diffraction studies.(14)

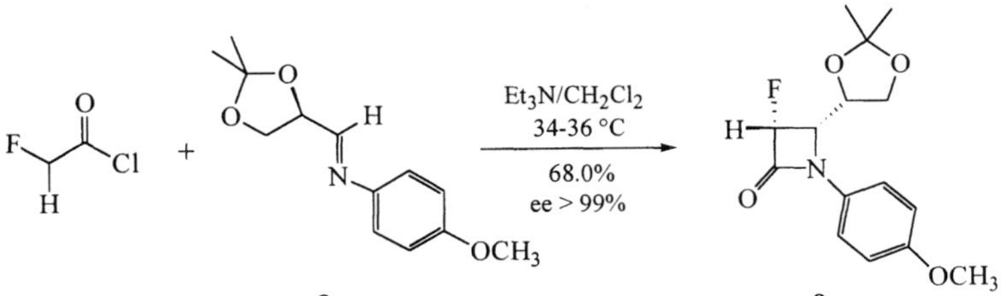

The high stereoselectivity observed, was previously discussed and may be rationalized as a consequence of the anti addition of the imine **8** to a single face of the fluoroketene.(14) The intermediate zwitterionic species **10** is postulated to collapse with 1,2-lk, ul topicity(15) i. e., antiperiplanar to the adjacent carbon-oxygen bond via a conrotatory ring closure forming **9** for stereoelectronic reasons.

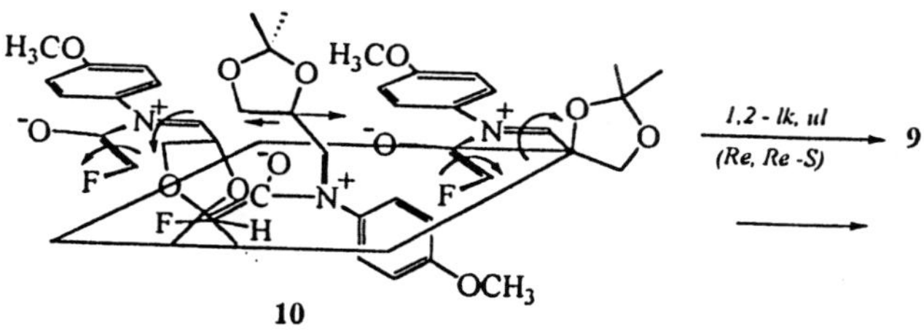

10

The chiral 3-fluoroazetidinone **9** was deprotonated with LDA at -90 °C and methylated without loss of stereochemical integrity at the fluorinated carbon as determined by ^{19}F NMR spectroscopy of the reaction mixture.(14) The very high stereoselectivity at C-3 in the ring could be explained by the steric demand of the bulky dioxolane ring at the C-4 position which directs electrophilic attack.

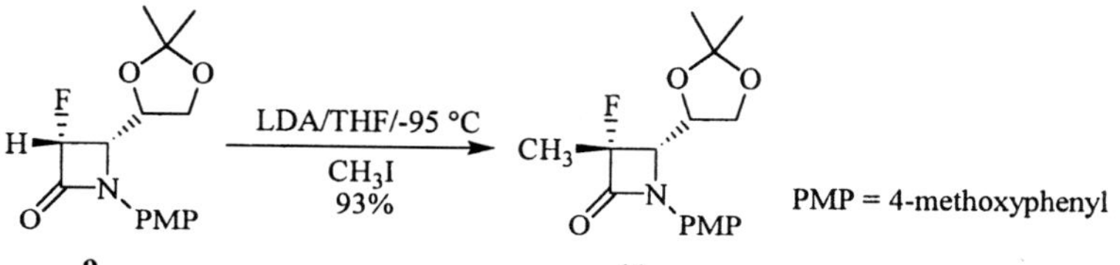

9 → LDA/THF/-95 °C, CH₃I, 93% → **11**   PMP = 4-methoxyphenyl

The transformation of **11** to the aldehyde **12** can proceed via either of two pathways. Cleavage of the acetal group with 70% acetic acid(16) followed by sodium periodate oxidative cleavage(17-18) of the formed diol **13**, furnished the aldehyde **12** in 88% yield. Alternatively, this conversion was realized in a single flask by simple treatment of **11** with periodic acid in dry ether.(19) The isolated yield, 83%, is comparable with the 88% yield of the two-step process.

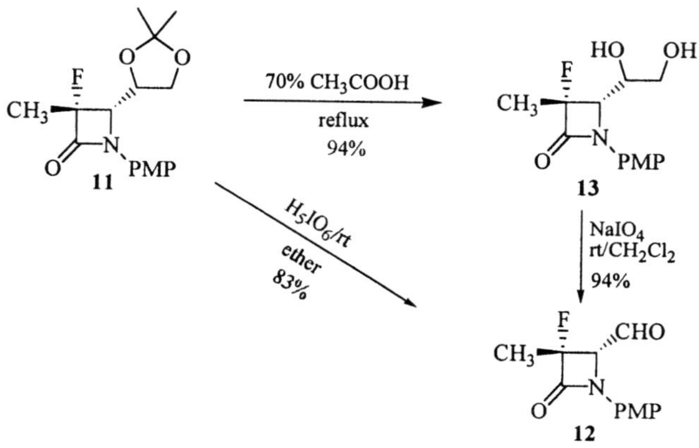

11 → 70% CH₃COOH, reflux, 94% → **13**

H₅IO₆/rt, ether, 83%

NaIO₄, rt/CH₂Cl₂, 94%

12

The aldehyde **12** was easily oxidized by potassium permanganate to the corresponding carboxylic acid(20) **13** in nearly quantitative yield. As mentioned previously, the 4-acetoxy derivatives of β-lactams are the key intermediates for the synthesis of thienamycin and related penems. In our case the conversion of the acid **13** to O-acetyl derivative **14** was performed by Hunsdiecker oxidative decarboxylation.(20)

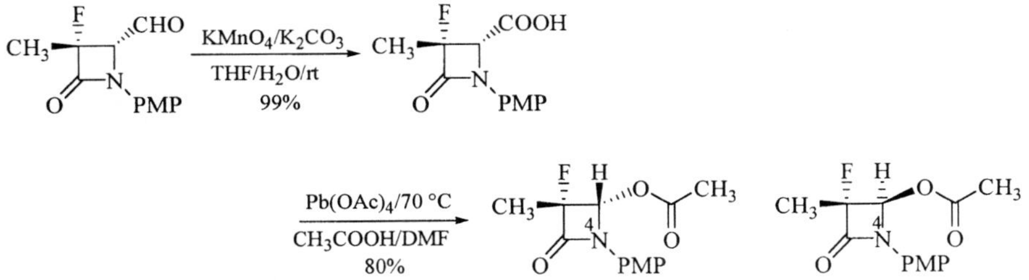

(4S)-**14**   (4R)-**14**

The reaction presumably proceeds via a radical mechanism with complete loss of the optical activity at C-4 to afford a 1:1 mixture of two diastereoisomers. Two resonances are observed in the ^{19}F NMR spectrum of this mixture: a quartet, corresponding to the (S) configuration at C-4, and a second resonance, a doublet of quartets that corresponds to the (R) configuration at C-4. Deprotection of the β-lactam nitrogen by treatment with ceric ammonium nitrate(21) (CAN) forms the N-dearylated product **15**.

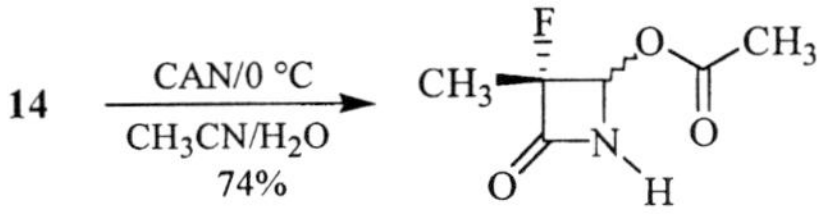

15

The introduction of the cyclohexanone ring at C-4 was affected by an addition-elimination under the two different reaction conditions. Initially, the Lewis acid catalyzed process, which requires TBDMS protection(22) of the β-lactam nitrogen, was explored. The protected product 16 was obtained smoothly in 90% yield from **15**.

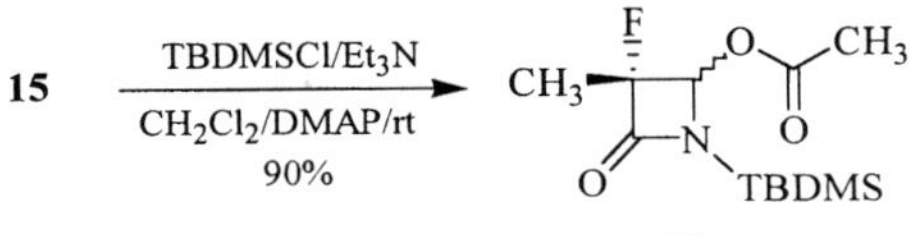

16

The Lewis acid (SnCl4)(23) catalyzed addition of the enol silyl ether(24-26) **17** to the protected β-lactam 16 was predicted to be a diastereoselective process after the literature.(23) After stirring the reaction mixture at 0 °C for 30 min, the starting material was decomposed under the published reaction conditions.

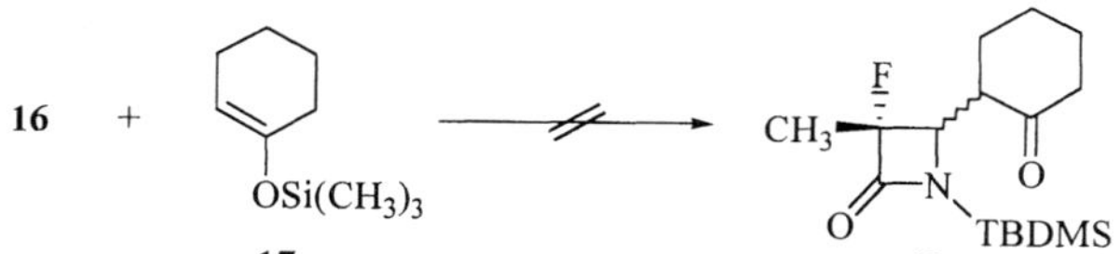

Failure of the Lewis acid-promoted condensation prompted our efforts to utilize the addition of the lithium enolate of cyclohexanone. The reaction proceeded smoothly in our hands at 0 °C in contrast to -78 °C conditions suggested by the literature.(27-28)

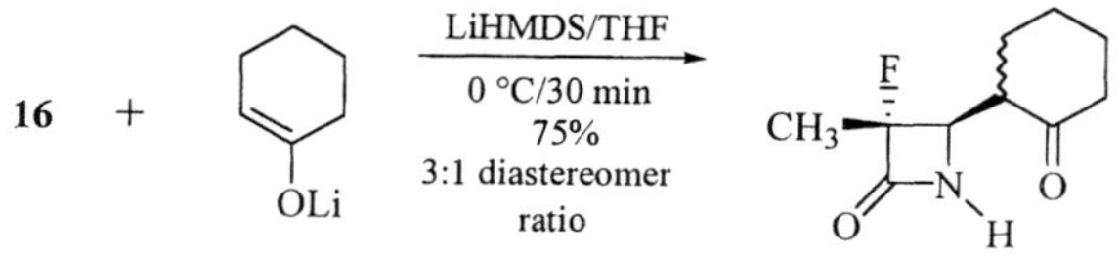

Stereochemistry of the Addition. In the published syntheses, enantiomerically pure 4-acetoxy **12**,(29-37) was employed in the addition reaction even though stereochemistry of the carbon atom at the position four is lost in the course of the reaction. The addition is thought to take place at the sp^2 hybridized carbon atom of the iminium ion. In our case, the lithium enolate of cyclohexanone must approach the re face of the double bond, trans to the fluorine.

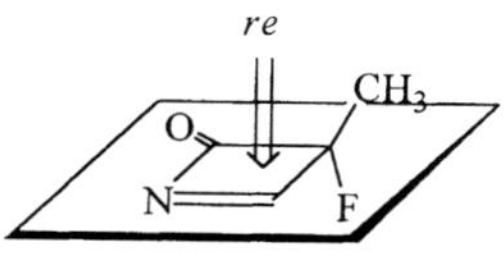

The result of the nonselective Hunsdiecker decarboxylation reaction of **13** is the loss of the optical purity the C-4 position and formation of the second diastereomer. In compound (4S)-**14**, the hydrogen on C-4 is trans to the fluorine on C-3 as it is in the starting material **9**. In compound (4R)-**14** the corresponding hydrogen is cis to the fluorine and appears as a doublet in ^{1}H-NMR spectrum with a coupling constant of 11.0 Hz. In (4S)-**14** this resonance is a singlet, due to the absence of coupling from the fluorine on the neighboring carbon.

The ^{1}H-NMR spectrum of **18** has resonance attributable to only two of the four possible diastereomers in a ratio of approximately 3:1. The resonances attributed to the pair of diastereomers are doublets of doublets with coupling constants of

$^3J_{F-H}$ = 9.2 Hz and $^3J_{H-H}$ = 4.0 Hz for δ = 3.77 ppm, and $^3J_{F-H}$ = 10.0 Hz $^3J_{H-H}$ = 4.5 Hz for δ = 3.54 ppm. The coupling from the fluorine on the adjacent carbon may indicate that the dihedral angles between C3-F and C4-H bonds are different from 90°. When the dihedral angle is 90°, the coupling constant vanishes as was observed in the case of β-lactam **9**.

Apparently, the stereoselectivity of the cyclohexanone addition is a consequence of electronic effects. By NMR analysis the cyclohexane moiety is most likely trans to the fluorine on C-3. The 3:1 ratio of the two diastereomers indicates selectivity at position 2' in contrast to the work previously reported by Glaxo on the formation of ((2'S)-**14** and (2'R)-**14**.

Construction of the B Ring. For construction of the penem nucleus the intramolecular Wittig-type cyclization developed by Woodward(38) as modified by Afonso(39) was employed. Compound **18** was N-acylated by treatment with methyl oxalyl chloride in the presence of triethylamine to oxalimide **19**.(29) This compound without purification was then heated under reflux in xylene in the presence of triethyl phosphite(40) to yield the tribactam **7**.

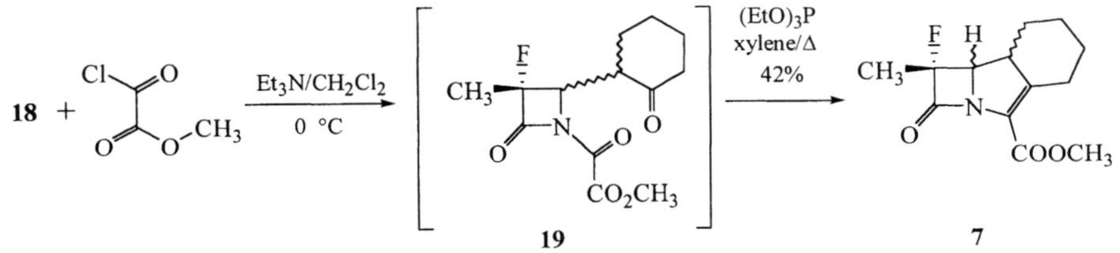

Characterization of the Product.

While cyclohexanone addition afforded only two diastereomers are observed after cyclization, fluorine and proton NMR resonances attributable to more than two diastereoisomers. Possibly, epimerization of position four occurs during the acylation of the lactam nitrogen. The presence of fluorine on the adjacent carbon atom may make this hydrogen sufficiently acidic to be abstracted by triethylamine, in contrast to the original trinem.

The final product was characterized using 500 MHz NMR spectroscopy. The individual methylene resonances of the cyclohexane ring as well as the remaining resonances were identified by two-dimensional proton-proton chemical shift correlation spectroscopy (COSY), which provided a map of coupling networks between protons in the molecule. The coupling pathway was determined by correlating the chemical shifts of coupled spins in a stepwise manner.

Three resonances can be assigned from the 1D-^{1}H-NMR. The first two of the three resonances can be attributed to methyl groups. One methyl is attached to C-10, the other is the methyl group of the ester function attached at C-2. The third resonance is derived from the proton attached to C-9. However, the information from the first two sets of hydrogens is limited since they do not have any cross peaks in the COSY spectrum. The proton on C-9 was the starting point for determining which protons are located on the six-membered ring. The only cross peak corresponding to H-9 is at δ 2.90-2.70 ppm. This chemical shift belongs to H-8. Continuing the connectivity analysis, the protons from H-7 through H-4 were identified (Table I).

Table I. NMR[a] spectroscopic data for compound 7 as a mixture of diastereomers

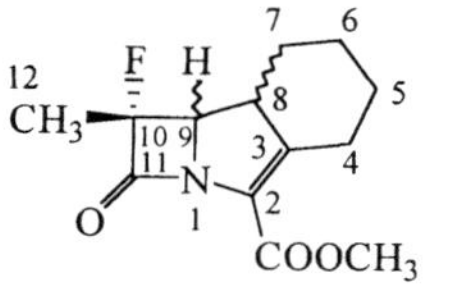

7

Position#	^{1}H δ (ppm)	^{1}H-^{1}H COSY	^{1}H-^{1}H TOCSY
4	1.95-1.80, br	H-5	H-5, H-6, H-8, H-9
5	1.80-1.50, br	H-6	H-6, H-7
6	1.80-1.50, br	H-7	H-7, H-8, H-9
7	2.50-2.30, br	H-8	H-8, H-9
8	2.90-2.70, m	H-9	H-9, H-12
9	4.30-4.05, m	——	H-12
12	1.74, d, 23.1; 1.73, d, 23.2;	——	——
	1.71, d, 22.2; 1.69, d, 23.1;		
13	3.83, 3.81, 3.80, 3.78, s	——	——

[a]NMR spectra recorded at 500 MHz..

For the confirmation of the structure, 2D TOCSY (2D totally correlated spectroscopy) was performed as well. In this spectrum, the long range couplings of H-9 can be clearly seen to H-7, H-4, H-6 and H-12 (the methyl group on C-10). The methyl on C-10 is also coupled with H-8. The couplings of H-8 to H-4 and H-6 are also detectable from the spectrum. From the two dimensional NMR spectra it is possible to assign a resonance to each hydrogen in the mixture of four diastereomers that were formed in a ratio of 1:1:2:6.

Conclusion

A fluorinated version of a "trinem" β-lactam was synthesized bearing a methyl group at position three instead of a hydroxyethyl unit. The electronic effect of the fluorine was observed during nucleophilic addition to the adjacent in situ -formed sp^2 carbon which resulted in exclusively trans addition to the carbon-fluorine bond. The electronic effect of fluorine dominates the steric influences of the methyl group at the same site. The addition of methyl oxalyl chloride in the cyclization process under basic conditions promotes epimerization of the C-4 stereocenter. The biological testing of the final compound is under investigation.

Acknowledgments. Financial Support of this work by NSF Grant Number CHE-9413004 and NIH Grant Number AI4097201 is gratefully acknowledged

Literature Cited

1.   Perboni, A.; Tamburini, B.; Rossi, T.; Tarzia, G.; Gaviraghi, G. In Recent Advances in the Chemistry of Anti-infective Agents; Bentley, P.H.; Ponsford, R., Eds.; RSC: Cambridge, 1993; pp 21-35.
2.   Wollmann, T.; Gerlach, U.; Horlein, R.; Krass, N.; Lattrell, R.; Limbert, M.; Markus, A. In Recent Advances in the Chemistry of Anti-infective Agents; Bentley, P.H.; Ponsford, R., Eds.; RSC: Cambridge, 1993; pp 50-66.
3.   Tamburini, B.; Perboni, A.; Rossi, T.; Donati, D.; Andreotti, D.; Gaviraghi, G.; Carlesso, R.; Bismara, C. Eur. Pat. Appl. EP0416953 A2, 1991; Chem. Abstr. 1992, 116, 23533t.
4.   Perboni, A.; Rossi, T.; Gaviraghi, G.; Ursini, A.; Tarzia, G. WO 9203437, 1992; Chem. Abstr. 1992, 117, 7735m.
5.   Biondi, S.; Rossi, T.; Contini, S. A. EP 617017, 1994; Chem. Abstr. 1994, 121, 280531r.
6.   Brown, A. G.; Pearson, M. J.; Southgate, R. In Agents Acting on Cell Walls; Wiley and Sons: New York, 1995; p 655.
7.   Matthew, M. J. Antimicrob. Chemother. 1979, 5, 349.
8.   Roy, C.; Foz, A.; Segura, C.; Tirado, M.; Fuster, C.; Reig, R. J. Antimicrob. Chemother. 1983, 12, 507.
9.   Skyes, R. B.; Matthew, M. J. Antimicrob. Chemother. 1976, 2, 115.
10.   Rasmussen, B. A.; Bush, K. Antimicrob. Agents Chemother. 1997, 223.
11.   Bush, K.; Jacoby, G. A.; Medeiros, A. A. Antimicrob. Agents Chemother. 1995, 39, 1211.
12.   Yocum, R. R.; Waxman, D. J.; Rasmussen, J. R.; Stromiger, J. L. Proc. Natl. Acad. Sci. U.S.A. 1979, 76, 2730.
13.   Fisher, J. F.; Belasco, J. G.; Khosla, S.; Knowles, J. R. Biochemistry 1980, 19, 2895.
14.   Welch, J. T.; Araki, K.; Kawecki, R.; Wichtowski, J. A. J. Org. Chem. 1993, 58, 2454.
15.   Seebach, D.; Prelog, V. Angew. Chem., Int. Ed. Engl. 1982, 21, 654.
16.   Ho, P-T. Tetrahedron Lett. 1978, 37, 1623.
17.   Bunton, C. A. In Oxidation in Organic Chemistry; Wiberg, K.B., Ed.; Academic: New York, 1965; Part A. pp 367-388.
18.   Perlin, A. S. In Oxidation; Augustine, R. L., Ed.; Marcel Dekker: New York, 1969; Vol. 1, pp 189-204.
19.   Wu, Y-L.; Wu, W-l. J. Org. Chem. 1993, 58, 2760.
20.   Georg, G. I.; Kant, J.; Gill, H. S. J. Am. Chem. Soc. 1987, 109, 1129.
21.   Kronenthal, D. R.; Han, C. Y.; Taylor, M. K. J. Org. Chem. 1982, 47, 2765.
22.   Hart, D. J.; Lel, C. S.; Pirkle, W. H.; Hyon, M. H.; Tsipouras, A. J. Am. Chem. Soc. 1986, 108, 6054.
23.   Rossi, T.; Marchioro, A.; Paio, A.; Thomas, R. J.; Zarantonello, P. J. Org. Chem. 1997, 62, 1653.
24.   Tirpak, R. E.; Rathke, M. W. J. Org. Chem. 1982, 47, 5099.

25. Brownbridge, P. Synthesis 1983, 1.

26. Murata, S. Tetrahedron, 1988, 44, 4259.

27. Kowalski, C.; Creary, X.; Rollin, A. J.; Burke, C. M. J. Org. Chem. 1978, 43, 2601.

28. Andreotti, D.; Rossi, T.; Gaviraghi, G.; Donati, D.; Marchioro, C.; Di Modugno, E.; Perboni, A. Bioorg. Med. Chem. Lett. 1996, 6, 491.

29. Jackson, P. M.; Roberts, S. M.; Davalli, S.; Donati, D.; Marchioro, C.; Perboni, A.; Proviera, S.; Rossi, T. J. Chem. Soc., Perkin Trans. 1. 1996, 2029.

30. Pentassuglia, G.; Tarzia, G.; Andreotti, D.; Bismara, C.; Carlesso, R.; Donati, D.; Gaviraghi, G.; Perboni, A.; Pezzoli, A.; Rossi, T.; Tamburini, B.; Ursini, A. J. Antibiotics 1995, 48, 399.

31. Greenlee, M. L.; DiNinno, F. P.; Salzmann, T. N. Heterocycles 1989, 28, 195.

32. Yoshida, A.; Hayashi, T.; Takeda, N.; Oida, S.; Ohki, E. Chem. Pharm. Bull. 1983, 31, 768.

33. Guthikonda, R. N.; Cama, L. D.; Quesada, M.; Woods, M. F.; Salzmann, T. N.; Christensen, B. G. J. Med. Chem. 1987, 30, 871.

34. Hayashi, T.; Yoshida, A.; Takeda, N.; Oida, S.; Sugawara, S.; Ohki, E. Chem. Pharm. Bull. 1981, 29, 3158.

35. Yoshida, A.; Hayashi, T.; Takeda, N.; Oida, S.; Ohki, E. Chem. Pharm. Bull. 1983, 31, 768.

36. Battistini, C.; Scarafile, C.; Foglio, M.; Franceschi, G. Tetrahedron Lett. 1984, 25, 2395.

37. Yoshida, A.; Tajima, Y.; Takeda, N.; Oida, S. Tetrahedron Lett. 1984, 25, 2793.

38. Pfaendler, H. R.; Gosteli, J.; Woodward, R. B. J. Am. Chem. Soc. 1980, 102, 2039.

39. Afonso, A.; Hon, F.; Weinstein, J.; Ganguly, A. K. J. Am. Chem. Soc. 1982, 104, 6139.

40. Padova, A.; Roberts, S. M.; Donati, D.; Perboni, A.; Rossi, T. J. Chem. Soc., Chem. Commun. 1994, 441.

BIOORGANIC SYNTHESIS OF ASYMMETRIC FLUOROORGANIC COMPOUNDS

Chapter 14

Chemical and Biochemical Approaches to the Enantiomers of Chiral Fluorinated Catecholamines and Amino Acids

K. L. Kirk[1], B. Herbert[1], S.-F. Lu[1], B. Jayachandran[1], W. L. Padgett[1],
O. Olufunke[1], J. W. Daly[1], G. Haufe[2], and K. W. Laue[2]

[1]Laboratory of Bioorganic Chemistry, National Institute of Diabetes
and Digestive and Kidney Diseases, National Institutes of Health,
Bethesda, MD 20817
[2]Organish-Chemishes Institut, Universitaet Muenster,
D–48149 Muenster, Germany

Several routes to the enantiomers of FNEs (**5**) and FDOPSs (**9**) have been explored. From these studies two asymmetric routes to FNEs **5b** and **5d** have been developed. A catalytic enantioselective oxazaborolidine reduction and a chiral (salen)Ti^{IV} catalyzed asymmetric silocyanation were used in the key stereo-defining steps of the respective routes. The syntheses that we carried out to demonstrate efficacy are complimentary in that they provided the targets with opposite configuration. Excellent progress has also been made towards the synthesis of (2*S*,3*R*)-FDOPS (**9b** and **9d**) using an Evans aldol approach. Binding studies revealed that the *R*-configuration of FNEs was essential for marked effects on the affinity towards adrenergic receptors. These studies also revealed that fluorination at the 2-position of NE reduced activity at the α-receptors and enhanced activity at β-receptors, while fluorination at the 6-position reduced activity at the β-receptors.

The naturally occurring catecholamines—dopamine (DA) (**1**), *R*-norepinephrine (*R*-NE) (**2**), and *R*-epinephrine (*R*-EPI) (**3**)—have many important biological functions. These catecholamines are produced *in vivo* from L-tyrosine. Tyrosine is first converted to dihydroxyphenylalanine (DOPA) by aromatic hydroxylation. L-DOPA is then decarboxylated to give DA, which is subsequently converted to *R*-NE by β-hydroxylation. DA is a vital neurotransmitter in the central nervous system (CNS) and has actions on the kidneys and heart. Norepinephrine is also present as a neurotransmitter in the CNS, and is the principal neurotransmitter of the peripheral sympathetic nervous system. Epinephrine, which is elaborated from *R*-NE by *N*-methylation in the adrenal medulla, has potent actions on the heart, smooth muscle, and other organs (*1*).

NE, EPI, and other adrenergic agonists function through interactions with membrane-bound adrenergic receptors. Two classes of adrenergic receptors,

194

1

2

3

designated α– and β-adrenergic receptors, have been identified in classic pharmacological studies. These receptors have been further divided into various subclasses (α_1–, α_2–, β_1–, β_2–, β_3–, for example) with further refinements in pharmacology and molecular biology producing additional classifications and sub-classifications (1,2).

The search for specific compounds that will function as an agonist or antagonist at a given type or subtype of adrenergic receptor has represented a major field of medicinal chemistry. Given that the structural unit common to DA, NE, and EPI is a catechol ring, we reasoned that alterations in the electronic environment produced by ring-fluorination of NE and EPI (**5b-d** and **6b-d**, respectively) should have a profound effect on their biological activity (3,4). We anticipated that fluorination would also have marked effects on phenolic and catecholic amino acids, including fluorinated tyrosines (**7**) (5), dihydroxyphenylalanines (DOPA) (**8b-d**) (6), and dihydroxyphenylserines (DOPS) (**9b-d**) (7). Our early studies included the synthesis of an extensive series of fluorinated compounds including DAs and racemic NEs, EPIs, DOPAs and DOPS (3).

We discovered that fluorine at the 2-position of NE (2-FNE) greatly reduced activity at α-adrenergic receptors resulting in a β-selective agonist. In contrast, fluorine at the 6-position greatly reduced activity at the β-adrenergic receptor resulting in a α-selective agonist. Fluorine at the 5-position had little effect on adrenergic activity (8,9). The binding of DA to either dopaminergic or adrenergic receptors was relatively unaffected by fluorine substitution (10). The striking effects of ring-fluorination on the adrenergic agonist properties of catecholamines correlated directly to the binding of the compounds to adrenergic receptors. This prompted us to investigate the underlying cause(s) of the selectivities. However, numerous structural changes in the agonists failed to reveal an unambiguous mechanism by which fluorine perturbs interactions of catecholamines with α- and β-adrenergic receptors (3,4). A fundamentally different approach involves the study of binding interactions of selective agonists and antagonists with chimeric α- and β-adrenergic receptors (11). Using FNEs (**5a,c**) and fluoroepinephrines (FEPIs) (**6a,c**) in this approach, we noted trends that were suggestive of receptor regions that may represent important differences in α– vs. β–adrenergic receptors. However, as the efficiency of binding to the chimeric receptors decreased, so did the selectivity of binding of 2- and 6-FNE and 2- and 6-FEPI, to the extent that interpretation of results became very difficult. Agonists that are more selective clearly were desirable.

It is important to note that although 2-, 5-, and 6-FDA (**4b-d**) did bind to adrenergic receptors the addition of a 2-fluoro or 6-fluoro substituent had little effect on the activity at the α- and β-adrenergic receptors compared to dopamine (10). This implicated the benzylic OH group as an important factor in defining receptor selectivity. Classic pharmacological studies have shown that DA and S-NE have comparable affinities at both α- and β-adrenergic receptors (12,13,14). These studies suggested that the S-isomers of FNEs, like S-NE, might have weak affinities towards both α- and β-adrenergic receptors. Since FNEs and FEPIs show marked adrenergic selectivities and FDAs show no such selectivities, it seems likely that *a reduction in affinity for either α- or β-receptors requires both the presence of the ring fluorine substituent and the benzylic OH group in the R-configuration.* In other words, R-2-FNE should be highly active at the β-adrenergic receptor and perhaps virtually inactive at the α-adrenergic receptor whereas the reverse should hold true for R-6-FNE.

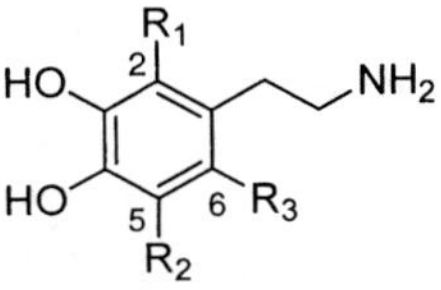

4

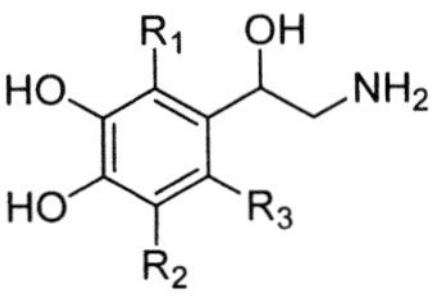

5

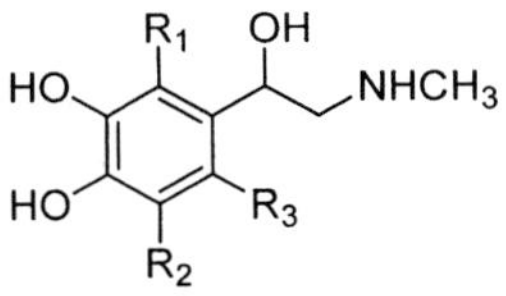

6

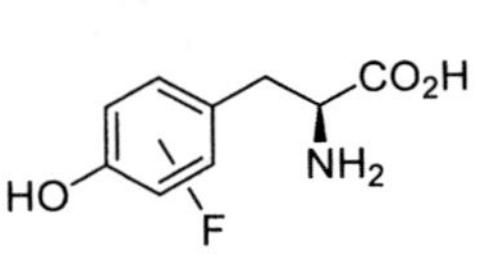

7

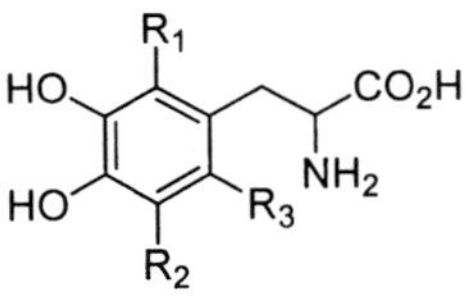

8

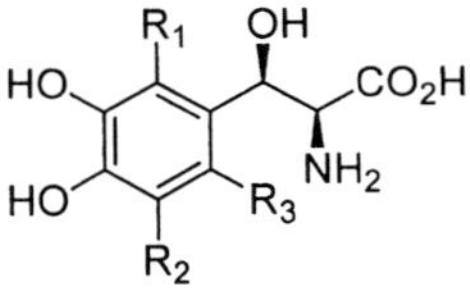

9

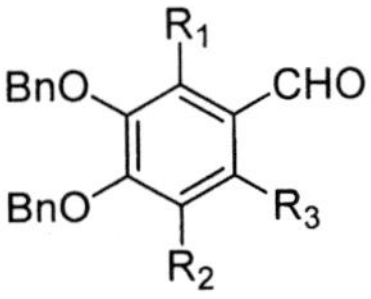

10

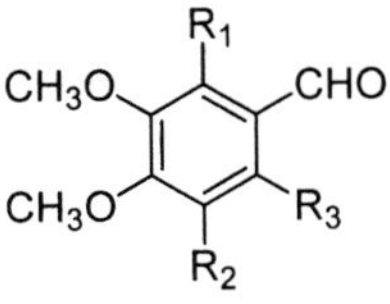

11

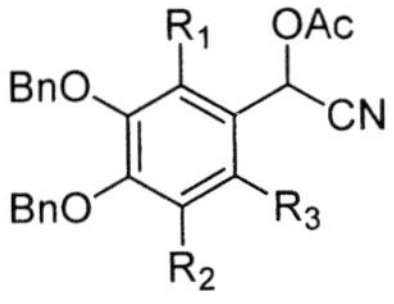

12

Series **a**: $R_1 = R_2 = R_3 = H$

b: $R_1 = F$, $R_1 = R_3 = H$

c: $R_2 = F$, $R_1 = R_2 = H$

d: $R_3 = F$, $R_1 = R_2 = H$

The elegance of fine-tuning NE and EPI to enhance their selectivity is not without limitation. To have beneficial clinical results it is imperative that the modified catecholamines reach all target areas. One particular problem is that NE cannot cross the blood-brain barrier. L-DOPA, the natural precursor to *R*-NE, can. It is known that fluorinated analogs of DOPA also possess the ability to cross the blood-brain barrier. These analogs are converted to *R*-FNEs by a pathway that is identical to that followed by L-DOPA (*15*). Several groups have reported results from clinical studies that indicate (2*S*,3*R*)-*threo*-DOPS (**9a**) can function *in vivo* as a direct biological precursor of NE as well. After crossing the blood-brain barrier, DOPS is enzymatically decarboxylated to provide *R*-NE. These findings implicate (2*S*,3*R*)-*threo*-DOPS as beneficial in treating disorders of both the sympathetic and CNS that are characterized by NE deficiencies, including certain manifestations of Parkinson's disease (*16*) and orthostatic hypotension (*17*). Given these findings it became clear that *both 2- and 6-fluoro-(2S,3R)-threo-DOPS (9b,d) have the potential of functioning in the CNS as direct biological precursors of the R-enantiomers of 2- and 6-FNE*. Generation of these selective α–and β–adrenergic agonists *in vivo* in the CNS could be beneficial, both clinically and for pharmacological studies.

Herein we detail our syntheses of the enantiomers of fluorinated norepinephrines. We also describe our synthetic approaches to 2- and 6-fluoro-(2*S*,3*R*)-*threo*-DOPS as alternative biological precursors of *R*-FNEs. Preliminary binding data for the enantiomeric FNEs are discussed.

Results and Discussion

1. Synthetic Approaches to the Enantiomers of Fluorinated Norepinephrines.

The above considerations made it clear that the development of an efficient enantioselective synthesis of *R*-FNEs and *R*-FEPIs should have high priority in our research program. At the same time, we recognized potential difficulties. The benzylic hydroxyl group of these catecholamines is very susceptible to solvolysis. For example, early attempts to recrystallize 6-FNE•HCl from methanol/ether led to extensive methyl ether formation at the benzylic position (*9*). In addition, Ding et al (*18*) reported that after separation of the enantiomers of 6-FNE by chiral HPLC, concentration of an acidic solution of the individual isomers led to extensive racemization. From this, it became clear that acid treatment as a final synthetic manipulation is not compatible with the high reactivity of the benzylic position. Another problem we encountered in our early work was the low electrophilicity of the carbonyl group of fluorinated 3,4-dibenzyloxybenzaldehydes (**10b-d**) and fluorinated veratraldehydes (**11b-d**). For example, attempts to prepare cyanohydrins by addition of HCN under usual conditions failed. In contrast, the ZnI$_2$-catalyzed addition of trimethylsilyl cyanide was facile and provided ready access to racemic FNEs by side-chain elaboration (*9*).

The lability of the benzylic OH group in fluorocatecholamino alcohols, and the low reactivity of precursor benzaldehydes, placed certain limitations on synthetic approaches. However, recent advances have provided powerful new procedures for asymmetric syntheses, and we have applied such new procedures to the preparation of the *R*- and *S*-enantiomers of FNEs.

1.1. Lipase-Catalyzed Hydrolysis of Cyanohydrin Acetates. Hydrolytic enzymes, especially lipases, are widely used for enantioselective transformations, and have been used to prepare optically active cyanohydrins. For example, the lipase-catalyzed kinetic resolution of racemic *m*-phenoxybenzaldehyde cyanohydrin acetate was an essential step in the synthesis of (1*R*,*cis*,α*S*)-cypermethrine (*19*). Another recent report described the lipase-catalyzed kinetic resolution of pentafluorobenzaldehyde cyanohydrin acetate (*20*). To examine this approach, 2- and 6-fluoro-3,4-dibenzyloxybenzaldehyde cyanohydrin acetates (**12b,d**) were prepared from the aldehydes **10b,d**. Preliminary attempts to carry out lipase-catalyzed kinetic resolutions of these cyanohydrin acetates were unsuccessful (unpublished results).

1.2. Oxynitrilase-Catalyzed Addition of HCN to Fluorinated Benzaldehydes. There have been similar advances in technology associated with the oxynitrilase-catalyzed addition of HCN to aromatic and aliphatic aldehydes and ketones (*21,22*). However, consistent with the documented low reactivity of the aldehydic carbonyl group of **10b-d** all attempts to achieve oxynitrilase-catalyzed cyanohydrin formation failed (unpublished results).

1.3. Sharpless Asymmetric Aminohydroxylation. The osmium-catalyzed asymmetric amino-hydroxylation (AA) of alkenes has emerged as a powerful and concise method for the enantioselective production of aminols. The reaction provides products with high enantioselectivities under admirable regio-control with most substrates. Prior to recent modifications, the reaction with styrene derivatives produced the benzylic amine as the predominant product (*23*). The regioselectivity was subsequently found highly dependent on both solvent and ligand, with formation of the benzyl alcohol favored by use of anthraquinone ligands and acetonitrile as the solvent. Modest yields of 2-aryl-2-hydroxyethylamines are now available using an improved procedure [(*24*), we are grateful to Professor Sharpless for providing details of this work prior to publication].

Fluorostyrene **13d** was prepared in 90% yield by Wittig olefination of aldehyde **10d**, Scheme 1. In the presence of 5 mol-% of (DHQD)₂AQN (**14**), 4 mol-% potassium osmate, *tert*-butylhypochlorite, sodium hydroxide, and benzylcarbamate in acetonitrile, alkene **13d** was transformed to a 3:2 mixture of benzylcarbamate **15d** and the desired benzyl alcohol **16d**. The regioisomers were readily separable by chromatography to provide benzyl alcohol **16d** in 45% yield. The benzyloxycarbonyl group was removed by basic hydrolysis to provide a sample of aminol **R-17d** for the determination of enantiopurity. We were disappointed with the modest enantioselectivity of 40% and we are continuing efforts to improve both regioselectivity and enantioselectivity.

1.4. Asymmetric Carbonyl Reduction. The enzyme-like CBS catalysts (e.g. **20**) developed by Corey (*25*) have been applied to the synthesis of arylethanolamines. Recent reports documenting successful enantioselective synthesis of denopamine **18** (*26*) in conjunction with the remarkably high enantioselectivities obtained with these catalysts provided the impetus for our successful approach to *R*-FNEs. The synthesis began with the production of chloroketones **19b,d** from aldehydes **10b,d** by a documented procedure (*27*) (Scheme 2). Reduction of **19b,d** with BH₃•THF in the

Scheme 1

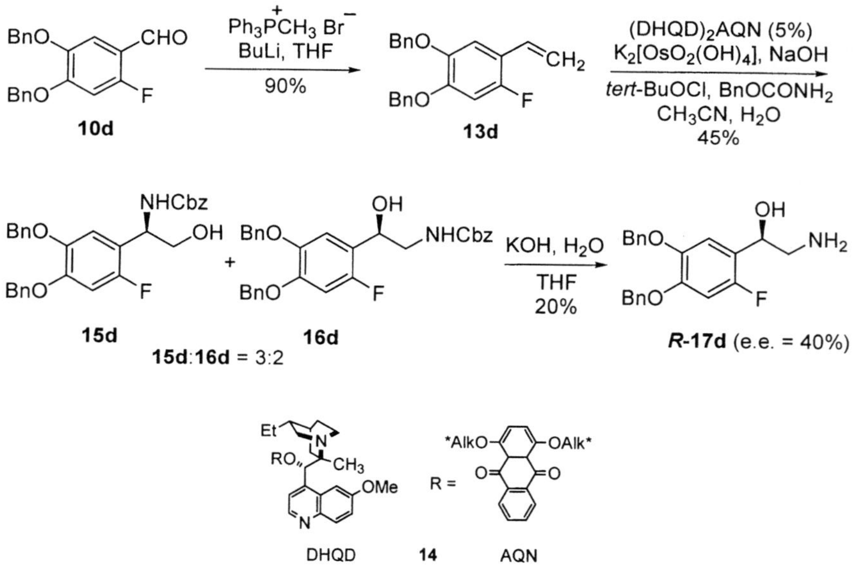

10d → (Ph₃PCH₃ Br⁻, BuLi, THF, 90%) → **13d** → ((DHQD)₂AQN (5%), K₂[OsO₂(OH)₄], NaOH, tert-BuOCl, BnOCONH₂, CH₃CN, H₂O, 45%)

15d + **16d** (15d:16d = 3:2) → (KOH, H₂O, THF, 20%) → **R-17d** (e.e. = 40%)

DHQD **14** AQN

Scheme 2

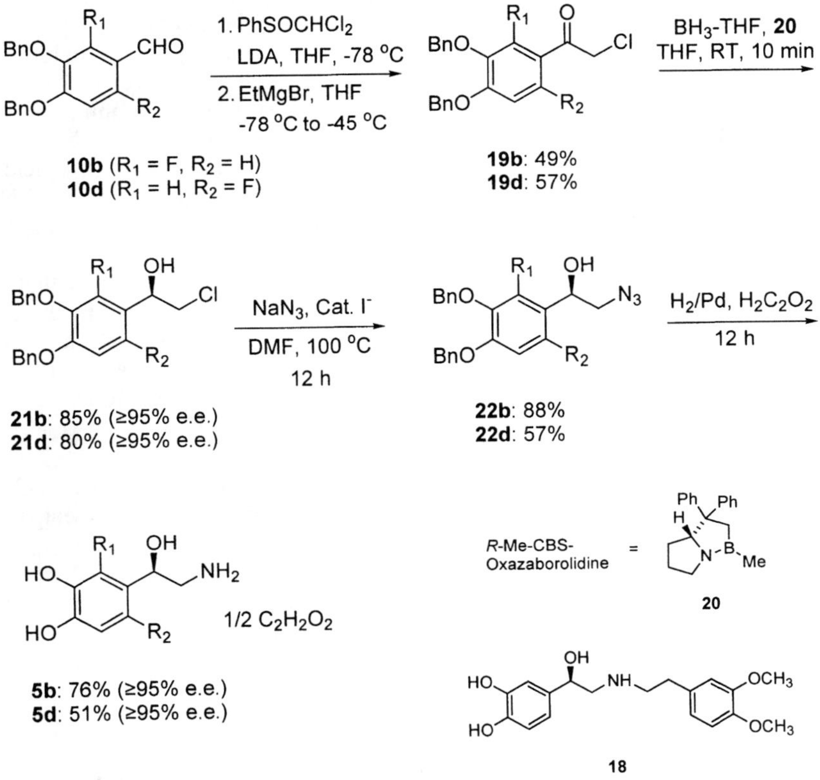

R1
BnO
CHO
BnO
R2
1. PhSOCHCl2
LDA, THF, -78 °C
2. EtMgBr, THF
-78 °C to -45 °C
10b (R1 = F, R2 = H)
10d (R1 = H, R2 = F)
O
Cl
19b: 49%
19d: 57%
BH3-THF, 20
THF, RT, 10 min
OH
21b: 85% (≥95% e.e.)
21d: 80% (≥95% e.e.)
NaN3, Cat. I-
DMF, 100 °C
12 h
N3
22b: 88%
22d: 57%
H2/Pd, H2C2O2
12 h
HO
NH2
1/2 C2H2O2
5b: 76% (≥95% e.e.)
5d: 51% (≥95% e.e.)
R-Me-CBS-
Oxazaborolidine
=
Ph
H
N
B
Me
20
NH
OCH3
18

presence of CBS catalyst **20** produced the *R*-chloroalcohols **21b** (80%) and **21d** (85%). A single recrystallization provided the chloroalcohols enriched to ≥ 95% e.e., as determined by chiral HPLC. After conversion to the azides, catalytic hydrogenation in the presence of oxalic acid gave the hemi-oxalates of *R*-2-FNE (*R*-**5b**) (76%) and *R*-6-FNE (*R*-**5d**) (51%). In both cases, the enantiomeric excesses of the final products were identical to that of chloroalcohols **21b** and **21d**.

1.5. Asymmetric Cyanohydrin Synthesis. The key step in our synthesis of racemic FNEs and FEPIs was the ZnI_2-catalyzed addition of trimethylsilyl cyanide to aldehydes **10b** and **10d**. Our desire to execute an asymmetric variant of this reaction led us to a report involving the use of chiral (salen)Ti^{IV} complexes as chiral Lewis acid catalysts (*28*). The (salen)Ti^{IV} complexes were successfully used to generate cyanohydrins in good yield and high enantioselectivity. The addition of TMSCN to dibenzyloxybenzaldehyde (**10a**) in the presence of (*R,R*)-catalyst **23** provided the crude trimethylsilyl cyanohydrin (**24a**), Scheme 3. Cyanohydrin **24a** was subsequently reduced with $LiAlH_4$ to provide a 25% over-all yield of amino alcohol **17a** ([α]$_D$ +16.4 °, CH_2Cl_2). In a similar fashion, fluorinated aldehydes **10b** and **10d** were converted to the corresponding phenethanolamines **17b** and **17d** in 25% yield (≥ 95% e.e.). Hydrogenolytic removal of the benzyl protecting groups provided FNEs **5b** and **5d** in 68% and 87% yield. Both of the final products were obtained in ≥ 95% e.e.

Through the use of the (*R,R*)-form of catalyst **23** we anticipated that the phenethanolamine generated would be enantiomeric to the series prepared by the CBS carbonyl reduction strategy. We were delighted to find that the amino alcohols **5b,d** prepared by both pathways were indeed enantiomeric, as determined by chiral HPLC and by comparison of their optical rotations. Thus, we now have two routes to both enantiomers of FNEs **5b** and **5d** in high enantiomeric excess (≥ 95%).

2. Synthetic Approaches to Enantiomers of Fluorinated *threo*-3,4-Dihydroxyphenylserine. Previous reports from these laboratories documented the successful preparation of both 2- and 6-F-*threo*-DOPS in racemic form (*7*). The compounds were prepared by the $ZnCl_2$-mediated addition of a protected silyl ketene acetal of glycine to aldehydes **10b** and **10d**. Unfortunately, neither of the fluorinated analogues were decarboxylated enzymatically (*29*). The possibility that the 2*R*,3*S*-*threo*-DOPS inhibits decarboxylation of its antipode (*30*) made it clear that an efficient asymmetric synthesis of fluorinated 2*S*,3*R*-*threo*-DOPS was critical to our goal of using these analogues as biological precursors of FNEs.

2.1. Ring-Opening of Chiral Aziridines. Recent attention has been given to the preparation of homochiral aziridines as intermediates for the synthesis of chiral amines and amino alcohols. In an elegant advance, Davis and coworkers (*31*) demonstrated that chiral sulfinimines (chirality resident on sulfur) derived from aldehydes condense with the lithium enolate of methyl bromoacetate to give, following *in situ* ring closure, chiral *Z*-2-substituted-1-carbomethoxy sulfinylaziridines. The reaction occurred with excellent diastereoselectivity with a variety of aldehydes. They further demonstrated that hydrolytic ring opening and removal of the sulfinate group occurred with minimal loss of stereochemical integrity

Scheme 3

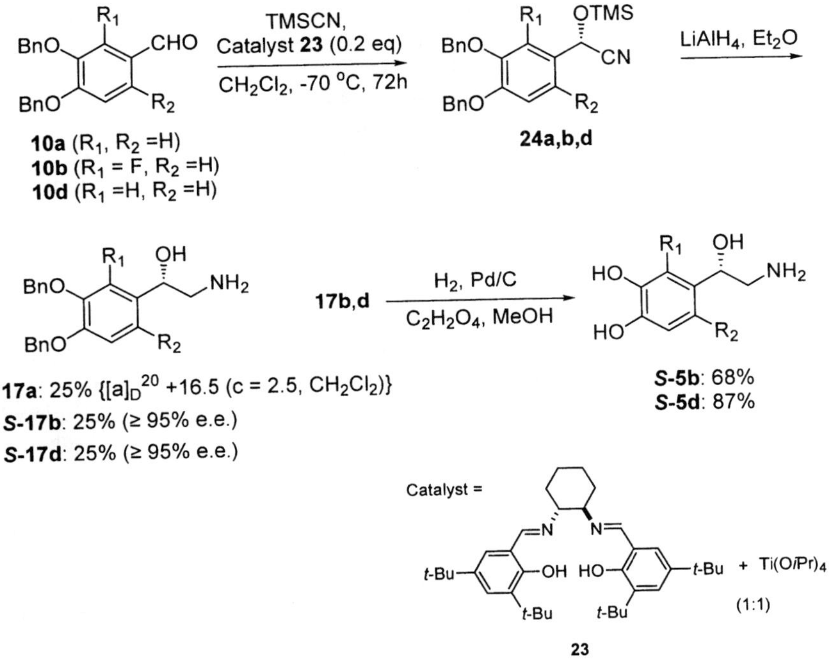

BnO, R$_1$, CHO, BnO   TMSCN, Catalyst **23** (0.2 eq) → CH$_2$Cl$_2$, -70 °C, 72h

10a (R$_1$, R$_2$ =H)
10b (R$_1$ = F, R$_2$ =H)
10d (R$_1$ =H, R$_2$ =H)

BnO, R$_1$, OTMS, CN, BnO, R$_2$ LiAlH$_4$, Et$_2$O →

24a,b,d

BnO, R$_1$, OH, NH$_2$, BnO, R$_2$

17a: 25% {[a]$_D^{20}$ +16.5 (c = 2.5, CH$_2$Cl$_2$)}
S-17b: 25% (≥ 95% e.e.)
S-17d: 25% (≥ 95% e.e.)

17b,d   H$_2$, Pd/C, C$_2$H$_2$O$_4$, MeOH →

HO, R$_1$, OH, NH$_2$, HO, R$_2$

S-5b: 68%
S-5d: 87%

Catalyst =

t-Bu ... OH HO ... t-Bu + Ti(OiPr)$_4$
t-Bu ... t-Bu   (1:1)

23

to give hydroxyamino esters (including the methyl ester of *2S,3R*-phenylserine). We initially carried out this sequence using tosyl sulfinimines derived in modest yield from 3,4-dibenzyloxy-6-fluorobenzaldehyde (**10d**). Condensation with the lithium salt of methyl bromoacetate produced the *p*-toluenesulfinyl aziridine, albeit in poor yield. Hydrolytic ring opening (water, trifluoroacetic acid, acetonitrile) produced a dibenzyl protected 6-fluoro-*threo*-DOPS ethyl ester, but poor overall conversions made final characterization difficult (*29*).

In a later development, Ellman and coworkers reported that chiral *t*-butylsulfinimides were both more reactive and gave better stereochemical control than the *p*-toluenesulfinimides (*32*). In our exploratory work, we used the chiral *t*-butylsulfinimide **25** derived from the less expensive *S-tert*-leucinol, aware that this would give us opposite absolute stereochemistry than that desired. Both sulfinimines **26a** and **26d** were prepared according to the literature procedure (*32*) in 74% and 34% yields, respectively. As previously noted, the presence of fluorine on the aromatic ring seemed to have a deleterious effect on the reactivity of the carbonyl. Sulfinimine **26a** was allowed to react with the lithium anion of methyl bromoacetate (5 eq) to provide a 9:1 ratio of *cis/trans*-aziridines, Scheme 4. Both isomers were readily separated by chromatography to provide aziridine **27a** (75%) and the *trans*-isomer in 9% yield. In contrast, the reaction with sulfinimine **26d** provided a 4:1 ratio of unstable aziridine isomers. Mindful of the acid lability of the incipient benzylic hydroxyl group, we anticipated that acid-catalyzed ring opening might prove problematic. Indeed, 1:1 to 1:2 mixtures of syn- and anti-products (**28** and **29**, respectively) were obtained during all attempts to carry out this transformation on aziridine **27a**. The same fate was met when a sulfonamide analog of **27a** was used.

2.2. Enantioselective Aldol Condensations.

The enantioselective addition of chiral Ni^{II} complexed glycine equivalents to aldehydes was reported by Belekon' (*33*) to proceed with good diastereoselectivity and enantioselectivity. Conditions were reported that gave predominantly (*2S,3R*)-*threo*-serine derivatives. However, our attempts to use this procedure were thwarted again by racemization of the benzylic OH group during the acid treatment required for disassembling the Ni^{II} complex of the final product (*7*).

Evans and coworkers have developed isothiocyanate **30** as a chiral glycine equivalent, and have used this for the synthesis of β-hydroxy-α-amino acids (*34,35*). With certain modifications, we adapted this strategy for the synthesis of the enantiomers of fluoro-*threo*-DOPS. Using two equivalents each of oxazolidinone **30**, stannous triflate, and lithium bis(trimethylsilyl)amide, aldehyde **10d** was converted to thiocarbamate **31d** (d.r. 100:1), Scheme 5. The major isomer was obtained as a single diastereoisomer by HPLC analysis in 89% yield after purification by chromatography and recrystallization. The chiral auxiliary was removed with methoxymagnesium bromide to provide methyl ester **32d** in 95% yield. The amide was converted to the *tert*-butyl carbamate **33d** (91%) and then subjected to the sulfur to oxygen exchange procedure (71%). The cyclic carbamate was cleaved (Cs$_2$CO$_3$ in MeOH. 82%) and the methyl ester saponified (LiOH in H$_2$O-THF, 85%) to provide the *N*-Boc protected (3,4-dibenzyloxy-6-fluorophenyl)serine (**35d**). The reaction of carbamate **35d** with gaseous HCl in ethyl acetate rapidly furnished amino acid **36d** in 64% yield. Further work on this route includes optimization for large-scale production of amino acid **9d** and the preparation of (*2S,3R*)-2-FDOPS (**9b**).

Scheme 4

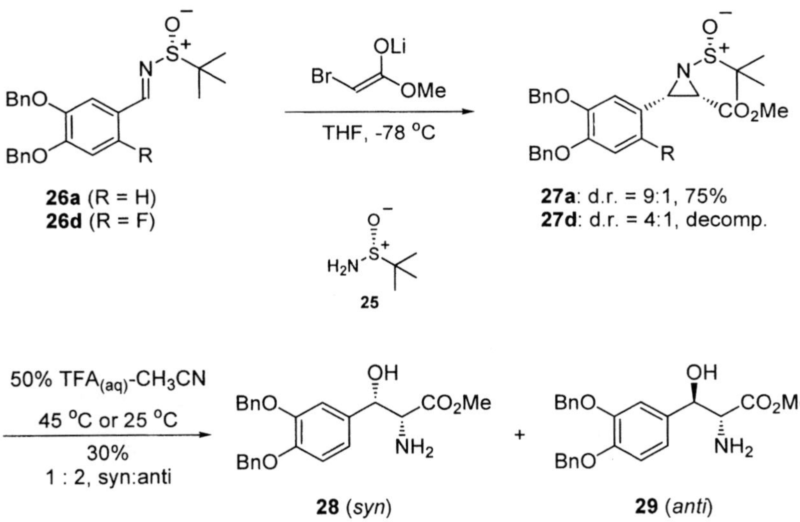

26a (R = H)
26d (R = F)

25

27a: d.r. = 9:1, 75%
27d: d.r. = 4:1, decomp.

50% TFA$_{(aq)}$-CH$_3$CN

45 °C or 25 °C

30%

1 : 2, syn:anti

28 (*syn*)

29 (*anti*)

Scheme 5

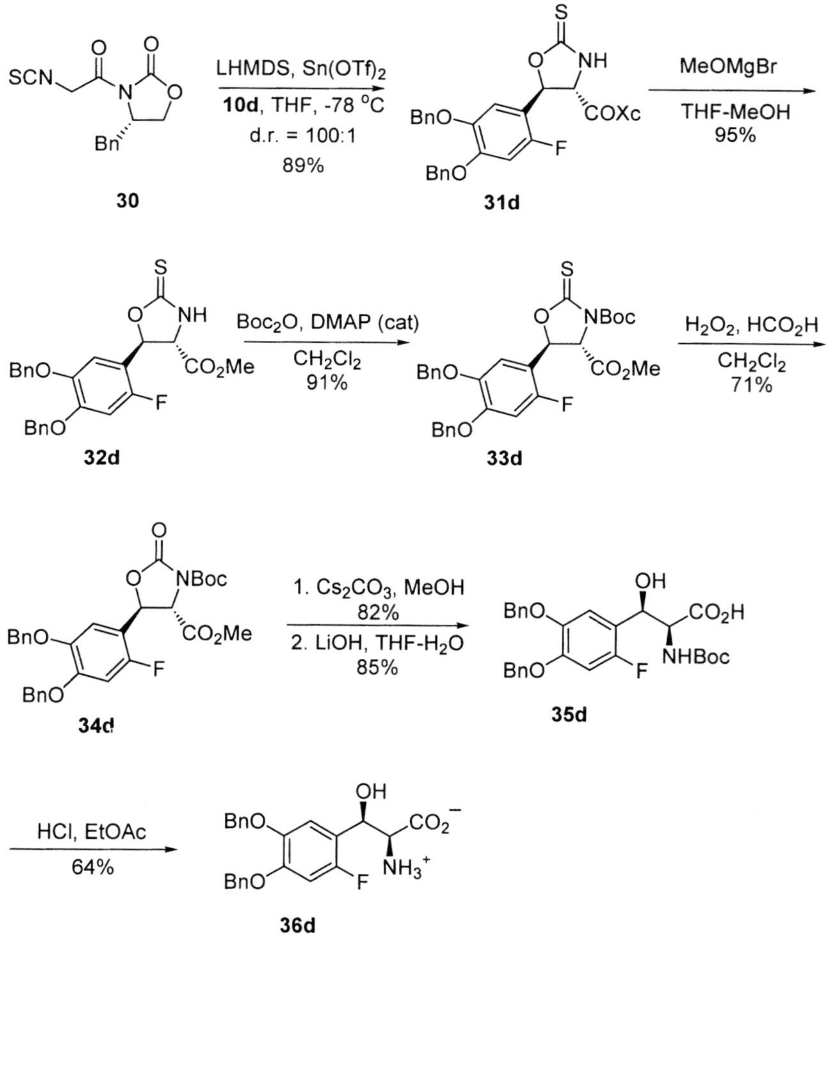
SCN
Bn
30
LHMDS, Sn(OTf)2
10d, THF, -78 °C
d.r. = 100:1
89%
S
O
NH
BnO
COXc
F
BnO
31d
MeOMgBr
THF-MeOH
95%
CO2Me
32d
Boc2O, DMAP (cat)
CH2Cl2
91%
NBoc
33d
H2O2, HCO2H
CH2Cl2
71%
34d
1. Cs2CO3, MeOH
82%
2. LiOH, THF-H2O
85%
OH
CO2H
NHBoc
35d
HCl, EtOAc
64%
CO2−
NH3+
36d

3. Binding Affinities of FNEs. Binding affinities of R-FNEs, S-FNEs, racemic FNEs, and fluorodopamines were carried out using procedures described (*36*). The natural R-NE, as expected, was 10 to 20-fold more active at the α_1-, α_2-, β_1-, and β_2-adrenergic receptors than the unnatural S-enantiomer. At β_1- and β_2-adrenergic receptors the β-selective R-2-FNE was over 100-fold more active than the S-enantiomer. At the α_1- and α_2-adrenergic receptor, R-2-FNE was more active or had similar activity, respectively, compared to the S-enantiomer. At α_1- and α_2-adrenergic receptors, the α-selective R-6-FNE was about 20-fold more active than the S-enantiomer. R-6-FNE was also several fold more active than the S-enantiomer at the β-adrenergic receptor. Thus, the presence of a ring fluorine has marked effects on the relative activity of the R- and S-enantiomers at the different adrenergic receptors.

Earlier studies with racemic 2-FNE and 6-FNE had suggested that the main effect of the 2-fluorine was to reduce activity at α-adrenergic receptors while the main effect of the 6-fluorine was to reduce activity at β-adrenergic receptors (*8,9*). The present preliminary results with the active R-enantiomers confirm that the 2-fluorine does reduce activity compared to R-NE at α-adrenergic receptors by about 20-fold. However, the 2-fluorine also enhances activity at β-receptors by at least 10-fold compared to R-NE. The 6-fluorine, as expected from earlier studies with racemates, has little effect on activity at α-adrenergic receptors but reduces activity compared to (-)-NE at β-adrenergic receptors by 10- to 20-fold. For the less active (+)-enantiomers, the results were as predicted. The 2-fluorine or 6-fluorine either had no effect or slightly increased or decreased activity. The preliminary results demonstrate that the natural R-configuration of the β-hydroxyl group of NE is essential for 2- or 6-fluorine substitution to have marked effects on the affinity of NEs at α- and β-adrenergic receptors.

Summary

Several chemical and biochemical approaches to the synthesis of non-racemic fluorinated analogs of norepinephrine and dihydroxyphenylserine have been examined. An asymmetric carbonyl reduction and an enantioselective addition of trimethylsilyl cyanide to aldehydes, followed by side-chain elaboration, provided two routes to enantiomeric fluorinated norepinephrine analogues. Of the several strategies explored, an Evans aldol strategy was most effective for the preparation of fluorinated dihydroxyphenylserine derivatives. Preliminary binding studies of the norepinephrine analogs revealed the sometimes-striking effects that fluorine substitution has on selectivities and metabolism. In all cases, NEs bearing the natural R-configuration were more active than their enantiomers. 6-FNE was found comparable in activity to NE at α-adrenergic receptors and less active at β-receptors. In contrast, 2-FNE had reduced activity at α-receptors and enhanced activity at β-receptors.

Literature Cited

1.   Cooper, J. R.; Bloom, F. E.; Roth, R. H. *The Biochemical Basis of Neuropharmacology*; Oxford University Press: New York, NY, 1996; pp 226-351.

2. Weiner, N. In *The Phrmacological Basis of Therapeutics*; Gilman, A. G.; Goodman, L. S.; Rall, T. W.; Murad, F., Eds.; MacMillan: New York, NY, 1985; pp 145-180.

3. Kirk, K. L. *J. Fluorine Chem.* **1995**, *72*, 261-266.

4. Kirk, K. L. In *Selective Fluorination in Organic and Bioorganic Chemistry*; Welch, J. T., Ed.; ACS Symposium Series 456; American Chemical Society: Washington, DC, 1991; pp 136-155.

5. Kirk, K. L. *J. Org. Chem.* **1980**, *45*, 2015-2016.

6. Creveling, C. R.,; Kirk, K. L. *Biochem. Biophys. Res. Commun.* **1986**, *136*, 1123-1131.

7. Chen, B.-H.; Nie, J.-y.; Singh, M.; Pike, V. W.; Kirk, K. L. *J. Fluorine Chem.* **1995**, *75*, 93-101.

8. Kirk, K. L.; Cantacuzene, D.; Creveling, C. R. In *Biomedicinal Aspects of Fluorine Chemistry*; Filler, R.; Kobayashi, Y., Eds.; Elsevier: New York, NY, 1982; pp 75-91.

9. Kirk, K. L.; Cantacuzene, D.; Nimitkitpaisan, Y.; McCulloh, D.; Padgett, W. L.; Daly, J. W.; Creveling, C. R. *J. Med. Chem.* **1979**, *22*, 1493-1497.

10 Nimit, Y.; Cantacuzene, D.; Kirk, K. L.; Creveling, C. R.; Daly, J. W. *Life Sciences*, **1980**, *27*, 1577-1585.

11. Kobilka, B.K.; Kobilka, T.S.; Daniel, K.; Regan, J.W.; Caron, M.G.; Lefkowitz, R.J. *Science* **1988**, *240*, 1310-1316

12. Pati, P. N.; LaPidus, J. B.; Tye, A. *J. Pharmacol. Exp. Therapeut.*, **1967**, *155*, 1-12.

13. Schonk, R. F.; Miller, D. D.; Feller, D. R. *Biochem. Pharmacol.* **1972**, *20*, 3403-3412.

14. Triggle, D. J. In *Burger's Medicinal Chemistry, Part III*; Wolf, M. E., Ed.; John Wiley & Sons: New York, NY, 1981; pp 246-251.

15. Fowler, J. S. In *Organofluorine Compounds in Medicinal Chemistry and Biomedical Applications*; Filler, R.; Kobayashi, Y.; Yagupolskii, L. M., Eds.; Elsevier Science Publishers B.V.: Amsterdam, 1993; pp 309-338.

16. Kondo, T.; *Adv. Neurology* **1993**, *60*, 660-665.

17. Suzuki, T.; Sakoda, S.; Hayashi, A.; Yamamura, Y.; Takaba, Y.; Nakajima, A. *Neurology*, **1981**, *34*, 1323-1326.

18. Ding, Y.S.; Fowler, J. S.; Gatley, S. J.; Dewey, S. L.; Wolf, A. P. *J. Med. Chem.* **1991**, *34*, 767-771.

19. Roos, J.; Stelzer, W.; Effenberger, F. *Tetrahedron Asymm.* **1998**, *9*, 1043-1049.

20. Sakai, T.; Miki, Y; Nakatani, M.; Ema, T.; Uneyama, K.; Utaka, M. *Tetrahredon Lett.* **1998**, 5233-5236.

21. Kanerva, L. T. VTT Symp. **1996**, *163*, 63-68.

22. Warmerdam, E.G.J.C.; van Rijn, R.D.; Brussee, J.; Kruse, C.G.; van der Gen, A. *Tetrahedron Asymm.* **1996**, *7*, 1723-1732.

23. Reddy, K. L.; Sharpless, B. *J. Am. Chem. Soc.* **1998**, *120*, 1207-1217, and references therein.

24. Bruncko, M.; Schlinghloff, G.; Sharpless, K. B. *Angew. Chem. Int. Ed. Eng.* **1997**, *36*, 1483-1486.

25. Corey, E. J.; Link, J. O. *Tetrahedron Lett.* **1989**, *30,* 6225-6278.
26. Corey, E. J.; Link, J. O. *J. Org. Chem.* **1991**, *56*, 442-444.
27. Satoh, T.; Mizu, Y.; Kawashima, T.; Yamakawa, K. *Tetrahedron* **1995**, *51*, 703-710.
28. Belekon', Y.; Flego, M.; Ikonnikov, N.; Moscalenko, M.; North, M.; Orizu, C.; Tararov, V.; Tasinazzo, M. *J. Chem. Soc. Perkin I* **1997**, 1293-1295.
29. Chen, B.-H.; Nie, J-y.; Singh, M.; Davenport, R.; Pike, V. W.; Kirk, K. L. In *Catecholamines, Bridging Science with Clinical Medicine*; Goldstein, D.; Eisenhofer, G.; McCarty, R., Eds.; Advances in Pharmacology; Academic Press: San Diego, CA, 1998, Vol. 42; pp 862-865.
30.. Inagaki, C.; Tanaka, C. *Biochem. Pharmacol.* **1978**, *27*, 1081-1086.
31. Davis, F. A.; Zhou, P.; Reddy, G. V. *J. Org. Chem.* **1994**, 3243-3245.
32. Liu, G.; Cogan, D. A.; Ellman, J. A. *J. Am. Chem. Soc.* **1997**, *119*, 9913-9914.
33. Belekon', Y. N.; Bulychev, A. G.; Vitt, S. V.; Struchkov, Y. T.; Batsanov, A. S.; Timofeeva, T.V.; Tsyryapkin, V. A.; Ryzhov, M. G.; Lysova, L. A.; Bakmutov, V. I.; Belikov, V. M. *J. Am. Chem. Soc.* **1985**, *107*, 4252.
34. Evans, D. A.; Weber, A. E. *J. Am. Chem. Soc.* **1986**, *108*, 6757-6761.
35. Evans, D. A.; Weber, A. E. *J. Am. Chem. Soc.* **1987**, *109*, 7151-7157.
36. Nie, J.-Y.; Shi, D.; Daly, J. W.; Kirk, K. L.; *Med. Chem. Res.* **1996** 318-331, and references therein

Chapter 15

Natural Products Containing Fluorine and Recent Progress in Elucidating the Pathway of Fluorometabolite Biosynthesis in *Streptomyces cattleya*

D. O'Hagan[1] and D. B. Harper[2]

[1]**Department of Chemistry, University of Durham, Science Laboratories, South Road, Durham DH1 3LE, United Kingdom**
[2]**School of Agriculture and Food Science, The Queen's University of Belfast, Newforge Lane, Belfast BT9 5PX, United Kingdom**

Nature appears to have made little progress in integrating fluorine into metabolic processes in living organisms. This is well illustrated by the restricted occurrence of the ability to biosynthesize the C-F bond. Little more than a handful of fluorinated natural products have been identified and these are confined to a few higher plant genera and a single bacterial genus. Although the subject of investigation and speculation for over half a century the enzymic mechanism of C-F bond formation remains a mystery. We have now traced the biosynthetic origin of fluoroacetate and 4-fluorothreonine in the bacterium *Streptomyces cattleya* and conclude that the substrate for fluorination is an intermediate in the glycolytic pathway or a closely related compound. The precise nature of the fluorination process has yet to be resolved.

Fluorine is the 13th most abundant element in the earth's crust and is the most abundant halogen. In this context it is perhaps remarkable that Nature has not developed a significant biochemistry exploiting fluorine (1). Chlorinated organic compounds are widespread as natural products and brominated compounds, whilst of more restricted occurrence, are common in the marine environment (2,3). Many fewer iodinated natural products have been isolated but this may be a reflection less of the rarity of such compounds in Nature than of the problems associated with chemical isolation of natural products with the relatively unstable C-I bond. Fluorinated organic compounds are extremely scarce in the biosphere and only five plus a series of long chain ω-fluoro-fatty acids have been identified. Several factors conspire to lead to this dearth of fluorinated natural products. The insoluble nature of many fluorine-containing minerals renders the element biologically unavailable. Thus, whereas seawater typically contains 1900 ppm of chloride, fluoride concentrations are around 1.3 ppm - four orders of magnitude less! Of course to become involved in biochemical processes the available fluoride must become reactive. Fluoride anion is however a relatively poor nucleophile and in water it is heavily hydrated which further compromises its nucleophilicity. The other halogens can be readily activated for electrophilic chemistry by haloperoxidase-catalysed oxidation with hydrogen peroxide to the corresponding halonium ions (Cl^+, Br^+ and I^+). However the high redox potential necessary for oxidation of F^- precludes incorporation of fluorine into natural products by this route. Thus the mechanism of C-F bond formation in the few organisms capable of this challenging biochemical feat is scientifically intriguing and also of potential biotechnological significance.

The Fluorinated Natural Products

The structures of all known fluorinated natural products are illustrated **1-6** apart from the series of long chain ω-fluoro fatty acids of which only the most abundant ω-fluoro-oleic acid is shown.

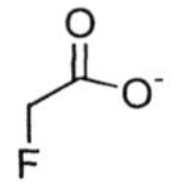

fluoroacetate **1**
principally *Dichapetalum*
Gastrolobium, Oxylobium spp.

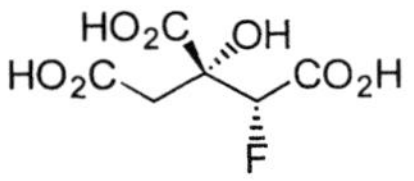

(2R, 3R)-fluorocitrate **2**
several higher plants

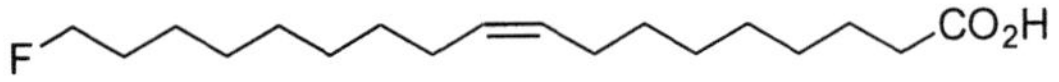

ω-fluoro-oleic acid **3**

Dichapetalum toxicarium

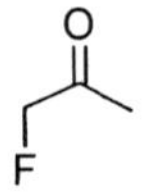

fluoroacetone **4**

Acacia georginea

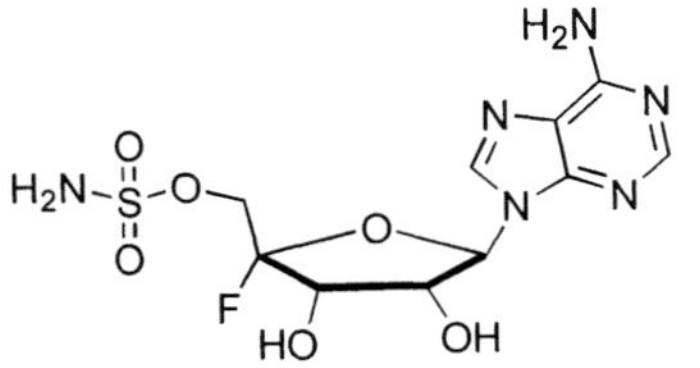

nucleocidin **5**

Streptomyces calvus

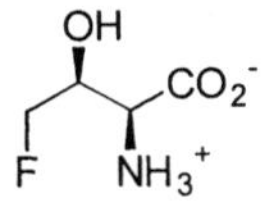

fluorothreonine **6**

Streptomyces cattleya

Fluoroacetate. Fluoroacetate **1** is the most ubiquitous fluorine-containing natural product and has been identified in over 30 plants, principally species from Africa and Australia, and also as a secondary metabolite in the bacteria *Streptomyces cattleya* (see below). The compound was first identified by Marais (4) in South Africa as the toxic principle in plants of *Dichapetalum cymosum* (gifblaar). The young leaves of this plant are particularly toxic in the spring when levels of fluoroacetate of up to

2500 ppm on a dry weight basis can kill livestock grazing on the veldt. Levels of fluoroacetate of up to 8000 ppm have been recorded in *Dichapetalum braunii* seeds from Tanzania (5) and the seeds and flower stalks of *Palicourea margravii* from Brazil have been shown to accumulate 5000 ppm of the toxin (6). The fluoroacetate-producing plants in Australia comprise about 20 leguminous species belonging to the genera, *Gastrolobium* and *Oxylobium* and also the small tree *Acacia georginae* (gidyea) (1,6).

Fluorocitrate. Fluoroacetate is toxic because it is converted *in vivo* to the (*2R, 3R*) stereoisomer of 2-fluorocitrate which is a potent inhibitor of citrate transport across the mitochondrial membrane (7) and is also an inhibitor of the citric acid cycle enzyme, aconitase (8,9). Accumulation of 2-fluorocitrate induces convulsions and ultimately heart failure and death. It is interesting that certain plants such as alfalfa (*Medicago sativa ssp sativa*) (10) soya bean (*Glycine max*) (11) and tea (*Thea sinensis*) (12) can accumulate low levels of 2-fluorocitrate when grown in the presence of fluoride suggesting that a wide range of plants possess the ability to biosynthesise fluoroacetate and convert it to 2-fluorocitrate in small quantities.

ω-Fluorinated Lipids. The shrubby plant *Dichapetalum toxicarium* from Sierra Leone is particularly interesting in that it accumulated ω-fluoro-oleic acid ($C_{18:1F}$) **3** and a number of other long chain ω-fluoro-fatty acids in its seeds (13,14). The fluorinated lipid component constitutes ~3% of the seed oil with ω-fluoro-oleic acid comprising 80% of the organic fluorine present. A comprehensive examination (15) of the other fatty acids in the fluorinated lipid fraction resulted in the identification of six other ω-fluoro compounds including both saturated and unsaturated acids namely $C_{16:0F}$, $C_{16:1F}$, $C_{18:0F}$, $C_{18:2F}$, $C_{20:0F}$ and $C_{20:1F}$. The pattern of ω-fluoro acids present in terms of both chain length and unsaturation mirrored that of their non-fluorinated analogues. Recently the sites of the double bonds in the unsaturated fluorofatty acids were established by GC/MS of the picolinyl esters (16). This study indicated that two isomers of the $C_{20:1F}$ acid were present with unsaturation at the 9 and 11 positions respectively. *Threo*-9,10-dihydroxystearic acid has also been isolated from *Dichapetalum toxicarium* seed oil (17) and appears to be derived metabolically from ω-fluoro-oleic *via* a 9,10-epoxide as an intermediate (15). The biosynthesis of ω-fluorofatty acids by *D. toxicarium* can be attributed to the activation of fluoroacetate as its CoA derivative which is then utilized as a starter unit in fatty acid biosynthesis. There is no evidence the fluorine is incorporated mid-chain into any of these lipids.

The presence of fluorine allows the distribution of fluorinated lipids on the seeds of *D. toxicarium* to be visualized *in situ* by ^{19}F magnetic resonance imaging (MRI). Organic fluorine is clearly shown to be located in the waxy seed kernel (Figure 1). As far as we are aware this is the only example of the location of a secondary metabolite in plant tissue by this technique.

Fluoroacetone. There is a single report from 1967 of the identification of fluoroacetone **4** in plant tissue homogenate (18). The volatiles from *Acacia georginae* homogenates incubated with inorganic fluoride were purged through a solution of 2,4-dinitrophenylhydrazine and the 2,4-dinitrophenylhydrazone derivative of fluoroacetone was identified. The metabolic origin of this compound has not been the subject of further investigation although a plausible route would be from 4-

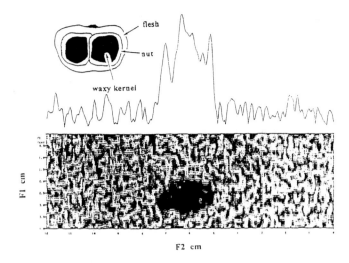

Figure 1 ^{19}F-MRI Image of a seed nut of *Dichapetalum toxicarium*. Only the waxy kernel contains ω-fluoro-oleate and other ω-fluorinated fatty acids.

fluoroacetoacetyl-ACP formed *via* condensation of fluoroacetyl CoA with malonyl-ACP. Hydrolysis and decarboxylation of 4-fluoroacetoacetyl-ACP would yield fluoroacetone.

Nucleocidin. Until comparatively recently, other than higher plants, the only organism known to elaborate a fluorinated compound was the bacterium *Streptomyces calvus*. The antibiotic nucleocidin **5** was first isolated from this organism in 1957 (19, 20) but it was not until twelve years later that the presence of fluorine in this compound was realised (21). The structure of the compound was confirmed by total synthesis in 1976 (22). Nucleocidin **5** is an intriguing compound as it is difficult to envisage a mechanism by which the fluorine in the molecule can be derived from fluoroacetate **1** and it therefore seems likely that the fluorination enzyme involved in biosynthesis of nucleocidin is unique to *S. calvus*. A recent attempt to re-isolate the compound from *S. calvus* failed and sadly it appears that the strain may have lost the capability of producing this fascinating secondary metabolite (23).

4-Fluorothreonine. In 1986 a second *Streptomyces* species capable of the biosynthesis of fluorinated compounds was identified (24). In the course of studies with *Streptomyces cattleya* to improve the yield of the ß-lactam antibiotic, thienamycin, it was discovered that when fluoride was present in the culture medium both fluoroacetate **1** and 4-fluorothreonine **6** accumulated in millimolar concentrations during fermentation. 4-fluorothreonine is the only naturally ocurring fluorinated amino acid known. When 4-fluorothreonine was initially isolated (24) the absolute stereochemistry of 4-fluorothreonine was predicted to be analogous to L-threonine. We have recently confirmed this by asymmetric synthesis (25) and demonstrated that natural 4-fluorothreonine has the (*2S, 3S*) configuration. The synthetic route used to prepare this metabolite is shown in the Figure 2 and exploited a modification of Seebach's imidazolidinone auxiliary methodology for amino acid synthesis (26). For the synthesis of L-threonine this involves a condensation between the auxiliary (enolate) and an aldehyde and a problem at the outset was the difficulty in obtaining fluoroacetaldehyde by a preparative method. In this event fluoroacetyl chloride was used in place of fluoroacetaldehyde to generate ketone **7**. X-ray crystal analysis of this ketone showed that it retained Td geometry at C-5, the stereogenic centre situated between the carbonyl groups. Reduction of **7** with NaBH$_4$ generated **8** in a highly stereoselective manner to give a single stereoisomer. Hydrolysis of **8** then released 4-fluorothreonine **6**, a compound which was crystallised and an X-ray structure solved to establish the relative and absolute stereochemistry as (2S, 3S). This compound was identical in all respects to the natural product.

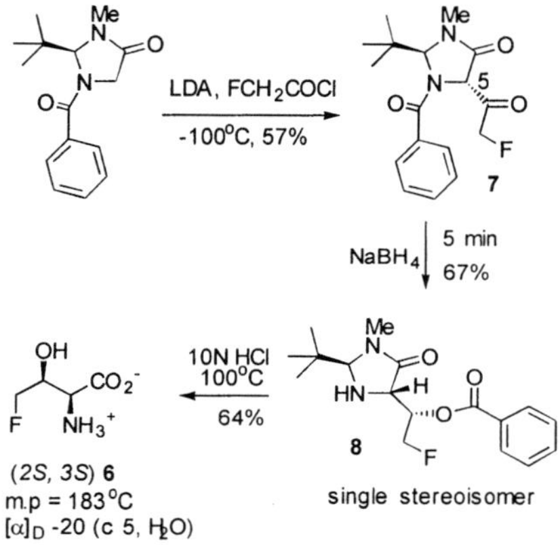

Figure 2   Asymmetric   synthesis   of   the   natural   product   (2S,   3S)-4-fluorothreonine **6** (25).

S. cattleya offers an excellent biological system in which to explore the biosynthesis of organo-fluorine metabolites and we have employed it over the last few years in a research programme directed towards elucidating the secrets of the biological fluorination process (27,28).

Biosynthesis of Fluoroacetate and 4-Fluorothreonine by *S. cattleya*.

Fluorometabolite production in relation to growth during batch culture (shake flasks) of *S. cattleya* on a defined glycerol/glutamate medium containing 2 mM fluoride is illustrated in Figure 3 (27). No uptake of flouride is observed prior to the growth maximum at 4 days but shortly thereafter, as the cells enter the stationary phase, fluoride concentration in the culture medium begins to fall and fluormetabolite production is initiated. When the concentration of fluorometabolites stabilises after about 28 days their combined concentration is about 1.8 mM with normally a 60:40 ratio of fluoroacetate to 4-fluorothreonine. Our investigations of the biogenesis of these metabolites have employed resting cell suspensions prepared from 6 to 8 day old cultures of *S. cattleya* as such cells had the highest rate of fluorometabolite biosynthesis. Washed cells suspended in sodium 2(N-morpholino)-ethane sulphonate (MES) buffer pH 6.5 at 29°C were supplied with isotopically labelled precursors and the incorporation of label into the fluorometabolites was assessed after 24 and 48 hours by $^{19}F\{^1H\}$-NMR and GC/MS using the selected ion monitoring mode. Such cell suspensions usually yielded approximately equal concentrations (~0.5 mM) of each fluorometabolites in the supernatant after 48 hours incubation.

A clear advantage in the investigation of the biosynthesis of fluorinated natural products is that isotopic incorporations are readily identified by $^{19}F\{^1H\}$-NMR. Carbon-13 (I=½) couples to fluorine-19 (I=½) with large J^1 (~165Hz) and J^2 (~18Hz) coupling constants and ^{13}C enrichments can be directly determined by examination of the coupling patterns in the resultant $^{19}F\{^1H\}$-NMR spectrum. Incorporation of deuterium into fluorometabolites is also readily ascertained. Deuterium bonded geminal to a fluorine in an FCDH group induces a large shift (~0.6ppm) in the fluorine signal; two deuteriums in an FCD_2 group lead to a shift of twice the magnitude (~1.2ppm). Deuterium bonded vicinal to the fluorine atom in a FC-CD relationship induces a smaller but detectable shift (0.2ppm). These shifts are clearly displayed in Figure 4 which is a $^{19}F\{^1H\}$-NMR spectrum of the supernatant of a resting cell suspension incubated in 30% D_2O where the deuteriums have exchanged into the fluorometabolites in a random manner (30). Both populations of labelled fluoroacetates ($FCDHCO_2H$ and FCD_2CO_2H) are apparent as signals shifted downfield from the parent fluoroacetate peak at -218ppm. Similarly the extent of deuterium incorporation into 4-fluorothreonine can be deduced by analysis of the signals of the various different isotopically labelled populations which are shifted downfield from the parent 4-fluorothreonine peak at -233ppm. The additive effects of multiple labelling are clearly evident and the incorporation of deuterium into C-3, the position vicinal to the fluorine atom, is well displayed. Thus the pattern and magnitude of the isotope incorporation into the fluorometabolites from labelled precursors can be readily determined. Some of the results obtained using this technique are described below.

Incorporation of Glycine, Serine and Pyruvate. Glycine was selected for initial studies as the amino acid could provide an attractive C_2 precursor for fluoroacetate. Incubation of $[1,2-^{13}C_2]$-glycine with resting cell suspensions of *S. cattleya* resulted in a high level of incorporation (34%) of both ^{13}C atoms as deduced from the doublet of doublets flanking the parent signals of fluoroacetate and 4-fluorothreonine in the resultant $^{19}F\{^1H\}$-NMR shown in Figure 5 (28,29). However when $[2-^{13}C]$-glycine was administered in a similar experiment an almost identical result (Figure 6) was obtained with 32% double labelling in fluoroacetate and also C-3 and C-4 of 4-fluorothreonine. This finding indicates that glycine is not incorporated as an intact unit but that C-2 of glycine is contributing both carbon atoms of fluoroacetate and also C-3 and C-4 of 4-fluorothreonine. The validity of this deduction was confirmed by the observation of negligible incorporation from $[1-^{13}C]$ glycine. This unexpected result can be rationalized if glycine is cleaved and C-2 is processed *via* N^5, N^{10}-methylenetetrahydrofolate and condensed with another molecule of glycine by the enzyme serine hydroxymethyl transferase to form L-serine. Indeed when $[3-^{13}C]$-serine was incubated with resting cell suspensions the isotope was incorporated in the expected regiospecific manner into the fluoromethyl group of each fluorometabolite

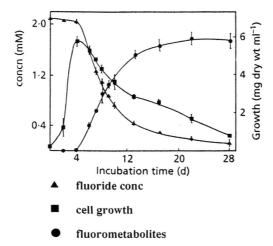

Figure 3 Fluoride consumption, cell growth and fluorometabolite production versus time relationships during production of **1** and **6** from batch cultures of *S. cattleya*.

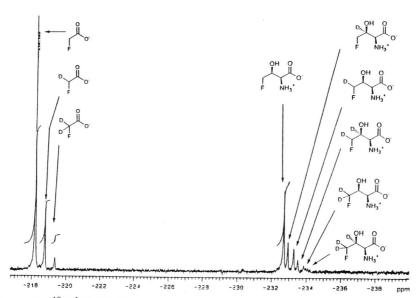

Figure 4 ^{19}F{^{1}H}-NMR spectra of fluoroacetate (-218 ppm) and 4-fluorothreonine (-233 ppm) after growth of a *S. cattleya* culture in a medium containing 30% D_2O.

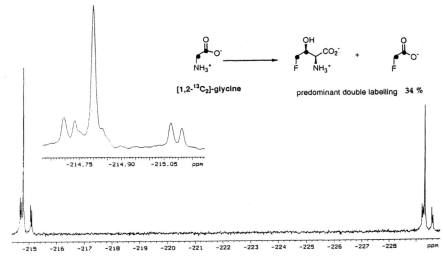

Figure 5 $^{19}F\{^1H\}$-NMR spectra of fluoroacetate (-214.8 ppm) and 4-fluorothreonine (-229 ppm) after 48h incubation of $[1,2-^{13}C_2]$-glycine with a *S. cattleya* cell suspension.

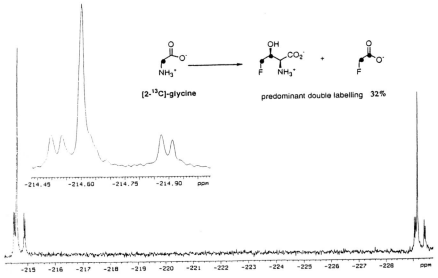

Figure 6 $^{19}F\{^1H\}$-NMR spectra of fluoroacetate (-214.6 ppm) and 4-fluorothreonine (-229 ppm) after 48h incubation of $[2-^{13}C]$-glycine with a *S. cattleya* cell suspension.

as indicated by the doublets in the ^{19}F{^{1}H}-NMR in Figure 7 (28). Since L-serine can be converted to pyruvate by the action of serine dehydratase [3-^{13}C]-pyruvate was tested as a precursor and was shown to be efficiently incorporated (Figure 8) into the fluorometabolites labelling the fluoromethyl group of each fluorometabolite in a similar manner to [3-^{13}C]-serine (28,29). It is significant that with each precursor the labelling pattern observed in C-1 and C-2 of fluoroacetate always mirrored that into C-3 and C-4 of 4-fluorothreonine indicating that these units have a common biosynthetic origin.

Incorporation of Succinate. Various metabolic fates for pyruvate are possible. The compound could undergo decarboxylation to yield acetyl-CoA but this metabolic option would seem an unlikely pathway to the fluorometabolites as the methyl group is no longer activated for a potential fluorination reaction. Alternatively pyruvate could be converted to oxaloacetate prior to activation to phosphoenolpyruvate. Entry to the glycolytic cycle by this route would result in the potential fluoromethyl carbon being activated as a phosphomethyl carbon, a feature which is intuitively attractive. In order to establish the status of oxaloacetate as an intermediate we studied the incorporation of [^{2}H$_4$]-succinate, a metabolite preceding oxaloacetate in the citric acid cycle. Efficient introduction of a single deuterium atom into the fluoromethyl groups of both fluoroacetate and 4-fluorothreonine was observed (28). This single incorporation is readily deduced from the ^{19}F{^{1}H}-NMR spectrum in Figure 9 where there is an isotope induced shift (0.6ppm) associated with each of the fluorometabolite signals. There is no evidence for populations of fluorometabolite molecules containing two deuterium atoms. The incorporation of just one deuterium atom is entirely consistent with the processing of succinate around the citric acid cycle *via* fumarate and malate to oxaloacetate prior to conversion to phosphoenolpyruvate as detailed in Figure 10. The foregoing experiments thus establish a link between the fluorometabolites and phosphoenolpyruvate, a metabolic intermediate situated at one end of the glycolytic pathway.

Incorporation of Glycerol. In 1995 Tamura and co-workers demonstrated that [2-^{13}C] glycerol was incorporated into the carboxyl group of fluoroacetate by resting cell suspensions of *S. cattleya* (31). This observation lends support to the idea that the glycolytic pathway may provide the carbon substrate for the fluorination enzyme as glycerol is routed into glycolysis *via* the glycolytic intermediate dihydroxyacetone phosphate. Glycerol is prochiral and its entry into the glycolytic pathway is controlled by the action of glycerol kinase, an enzyme which phosphorylates the pro-(R) hydroxymethyl group of glycerol to generate *sn*-glycerol-3-phosphate.

To explore the role of glycerol in providing a substrate for fluorination we prepared by synthesis the (R) and (S)-[1-^{2}H$_2$]-glycerols which are chiral by virtue of deuterium labelling. The compounds were incubated in separate experiments with cell suspensions of *S. cattleya* but only the (R)-isomer gave deuterium incorporation in the fluoromethyl groups of fluoroacetate and 4-fluorothreonine (32) as indicated in Figure 11. This is a most revealing finding as it demonstrates unequivocally that the hydroxymethyl group of glycerol that becomes phosphorylated by the action of the glycerol kinase is also the hydroxymethyl group that becomes fluorinated. Furthermore both deuterium atoms were retained in the fluorometabolites indicating that there was no change in the oxidation state of the hydroxymethyl group between

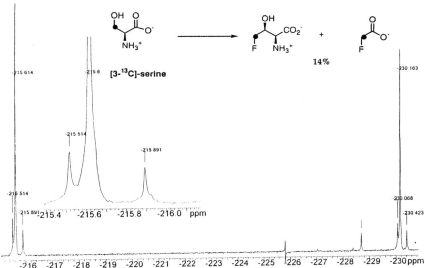

Figure 7 $^{19}F\{^1H\}$-NMR spectra of fluoroacetate (-215 ppm) and 4-fluorothreonine (-230 ppm) after 48h incubation of [3-^{13}C]-serine with a *S. cattleya* cell suspension.

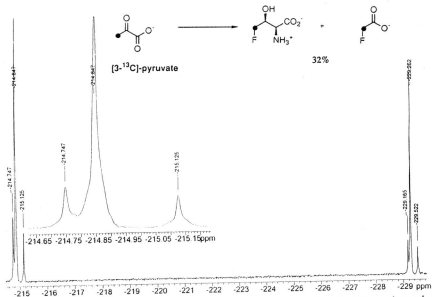

Figure 8 $^{19}F\{^1H\}$-NMR spectra of fluoroacetate (-214.8 ppm) and 4-fluorothreonine (-229ppm) after 48h incubation of [3-^{13}C]-pyruvate with a *S. cattleya* cell suspension.

220

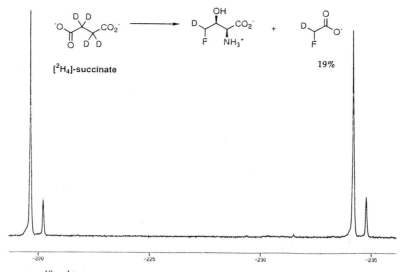

Figure 9 $^{19}F\{^1H\}$-NMR spectra of fluoroacetate (-220 ppm) and 4-fluorothreonine (-234 ppm) after 48h incubation of $[^2H_4]$-succinate with a *S. cattleya* cell suspension.

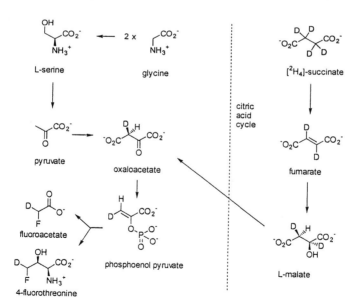

Figure 10 Rationale for the incorporation of a single deuterium atom from $[^2H_4]$-succinate into the fluorometabolites in *S. cattleya*.

phosphorylation and fluorination. This observation would imply that phosphorylation is part of the activation process, perhaps for a fluoride displacement reaction, although the mechanism of the process remains elusive.

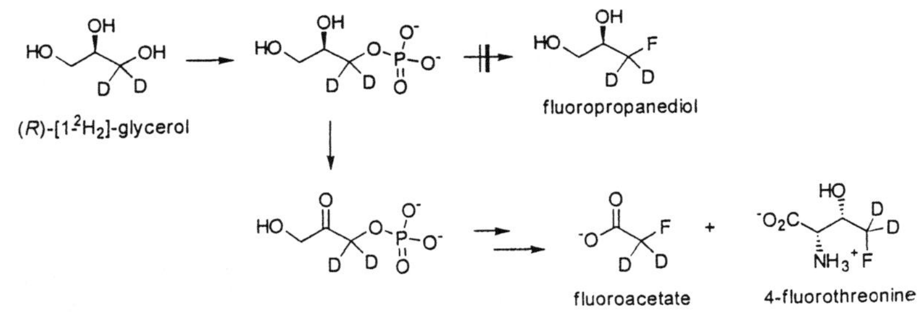

Figure 11 Labelling rationale from (R)-[1-^{2}H$_2$]-glycerol. Fluoropropanediol is not an intermediate in fluorometabolite biosynthesis.

In a further attempt to identify the point at which fluorination occurs we have synthesised [1-^{2}H$_2$]-fluoropropane-2,3-diol which would be the expected product if the phosphate group of sn-glycerol-3-phosphate is displaced by fluoride in a substitution reaction. This immediate post fluorination product may then be further metabolized to fluoroacetate and 4-fluorothreonine. However, on incubation of cell suspensions will [1-^{2}H$_2$]-fluoropropane-2,3-diol, deuterium was not incorporated into the fluoromethyl groups of the fluorometabolites and we have now eliminated this compound as a possible intermediate (Figure 11). [3-^{2}H$_2$]-3-fluorolactate also does not label the fluorometabolites implying that the various phosphorylated glycerates involved in the glycolytic pathway are not substrates for fluorination (Figure 13). Other possibilities currently under investigation in our laboratories are (a) that either dihydroxyacetone phosphate or glyceraldehyde-3-phosphate is the substrate for fluorination and (b) that a C$_3$ glycolytic intermediate is oxidised to a C$_2$ phosphate e.g. glycoaldehyde phosphate prior to fluorination.

A Role for Fluoroacetaldehyde

We have established by experiments using isotopically labelled fluoroacetate and 4-fluorothreonine that neither fluorometabolite is derived by metabolism of the other. Also our experiments indicate that the C$_2$ unit of fluoroacetate and C-3 and C-4 of 4-fluorothreonine have a common biosynthetic origin. It therefore seems likely that a C$_2$ fluorinated compound must act as a common precursor of both fluorometabolites. Some recent results suggested that fluoroacetaldehyde may fulfil this role. We have shown (32) that a cell-free extract of S. cattleya can efficiently oxidise fluoroacetaldehyde to fluoroacetate as illustrated in Figure 12. The enzyme responsible is a NAD dependant aldehyde dehydrogenase. It is not clear at present whether the metabolic role of this enzyme is to mediate the final step in fluoroacetate biosynthesis or whether it has a more general function in cellular metabolism and a sufficiently relaxed substrate specificity to oxidise fluoroacetaldehyde.

The extremely rapid oxidation of fluoroacetaldehyde by cell suspensions renders investigation of its possible utilization in 4-fluorothreonine biosynthesis difficult. However if cell suspension of *S. cattleya* are incubated with 4 mM fluoroacetaldehyde in the absence of fluoride ion and 4-fluorothreonine formation is monitored during the period (approximately 3 hours) before total oxidation of the aldehyde has occurred, the amino acid is found to accumulate in the supernatant in quantities comparable to those normally formed in the presence of fluoride ion. This suggests that fluoroacetaldehyde is contributing to 4-fluorothreonine biosynthesis and is a common intermediate to both metabolites.

Conclusion.

We have traced the metabolic pathways channelling metabolites towards the fluorometabolites in *S. cattleya* and an overview of these relationships is shown in Figure 13. We believe that the substrate for the fluorination process is either a C_3 phosphorylated glycolytic intermediate such as dihydroxyacetone phosphate or glyceraldehyde-3-phosphate, or alternatively a C_2 metabolite such glycoaldehyde phosphate, derived from these compounds. Fluoroacetaldehyde appears to be a strong candidate for the role of common intermediate to both fluoroacetate and 4-fluorothreonine.

Acknowledgments. It is a great pleasure to thank our enthusiastic coworkers Muhammad Amin, Jack Hamilton, Steven Moss, Cormack Murphy, Jens Nieschalk and Karen Reid for their commitment to the project and to thank Peter Binks, Roy Bowden and Harry Eccles, from BNFL Central Research Labs. Preston, UK, for financial support, for many stimulating discussions and for taking a close interest in biological fluorination. We also thank Dr Ian Rowland of the Royal Marsden Hospital, London for obtaining the ^{19}F-MRI image of the *D. Toxicarium* seed.

Literature Cited.

1. Harper, D. B.; O'Hagan, D. *Nat. Prod. Rep.* **1994**, *11*, 4.
2. Gribble, G. W. *Acc. Chem. Res.* **1998**, *31*, 141.
3. Neidleman, S. L.; Geigert, J. *Biohalogenation: Principles, Basic Roles and Applications,* Ellis Horwood Ltd., Chichester, **1986**.
4. Marais, J. S. C. *Onderstepoort J. Vet. Sci. Anim. Ind.* **1944**, *20*, 67.
5. O'Hagan, D.; Perry, R.; Lock, M. J.; Meyer, J. J. M.; Dasaradhi, L.; Hamilton, J. T. G.; Harper, D. B. *Phytochemistry,* **1993**, *33*, 1043.
6. Hall, R. J. *New Phytologist,* **1972**, *71*, 855.
7. Kun, E.; Kirsten, E.; Sharma, M. L. *Proc. Nat. Acad. Sci.* **1977**, *74*, 4942.
8. Peters, R. A.; Wakelin, R. W.; Buffa, P.; Thomas, L. C. *Proc. Roy. Soc. B.,* **1953**, *140*, 497.
9. Lauble, H.; Kennedy, M. C.; Emptage, M. H.; Beinert, H.; Stout, C. D. *Proc. Nat. Acad. Sci.* **1996**, *93*, 13699.
10. Lovelace, C. J.; Miller, G. W.; Welkie, G. W. *Atmos. Environ.* **1968**, *2*, 187.
11. Yu, M. H.; Miller, G. W. *Environ. Sci. Technol.* **1970**, *4*, 492.
12. Peters, R. A.; Shorthouse, M. *Phytochemistry,* **1972**, *11*, 1337.
13. Peters, R. A.; Hall, R. J.; Ward, P. F. V.; Sheppard, N. *Biochem. J.,* **1960**, *77*, 17.

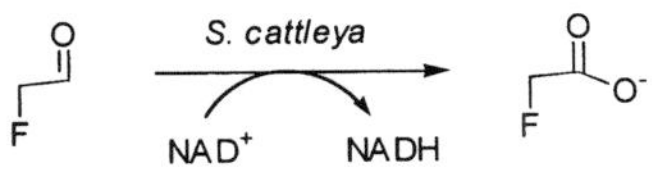

Figure 12   Cell free extracts of *S. cattleya* posess an NAD dependant aldehyde dehydrogenase activity which oxidises fluoroacetaldehyde to fluoroacetate

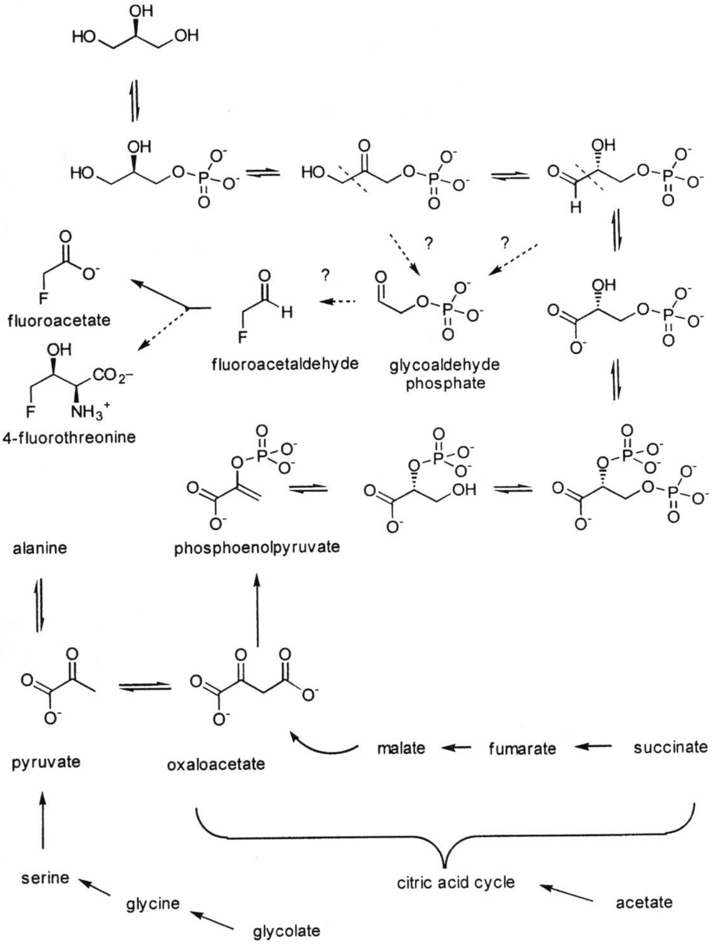

Figure 13   Overview of the metabolic relationships involved in fluoroacetate and 4-fluorothreonine biosynthesis in *S. cattleya*.

14. Ward, P. F. V.; Hall, R. J. ; Peters, R. *A. Nature*, **1964**, *201*, 611.

15. Hamilton, J. T. G.; Harper, D. B. *Phytochemistry*, **1997**, *44*, 1179.

16. Christie, W. W.; Hamilton, J. T. G.; Harper, D. B. *Chem. Phys. Lipids* (in press).

17. Harper, D. B.; Hamilton, J. T. G.; O'Hagan, D. *Tetrahedron Lett.*, **1990**, *31*, 7661.

18. Peters, R. A.; Shorthouse, *M. Nature*, **1971**, *231*, 123.

19. Thomas, S. O.; Singleton, V. L.; Lowery, J. A.; Sharpe, R. W.; Pruess, L. M.; Porter, J. N.; Mowatt, J. H.; Bohonas, N. *Antibiotics Ann.* **1957**, *1956-7*, 716.

20. Waller, C. W.; Patrick, J. B.; Fulmor, W.; Meyer, W. E. *J. Am. Chem. Soc.*, **1957**, *79*, 1011.

21. Morton, G. O.; Lancaster, J. E.; Van Lear, G. E.; Fulmor, W.; Meyer, W. E. *J. Am. Chem. Soc.*, **1969**, *91*, 1535.

22. Jenkins, I. D.; Verheyden, J. P. H.; Moffat, J. G. *J. Am. Chem. Soc.*, **1976**, *98*, 3346.

23. Maguire, A. R.; Meng, W.-d.; Roberts, S. M.; Willets, A. J. *J. Chem. Soc. Perkin Trans. 1*, **1993**, 1795.

24. Sanada, M. Miyano, T.; Iwadare, S.; Williamson, J. M.; Arison, B. H.; Smith, J. L.; Douglas, A. W.; Liesch, J. M.; Inamine, E. *J. Antibiot.* **1986**, *39*, 259.

25. Amin, M. R.; Harper, D. B.; Moloney, J. M.; Murphy, C. D.; Howard, J. A. K.; O'Hagan, D. *Chem. Commun.* **1997**, 1471.

26. D. Seebach, E. Juaristi, D. D. Miller, C. Schickli, T. Weber. *Helv. Chim. Acta.* **1987**, *70*, 237

27. Reid, K. A.; Hamilton, J. T. G.; Bowden, R. D.; O'Hagan, D.; Dasaradhi, L.; Amin, M. R.; Harper, D. B. *Microbiology* **1995**, *141*, 1385.

28. Hamilton, J. T. G.' Murphy, C. D.; Amin, M. R.; O'Hagan, D.; Harper, D. B. *J. Chem. Soc. Perkin Trans. 1*, **1998**, 759.

29. Hamilton, J. T. G.; Amin, M. R.; Harper, D. B.; O'Hagan, D. *Chem. Commun.* **1997**, 797.

30. Amin, M. R.; Harper, D. B.; O'Hagan, D. *J. Labelled Cpds, Radiopharms.* (in press).

31. Tamura, T.; Wada, M.; Esaki, N.; Soda, K. *J. Bacteriol.* **1995**, *177*, 2265.

32. Nieschalk, J.; Hamilton, J. T. G.; Murphy, C. D.; Harper, D. B.; O'Hagan, D. *Chem. Commun.* **1997**, 799.

Asymmetric Fluoroorganic Chemistry in Materials Chemistry

Chapter 16

Synthetic Aspects of Fluorine-Containing Chiral Liquid Crystals

Tamejiro Hiyama[1], Tetsuo Kusumoto[2], and Hiroshi Matsutani[2]

[1]Department of Material Chemistry, Kyoto University, Yoshida, Sakyo-ku, Kyoto 606–8501, Japan
[2]Sagami Chemical Research Center, 4–4–1 Nishiohnuma, Sagamihara, Kanagawa 229–0012, Japan

Synthesis of fluorine-containing chiral molecules useful as a part or a whole of liquid crystalline materials is described with the emphasis on (1) asymmetric hemiacetal synthesis of trifluoroacetaldehyde using a BINOL-titanium catalyst, (2) first stereospecific carbon-carbon bond formation through a nucleophilic substitution at the chiral acetal carbon bearing a trifluoromethyl group using organoaluminium reagents, (3) synthesis of diastereomeric dichiral liquid crystals containing two fluorine atoms at two chiral centers, and (4) unusual transition behaviour of the dichiral liquid crystals.

Since the display using liquid crystals (LCs) is characterized by the features of light weight, space- and energy-saving, low-voltage-drive, and lack of flickering, the LC display is now commonly used for lap-top computers and portable TV and will be used in the near future for most of the office automation (OA) display and the home TV. Although the display using a thin-layer-film transistor (TFT) has been put on the market at present, the ones of large size are less accessible and extremely expensive. In view that high definition TV is growing to be common, the LC display of larger size with high quality and high capacity is yet to be developed. The promising materials are ferroelectric or antiferroelectric liquid crystals (FLCs or AFLCs) which consist of chiral molecules and show chiral smectic C (SmC*) phase or chiral smectic C_A (SmC$_A$*) phase.

Typical LC molecules that exhibit SmC* and/or SmC$_A$* phase are shown in Figure 1. As readily seen, the LC compounds consist of optically active 1,1,1-trifluoro-2-alkanols which contribute to the stabilization of the SmC$_A$ phase and thus play key roles of these LC materials. Accordingly methods for the 1,1,1-trifluoro-2-alkanols have been the target of organic synthesis. Conventional retrosynthetic analysis of the trifluoro alcohols leads to (1) resolution of 1,1,1-trifluoro-2-alkanols by chemical or enzymatic methods, (2) asymmetric reduction of 1,1,1-trifluoro-2-alkanones, or (3) asymmetric carbonyl addition of carbonaceous nucleophiles. Indeed, at present, the enzymatic resolution through the hydrolysis of the carboxylates of 1,1,1-trifluoro-2-alkanols prevails. Other methods remain yet to be studied.

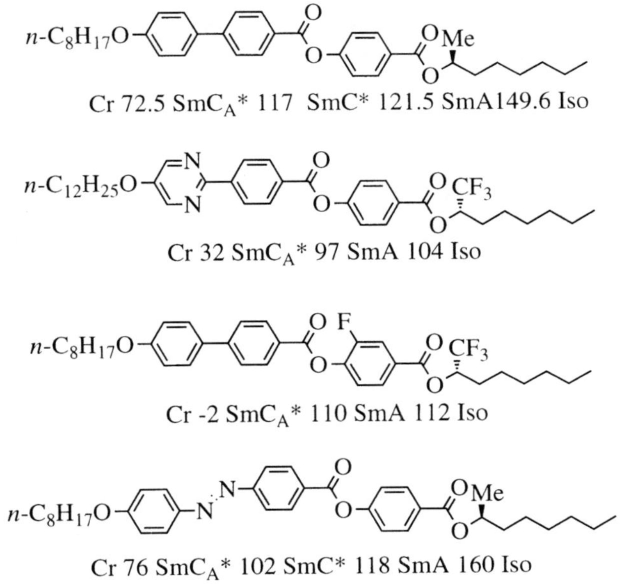

$n\text{-}C_8H_{17}O$—〈〉—〈〉—

Cr 72.5 SmC$_A$* 117 SmC* 121.5 SmA149.6 Iso

$n\text{-}C_{12}H_{25}O$—

Cr 32 SmC$_A$* 97 SmA 104 Iso

$n\text{-}C_8H_{17}O$—

Cr -2 SmC$_A$* 110 SmA 112 Iso

$n\text{-}C_8H_{17}O$—

Cr 76 SmC$_A$* 102 SmC* 118 SmA 160 Iso

Figure 1. Typical AFLCs

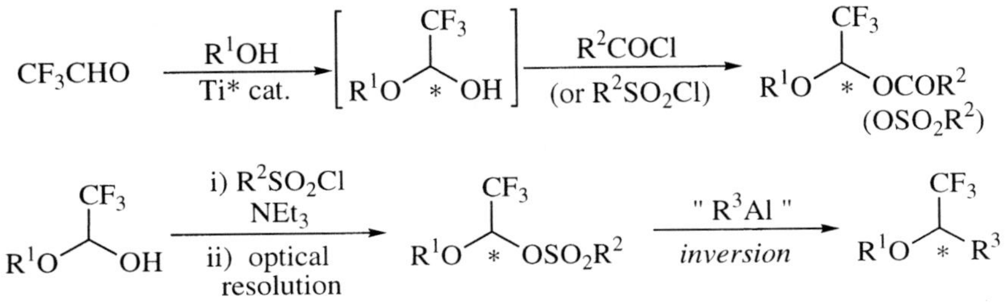

$CF_3CHO \xrightarrow[\text{Ti* cat.}]{R^1OH} \left[R^1O \overset{CF_3}{\underset{*}{\diagup}} OH \right] \xrightarrow[\text{(or } R^2SO_2Cl)]{R^2COCl} R^1O \overset{CF_3}{\underset{*}{\diagup}} OCOR^2 \; (OSO_2R^2)$

$R^1O \overset{CF_3}{\underset{OH}{\diagup}} \xrightarrow[\substack{\text{ii) optical} \\ \text{resolution}}]{\substack{\text{i) } R^2SO_2Cl \\ NEt_3}} R^1O \overset{CF_3}{\underset{*}{\diagup}} OSO_2R^2 \xrightarrow[inversion]{\text{" } R^3Al \text{ "}} R^1O \overset{CF_3}{\underset{*}{\diagup}} R^3$

Scheme 1. Research Plan

Asymmetric Synthesis of 1-Alkoxy-2,2,2-trifluoroethanol Derivatives

Trifluoroacetaldehyde, generated from 1-ethoxy-2,2,2-trifluoroethanol (4 mmol) by treatment with polyphosphoric acid (PPA), was passed into a toluene solution of (R)-BINOL-Ti(O-i-Pr)₂ (1 mmol) at -78 °C in the presence of 2-propanol (2 mmol) co-produced during the preparation of the catalyst from Ti(O-i-Pr)₄ and (R)-binaphthol (BINOL). The resulting hemiacetal was treated with benzoyl chloride, triethylamine and 4-dimethylaminopyridine (DMAP) at -78 °C to afford 2,2,2-trifluoro-1-isopropoxyethyl benzoate (80% ee) in 85% yield. When the esterification was carried out at higher temperatures, % ee of the product considerably decreased due probably to racemization of the intermediate hemiacetal.

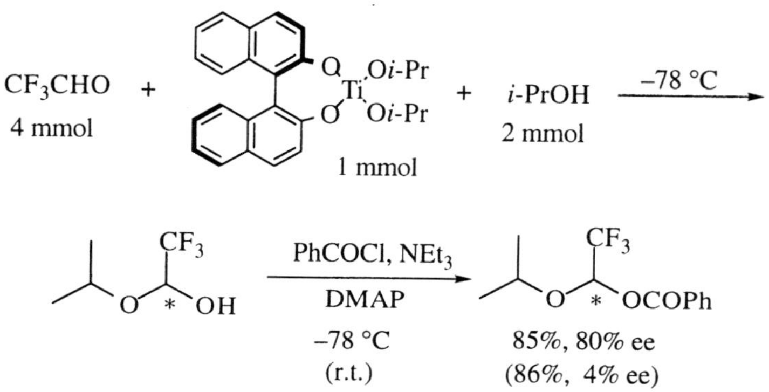

Scheme 2. Asymmetric Hemiacetal Synthesis

Optimization of the rate and order of addition of substrates suggested that slow and *simultaneous* addition of liquid 2-propanol *and* gaseous trifluoroacetaldehyde to a toluene solution of (R)-BINOL-Ti(O-i-Pr)₂ gave better results. To extend the reaction to various alcohols, it was necessary to prepare the chiral catalyst by treatment of dilithium (R)-binolate with Ti(O-i-Pr)₂Cl₂ in toluene at room temperature for 2 h. The catalyst solution was then cooled to -78 °C, and a toluene solution of 2-propanol (2 mmol) was added slowly by a syringe pump over 20 min while simultaneously passing trifluoroacetaldehyde (4 mmol) into the solution. This procedure improved the ee of the benzoate to 86% or 82% using 50 or 10 mol% of (R)-BINOL-Ti(O-i-Pr)₂, respectively as summarized in Table 1.

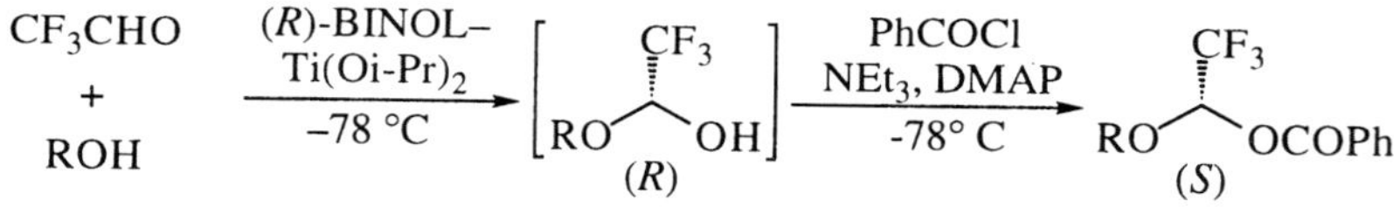

Scheme 3. Optimization of Asymmetric Hemiacetal Synthesis

The hemiacetal formation using benzyl alcohol was performed by decreasing the amount of trifluoroacetaldehyde under the same conditions, and the benzoate was attained with 74% ee. The ee was further improved to 91% by use of 20 mol% of the chiral BINOL-Ti catalyst. According to this procedure, 2-phenylethanol also was allowed to react with trifluoroacetaldehyde in the presence of (*R*)-BINOL-Ti(O-*i*-Pr)$_2$ (20 mol%). These resutls also are summarized in Table 1 (*1*).

Table 1. Asymmetric Hemiacetal Synthesis

(*R*)-BINOL–Ti(O*i*-Pr)$_2$ (mmol)	CF_3CHO (mmol)	ROH (mmol)	Yield (%)	% ee (sign of $[\alpha]_D$)
1	4	*i*-PrOH (2)	86	86 (+)
0.2	4	*i*-PrOH (2)	81	82 (+)
0.2	4	$PhCH_2OH$ (2)	61	25 (+)
0.2	2.2	$PhCH_2OH$ (2)	48	74 (+)
0.4	2.2	$PhCH_2OH$ (2)	54	91 (+)
0.4	2.2	$PhCH_2CH_2OH$ (2)	57	78 (+)

Determination of the Absolute Configurations of $CF_3CH(OCH_2Ph)OCOPh$. The assignment of the absolute cofiguration was performed by the Mosher method as illustrated below. The sample of the resulting (*S*,*R*)-diastereomer was derivatized to the benzoate by treatment with HAl(*i*-Bu)$_2$ followed by benzoylation and then compared with the product of the asymmetric hemiacetal synthesis. In this manner the product was shown to be (*S*).

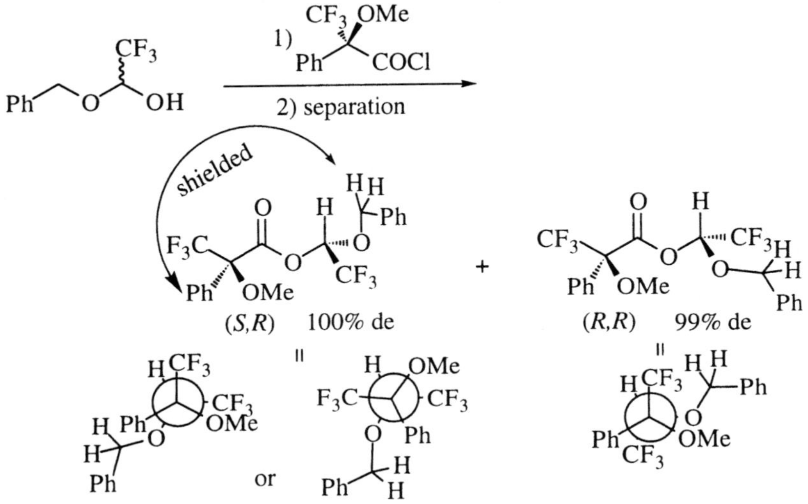

δ 4.62 (d, J = 12 Hz, 1 H)
δ 4.78 (d, J = 12 Hz, 1 H)

δ 4.79 (d, J = 12 Hz, 1 H)
δ 4.94 (d, J = 12 Hz, 1 H)

Scheme 4. Determination of Absolute Configurations of $CF_3CH(OCH_2Ph)OH$

The Trost method using (S)-(+)-α-methoxyphenylacetic acid also led to the same assignment.

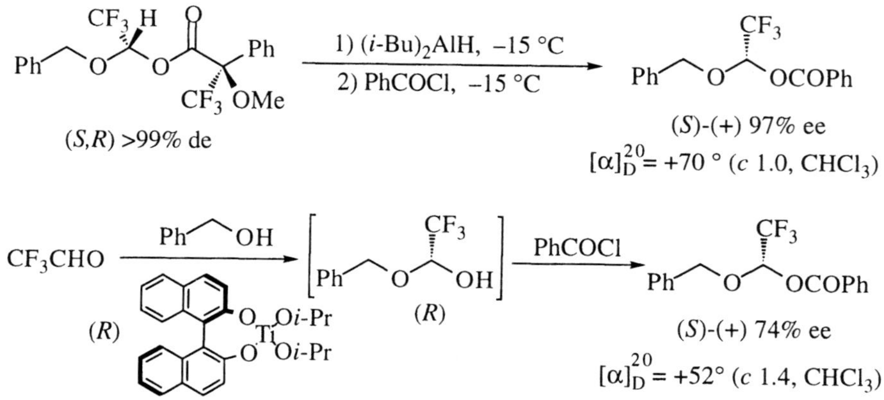

Scheme 5. Structural Correlation

Effect of Alkoxy Ligand. Since the modification of the chiral diol in the titanium complex affected the enantioselectivity, we studied the effect of the alkoxide ligand in (R)-BINOL-Ti(OR)$_2$ and prepared several complexes by treatment of lithium (R)-binolate with TiCl$_2$(OR)$_2$. Although a primary alkoxide ligand led to minimal asymmetric induction, a secondary alkoxide resulted in reasonable ee's. A tertiary butoxide or binolate ligand decreased the ee considerably. Thus, the bulk of the alkoxide ligand on the titanium complex appears to be extremely important to create an appropriate size of the reaction site.

Table 2. Effect of Alkoxy Ligand

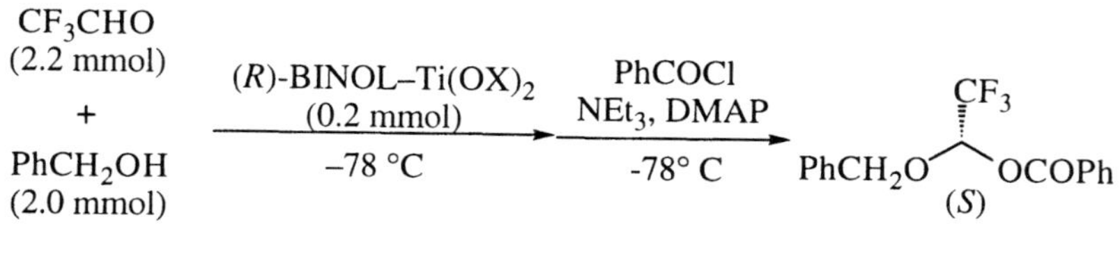

X	yield (%)	% ee
PhCH$_2$	63	8
i-Pr	48	74
(CF$_3$)$_2$CH	65	77
c-Hex	50	79
(PhCH$_2$)$_2$CH	67	63
t-Bu	49	32

Nucleophilic Substitution of Optically Active 1-Alkoxy(polyfluoro)alkyl Sulfonates

We have demonstrated that optically active 1-alkoxyalkyl carboxylates undergo the substitution reactions with lithium dialkylcuprates in the presence of boron trifluoride etherate to give optically active alkoxyalkanes with inversion of configuration (2). We have been attempting to expand this method to reaction using trifluoroacetaldehyde. Furthermore, the nucleophilic C–C bond-forming substitution reaction at the carbon bearing a trifluoromethyl group is unprecedented. In fact, we found that a nucleophilic substitution of optically active 1-alkoxy(polyfluoro)alkyl sulfonates with an organoaluminum reagent takes place smoothly to give optically active alkoxy(polyfluoro)alkanes with a high degree of inversion of configuration (3).

Optical Resolution of CF$_3$CH(OCH$_2$Ph)OSO$_2$R. Although optically active 1-alkoxy(polyfluoro)alkyl sulfonates were available by asymmetric synthesis followed by sufonation at 0 °C as discussed above (1), the ee's obtained by this procedure were not always high. Therefore, we resolved the sulfonates by preparative HPLC. We first liberated trifluoroacetaldehyde or heptafluorobutanal from the corresponding hemiacetal or hydrate, respectively, and treated the free aldehyde with benzyl alcohol to give the racemic hemiacetal, which was converted into the corresponding sulfonate with a sulfonyl chloride, triethylamine, and DMAP. The resulting hemiacetal derivatives were resolved by HPLC (Daicel CHIRALPAK AD or CHIRALCEL OD) to afford (+)- and (−)-tosylates, mesylates, and benzenesulfonates.

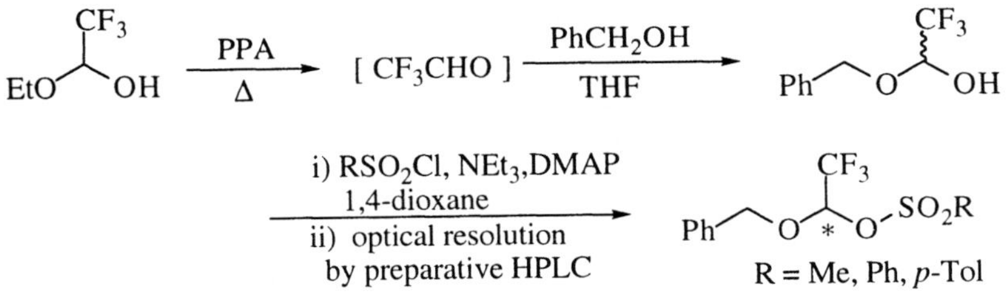

R = Me   (S)-(+) : 25%, 83% ee, $[\alpha]_D^{20} = +46°$ (c 1.0, CHCl$_3$)

         (R)-(−) : 36%, 61% ee, $[\alpha]_D^{20} = -35°$ (c 1.0, CHCl$_3$)

R = p-Tol   (S)-(+) : 27%, 100% ee, $[\alpha]_D^{20} = +38°$ (c 1.0, CHCl$_3$)

         (R)-(−) : 29%, >98% ee, $[\alpha]_D^{20} = -38°$ (c 1.0, CHCl$_3$)

Scheme 6. Resolution of Trifluoroacetaldehyde Hemiacetal Sulfonates

Nucleophilic Substitution of Hemiacetal Sulfonates. The resolved sulfonates were allowed to react with various organometallic reagents. In contrast to the corresponding acetaldehyde hemiacetal carboxylates, 1-benzyloxy-2,2,2-trifluoroethyl acetate did not react with Bu$_2$CuLi•LiI/BF$_3$•OEt$_2$. However, we were pleased to find that the treatment of (R)-(−)-mesylate of 61% ee with triethylaluminum in toluene at −20 °C gave (−)-2-benzyloxy-1,1,1-trifluorobutane in 66% yield with 46% ee. The absolute configuration of the product was determined by the synthesis of an authentic sample from (S)-1,1,1-trifluoro-1,2-epoxypropane.

$$\text{Ph} \sim \text{O} \overset{\text{CF}_3}{\underset{\text{O}}{\diagup}} \text{SO}_2\text{Me} \xrightarrow[\substack{\text{toluene} \\ -20\ ^\circ\text{C, 0.5 h} \\ \textit{inversion}}]{\text{AlEt}_3\ (2.5\ \text{equiv})} \text{Ph} \sim \text{O} \overset{\text{CF}_3}{\diagup}$$

(R)-(–), 61% ee   (S)-(–), 46% ee

                                       66% yield

Since the absolute configuration of the resulting ether is S and that of the mesylate was shown to be R, we conclude that the reaction proceeded with inversion of configuration. *This is the first observation of a nucleophilic substitution at a CF$_3$-substituted carbon under a C–C bond formation.* The stereochemistry also was shown to be inversion as observed before (2).

The observation that ee was fairly lost may suggest that AlEt$_3$ induces to small extent the formation of an oxocarbenium ion intermediate. In order to prevent such racemization, LiAlEt$_4$ was found to be a reagent of choice: the nucelophilic alkylation with this aluminate reagent gave the same product in 59% yield with 83% ee, starting with 83% ee of the mesylate. Noteworthy is that the stereochemical integrity of the substrate was completely maintained as shown in Table 3.

Table 3. Ethylation of CF$_3$CH(OCH$_2$Ph)OSO$_2$R with Aluminium Reagents

$$\text{Ph} \sim \text{O} \overset{\text{CF}_3}{\underset{* \ \text{O}}{\diagup}} \text{SO}_2\text{R} \xrightarrow[\substack{\text{toluene} \\ \textit{inversion}}]{\substack{"\text{AlEt}" \\ (2.5\ \text{equiv})}} \text{Ph} \sim \text{O} \overset{\text{CF}_3}{\underset{*}{\diagup}}$$

R	% ee	" AlEt "	Temp. (°C)	Time (h)	Yield (%)	% ee
Me	61 (R)-(–)	AlEt$_3$	–20	0.5	66	46 (S)
Me	83 (S)-(+)	LiAlEt$_4$	0	1.5	59	83 (R)
Ph	93 (R)-(–)	AlEt$_3$	–20	0.5	57	71 (S)
Ph	98 (R)-(–)	LiAlEt$_4$	0	1.5	61	95 (S)
p-Tol	100 (S)-(+)	AlEt$_3$	–20	0.5	75	82 (R)
p-Tol	100 (S)-(+)	LiAlEt$_4$	0	1.5	69	98 (R)

We also studied the feasibility of using triethylborane, lithium tetraethylborate, diethylzinc, and lithium triethylzincate (*vide infra*) and found that the borate and zincate also underwent the substitution reaction with high degree of inversion but with a formation of by-products as summarized in Table 4. Thus, lithium tetraethylaluminate is concluded to be the best ethylation reagent for the ethylation of hemiacetal tosylate CF$_3$CH(OCH$_2$Ph)OTs.

$$\text{Ph} \sim \text{O} \overset{\text{CF}_3}{\diagup} \text{OTs} \ + \ \substack{\text{MEt}_n\ \text{or} \\ \text{LiMEt}_{n+1}} \xrightarrow{\text{toluene}} \text{Ph} \sim \text{O} \overset{\text{CF}_3}{\diagup}$$

100% ee

Table 4. Comparison of Organometallic Reagents for Ethyl Substitution

MEt_n or $LiMEt_{n+1}$	Temp (°C)	Time (h)	Yield (%), % ee		By-product(s)
$AlEt_3$	-20	0.5	75	82	
$LiAlEt_4$	0	1.5	69	98	
BEt_3	0	overnight	no reaction		
$LiBEt_4$	r.t.	2.0	32	95	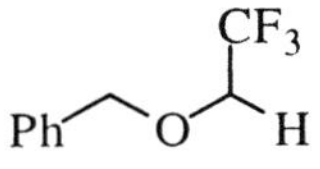 $Ph\diagup O \diagup CF_3 H$ >15%
$ZnEt_2$	0	overnight	no reaction		
$LiZnEt_3$	r.t.	2.0	30	96	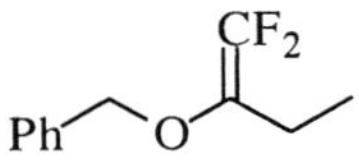 $Ph\diagup O \diagup CF_2$ 30%

To demonstrate the generality of the substitution reaction, the (R)- or (S)-tosylate was allowed to react with various commercially available organoaluminum reagents (Table 5). The degree of inversion was remarkable when lithium tetraalkylaluminates were used. The low yield of products when $LiAlMe_4$ was used may be ascribed to a poor solubility of the reagent in toluene.

Table 5. Alkylation of $CF_3CH(OCH_2Ph)OTs$ with Aluminium Reagents

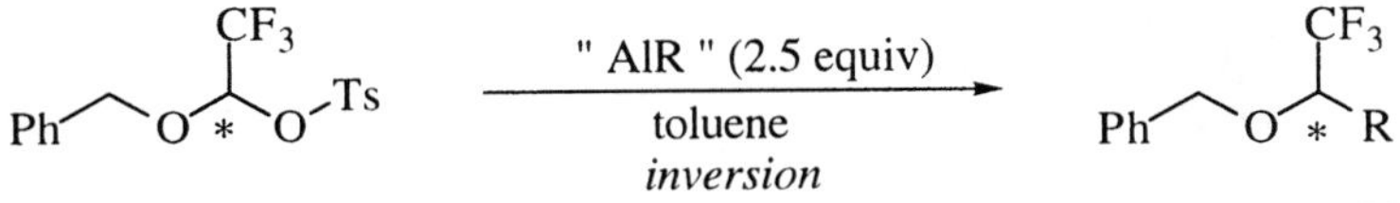

$$Ph\diagup O_* \diagup O^{-Ts}_{CF_3} \xrightarrow[\substack{\text{toluene} \\ \textit{inversion}}]{\text{" AlR " (2.5 equiv)}} Ph\diagup O_* R^{CF_3}$$

% ee	" AlR "	Temp. (°C)	Time (h)	Yield (%)	% ee
92 (R)-(−)	$Al(n\text{-}Bu)_3$	−20	0.5	67	72 (S)-(−)
100 (S)-(+)	$LiAl(n\text{-}Bu)_4$	r.t.	1.0	62	90 (R)-(+)
94 (R)-(−)	$Al(i\text{-}Bu)_3$	−20	0.5	76	61 (S)-(−)
100 (S)-(+)	$LiAl(i\text{-}Bu)_4$	0	1.5	46	93 (R)-(+)
91 (R)-(−)	$AlMe_3$	−20	0.5	66	69 (S)-(−)
90 (R)-(−)	$LiAlMe_4$	0 ~ 40	2.5	26	78 (S)-(−)

In addition to the trifluoroacetaldehyde hemiacetal tosylate, optically active 1-benzyloxy-2,2,3,3,4,4,4-heptafluorobutyl tosylate derived from heptafluorobutanal also reacted with LiAl(*i*-Bu)$_4$ stereospecifically.

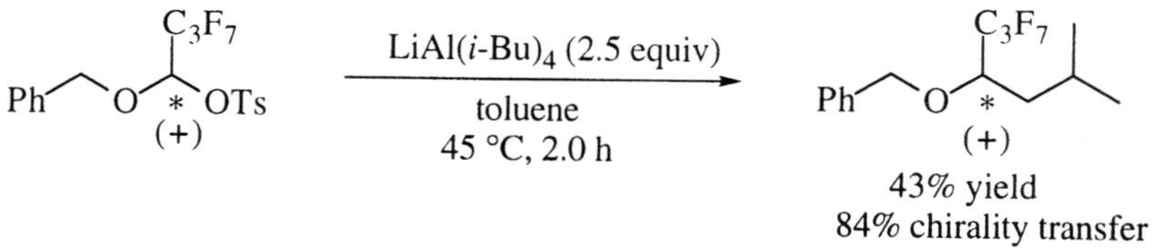

$$
\text{Ph}\diagup\text{O}\underset{(+)}{\overset{C_3F_7}{\diagdown}}\text{OTs} \quad \xrightarrow[\substack{\text{toluene}\\ \text{45 °C, 2.0 h}}]{\text{LiAl}(i\text{-Bu})_4 \ (2.5 \text{ equiv})} \quad \text{Ph}\diagup\text{O}\underset{\substack{(+)\\ 43\% \text{ yield}\\ 84\% \text{ chirality transfer}}}{\overset{C_3F_7}{\diagdown}}
$$

The versatility of the present reaction is demonstrated by the transformation of the products to optically active 1,1,1-trifluoro-2-alkanols by hydrogenolysis. For example, (*R*)-(+)-1,1,1-trifluoro-2-hexanol (*4*), a key chiral building block of AFLCs, is readily obtained by the sequence of these transformations.

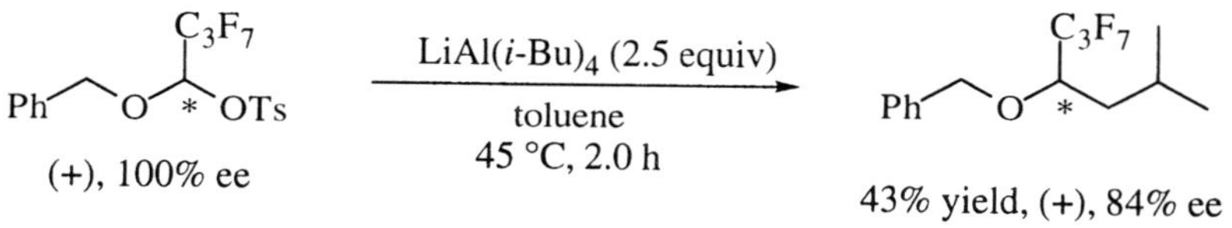

$$
\text{Ph}\diagup\text{O}\overset{C_3F_7}{\underset{\substack{* \ \text{OTs}\\ (+),\ 100\% \text{ ee}}}{\diagdown}} \quad \xrightarrow[\substack{\text{toluene}\\ \text{45 °C, 2.0 h}}]{\text{LiAl}(i\text{-Bu})_4 \ (2.5 \text{ equiv})} \quad \text{Ph}\diagup\text{O}\overset{C_3F_7}{\underset{\substack{*\\ 43\% \text{ yield, (+), 84\% ee}}}{\diagdown}}
$$

Use of Aluminate Reagents Derived from 1-Alkenes. The nucleophilic substitution was achieved using lithium tetraalkylaluminates produced by the Ti-catalyzed hydroalumination reaction of 1-alkenes with LiAlH$_4$ (*5*). The whole transformation is illustrated in Scheme 7, and the results are summarized in Table 6. Because the hydroalumination is site-selective, optically active phenyl-substituted or olefinic trifluoro alcohols are now readily available.

Synthesis of Dichiral Liquid Crystals Containing Two Fluorines at Two Chiral Centers

During the synthetic research of FLC materials for fast switching, we prepared 2-[4-[(*R*)-2-fluorohexyloxy]phenyl]-5-[4-(*S*)-2-fluoro-2-methyldecanoyloxy]phenyl]-pyrimidine shown in Figure 2 and its diastereomers and homologs.

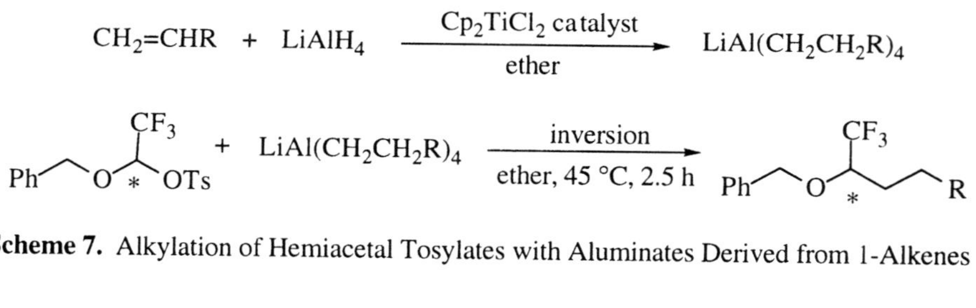

$$
\text{CH}_2=\text{CHR} + \text{LiAlH}_4 \xrightarrow[\text{ether}]{\text{Cp}_2\text{TiCl}_2 \text{ catalyst}} \text{LiAl(CH}_2\text{CH}_2\text{R)}_4
$$

$$
\text{Ph}\diagup\text{O}\overset{CF_3}{\underset{* \ \text{OTs}}{\diagdown}} + \text{LiAl(CH}_2\text{CH}_2\text{R)}_4 \xrightarrow[\text{ether, 45 °C, 2.5 h}]{\text{inversion}} \text{Ph}\diagup\text{O}\overset{CF_3}{\underset{*}{\diagdown}}\text{R}
$$

Scheme 7. Alkylation of Hemiacetal Tosylates with Aluminates Derived from 1-Alkenes

Table 6. Alkylation with the Aluminates Derived from 1-Alkenes

Tosylate (Abs. Config., % ee)	CH$_2$=CHR	Product (Abs. Config., % yield, % ee)
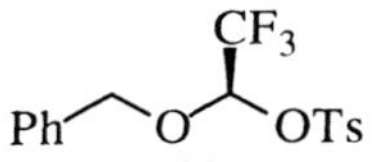 Ph–O–C(CF$_3$)–OTs (R, 99)	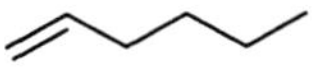 (1-hexene structure)	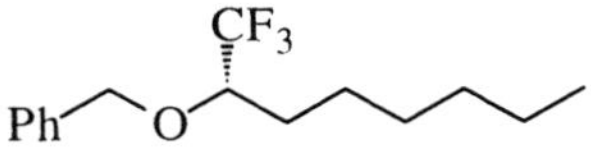 Ph–O–C(CF$_3$) chain (S, 54, 87)
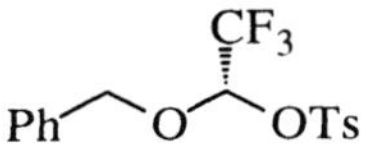 Ph–O–C(CF$_3$)–OTs (S, 100)	 (1-heptene structure)	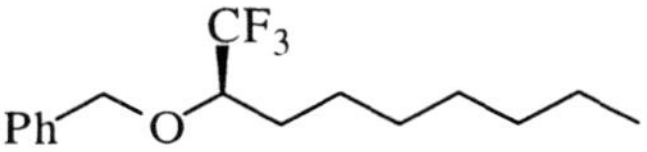 Ph–O–C(CF$_3$) chain (R, 58, 84)
(S, 100)	 (4-phenyl-1-butene) Ph	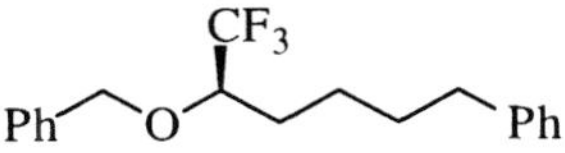 Ph–O–C(CF$_3$) chain Ph (R, 43, 77)
(R, 99)	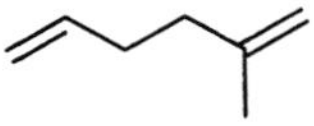 (4-methyl-1,4-pentadiene)	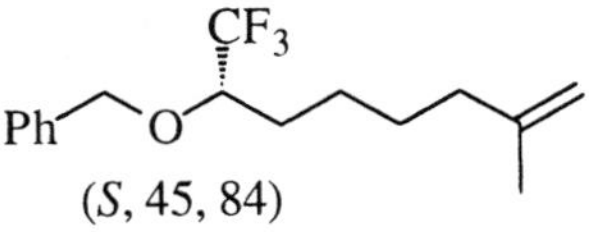 Ph–O–C(CF$_3$) chain (S, 45, 84)
(R, 99)	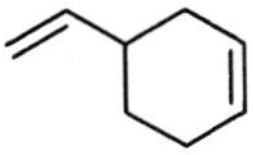 (vinylcyclohexene)	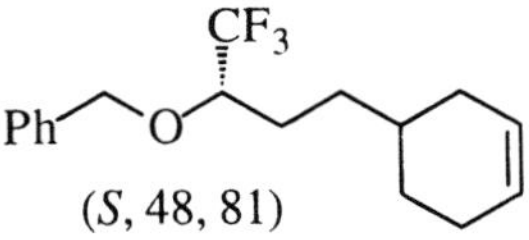 Ph–O–C(CF$_3$) chain cyclohexene (S, 48, 81)

Phase transition diagram:

Cr $\underset{68}{\overset{93}{\rightleftarrows}}$ SmX $\underset{97}{\overset{99}{\rightleftarrows}}$ IsoX $\overset{131}{\rightarrow}$ Iso Liq

SmC* — 107 (to IsoX), 117 / 116 (to Iso Liq)

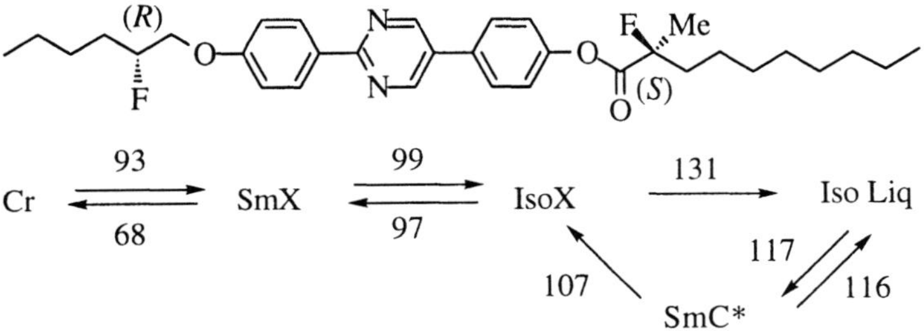

Figure 2. A Dichiral Liquid Crystalline Compound and Phase Transtion Temperatures

The phenolic part was prepared according to the route shown in Scheme 8; the carboxylic moiety by the sequence of reactions shown in Scheme 9. Ee of the chiral center of each substrate was determined to be over 98 % by HPLC. The two parts were combined by conventional esterification.

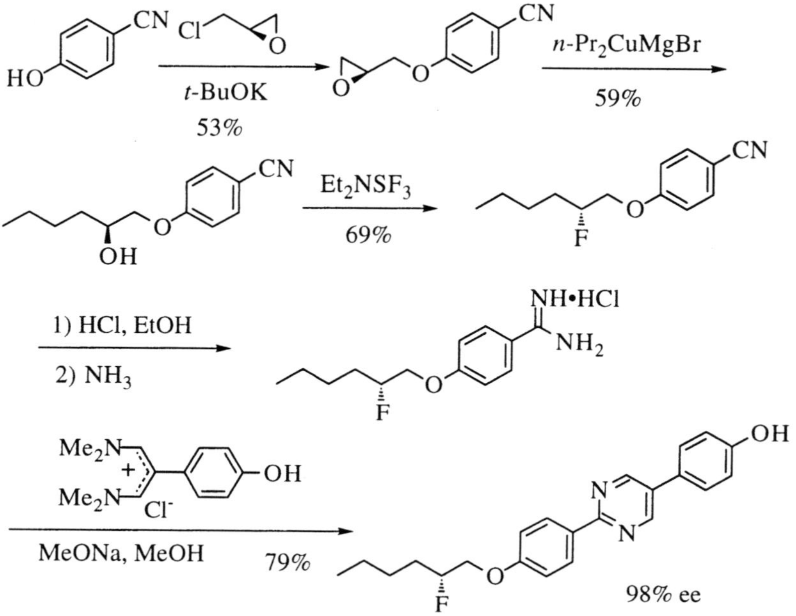

Scheme 8. Synthesis of a Phenolic Part

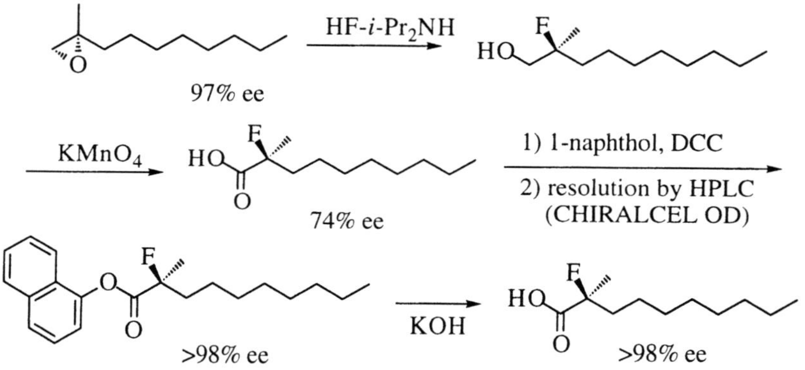

Scheme 9. Synthesis of a Carboxylic Acid Part

Unusual Endothermic Transition

The ester in Figure 2 showed an endothermic transition from SmC* to IsoX, where the apparently optically isotropic phase is denoted as IsoX. Unusual behaviors of IsoX are following: (1) All properties suggest isotropic liquid except for X-ray analysis which indicates layer ordering. (2) The sequence of the phase transitions on cooling differs from those on heating. (3) Phase transition of SmC* → IsoX taking place on cooling is endothermic!

In contrast, the diastereomer of the compound in Figure 2 showed a direct exothermic transition from the isotropic liquid to the IsoX phase. We have studied the transition behavior and the structure of the IsoX phase by means of differential scanning calorimetry (DSC), optical microscopy, solid-state NMR and X-ray diffraction measurements.

On cooling the compound, the Iso Liq phase changed at 117 °C into an SmC* phase, which slipped into an optically isotropic phase (IsoX) with a sharp endothermic peak at 107 °C. The transition was observed by optical microscopy to occur immediately. The texture of the IsoX phase was apparently different from that of a homeotropic smectic A (SmA) or SmC phase. On further cooling, an unidentified ferroelectric smectic modification (SmX) appeared gradually from the IsoX phase with a broad exothermic peak. Spontaneous polarization in the SmC* and SmX phases were -249 nC/cm^2 at 115 °C and -341 nC/cm^2 at 95 °C, respectively; spontaneous polarization and switching behavior were not observed in the IsoX phase. Viscosity of the IsoX phase was higher than that of the SmC* and SmX phases. A marked supercooling was observed for the appearance of the IsoX phase. Changing the cooling rate from 10 K/min to 0.1 K/min reproduced the endothermic transition behavior. Although the endothermic transition enthalpy can be explained in terms of a transition of a metastable SmC* phase to a stable IsoX phase, such a well reproducible endothermic transition is unprecedented in thermotropic liquid crystals (6).

Structural Elucidation of IsoX Phase. Spectral analysis by ^{13}C NMR and X-ray diffraction measurements suggests that upon the SmC* → IsoX transition, an interlayer correlation may be broken, whereas dimerization or tetramerization via a stereospecific intermolecular interaction through fluorines at chiral centers within each layer is assumed to be operating. The dimerization or tetramerization reduces the anisotropy of molecular motion to increase rotation around the short axis. This change in molecular dynamics may release entropy of the system to induce the endothermic transition. The IsoX phase is not a result of the competition between helical structure and mesophase ordering but a result of the chirality-dependent stereospecific interaction or chiral molecular recognition (7.

Acknowledgments

The authors sincerely thank co-workers listed in the following references for their sincere devotion.

238

References

1. Poras, H.; Matsutani, H.; Yaruva, J.; Kusumoto, T.; Hiyama, T. *Chem. Lett.* **1998**, 665.
2. Matsutani, H.; Ichikawa, S.; Yaruva, J.; Kusumoto, T.; Hiyama, T. *J. Am. Chem. Soc.* **1997**, *119*, 4541.
3. Matsutani, H.; Poras, H.; Kusumoto, T.; Hiyama, T. *Chem. Commun.* **1998**, 1259.
4. Yonezawa, T.; Sakamoto, Y.; Nogawa, K.; Yamazaki, T.; Kitazume, T. *Chem. Lett.*, **1996**, 855.
5. Matsutani, H.; Poras, H.; Kusumoto, T.; Hiyama, T. *Synlett*, in press.
6. Yoshizawa, A.; Umezawa, J.; Ise, N.; Sato, R.; Soeda, Y.; Kusumoto, T.; Sato, K.; Hiyama, T.; Takanishi, Y.; Takezoe, H. *Jpn. J. Appl. Phys.*, **1998**, *37*, L942.
7. Kusumoto, T.; Sato, K.; Katoh, M.; Matsutani, H.; Yoshizawa, A.; Ise, N.; Umezawa, J.; Takanishi, Y.; Takezoe, H.; Hiyama, T. *Mol. Cryst. Liq. Cryst.*, in press.

Chapter 17

Resolution of Racemic Perfluorocarbons through Self-Assembly Driven by Electron Donor–Acceptor Intermolecular Recognition

Maria Teresa Messina, Pierangelo Metrangolo, and Giuseppe Resnati[1]

**Dipartimento Chimica, Politecnico 7, via Mancinelli, I–20131 Milano, Italy
(e-mail: resnati@dept.chem.polimi.it)**

The attractive intermolecular interaction between iodine or bromine atoms of perfluorocarbon halides and anions or heteroatomic sites of hydrocarbons is robust enough to overcome the low affinity between perfluorocarbons and hydrocarbons and to drive their molecular recognition directed self-assembly. This strong and specific intermolecular interaction can be readily exploited in the design and manipulation of molecular aggregation processes. From a preparative point of view, this recognition enabled the resolution of chiral perfluorocarbon derivatives.

Introduction.

Perfluorocarbon (PFC) compounds show a unique combination of physical and chemical properties (1). As a result, synthetic methodologies commonly employed for the preparation of hydrocarbon (HC) derivatives cannot be directly extrapolated to the preparation of their PFC analogues. These considerations gain particular relevance in asymmetric synthesis where high enantiomer ratios derive from subtle and often unpredictable differences in the kinetic and/or thermodinamic parameters of diastereoisomeric reaction pathways. A specifically tailored approach has therefore to be developed if the obtainment of PFC derivatives in enantiopure forms is pursued. The well-established approach to obtain enantiopure compounds through resolution of their racemic forms *via* formation of diastereoisomeric and non-covalent adducts is particularly promising, the convenient intermolecular interaction between enantiopure HC-resolving agents and racemic PFCs being developed. Here we describe how the electron donor-acceptor complex formation between perfluoroalkyl halides and anions or heteroatom containing hydrocarbons (R_f - X + El $\rightleftharpoons$ R_f - X····El, X = Br, I; El =

---

[1]Corresponding author.

Br⁻, N, O, S), while being so far poorly characterized in the literature, is a general, robust, specific, and directional motif of PFC-HC self-assembly both in the liquid and the solid phase. This recognition motif fulfills the energetic and geometric requirements for effective and diastereoselective recognition processes and allowed the resolution of 1,2-dibromohexafluoropropane.

General Aspects of the R_f - X····El Interaction.

Numerous analytical techniques (microwave, infrared, and ultraviolet spectroscopy, nuclear magnetic resonance and nuclear quadrupolar resonance, X-ray analyses, etc.) consistently show that carbon-bound halogens (C—X, X = Cl, Br, I) can be involved as electron acceptors in attractive interactions (C—X····El) with electronegative atoms as donors (El = N, O, S) (2-5). The tendency to give short C-X····El contacts (namely deeply interpenetrating van der Waals volumes) increases on moving from chlorine to bromine to iodine and tends to be highly directional on the extended C-X bond axis (6-8). In the Table 1 we have reported some prototype examples from the literature showing how the interaction occurs both in the solid and in the liquid phase. The environment around the halogen having a strong electron-withdrawing effect, the resulting interaction becomes robust enough to influence the crystal packing in the solid. For instance, the X-ray structure of chlorocyanoacetilene (10) shows how a dimer is formed, the Cl····N distance is 2.984 (0.90) Å, and the maximum energy gain is about 10 kJ/mol, about half the magnitude of an average hydrogen bond.

Table 1. Electron donor-acceptor complexes formed by aliphatic halides and organic bases.

Acceptor	Donor	Interaction parameter
$CHI_3^{(1)}$	HMTA[3]	$d_{I\cdots N}$ 2.94 (3.65)[4]
	Quinoline	$d_{I\cdots N}$ 3.05 (3.65)[4]
	1,4-Dioxane	$d_{I\cdots O}$ 3.04 (3.55)[4]
	1,4-Dithiane	$d_{I\cdots S}$ 3.32 (4.00)[4]
	1,4-Diselenane	$d_{I\cdots Se}$ 3.51 (4.15)[4]
$CHCl_3^{(2)}$	Triethylamine	K_{eq} 0.4[5]
$CHBr_3^{(2)}$	Triethylamine	K_{eq} 0.9[5]
$CHI_3^{(2)}$	Triethylamine	K_{eq} 3.2[5]

[1] Prout, C. K.; Kamenar, B. *Molecular Complexes*, Foster, R. Ed., Elek Science, London **1973**, *1*, 151. [2] Geron, C.; Gomel, M. *J. Chim. Phys. Phys.-Chim. Biol.* **1978**, *75*, 241. [3] HMTA = hexamethylenetetramine. [4] Bond lenght established through X-ray analyses; sum of van der Waals radii in parenthesis. [5] Equilibrium costant determined through dielectric polarization measurements.

Scheme 1. Infinite 1D network **6a** formed by TMEDA (**2**) and 1,2-diiodotetrafluoro-
ethane (**1a**).

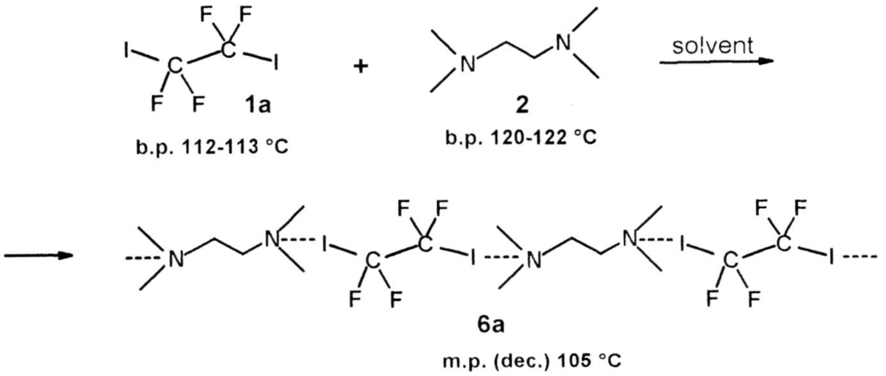

Iodotrifluoromethane has been reported to form discrete molecular complexes (1:1
stoichiometry) with various monoamines (11-23). These complexes have been studied
in the gas phase at room temperature and in the solid phase in the cold (80 K). The
interaction has been rationalised as an n → σ* donation (24) and a similar behaviour
has been described for perfluoroethyl- and *n*-perfluoropropyliodides. While no energy
data are available for this interaction, we expected it to be particularly strong due to the
powerful electron withdrawing ability of perfluoroalkyl residues. We also expected
that by using bidentate donors (*e.g.* diamines) and acceptors (*e.g.*
perfluoroalkyldihalides), the recognition process could be doubled at either ends of the
two self-assembling motifs. Here we report how this double lock results in a
particularly effective PFC-HC recognition process.

The R$_f$- X$\cdots$EI Interaction in the Solid Phase.

An equimolar solution of 1,2-diiodotetrafluoroethane (b.p. 112 - 113 °C, **1a**), and
N,N,N',N'-tetramethylethylediamine (m.p. 120 - 122 °C, **2**, TMEDA) in chloroform
affords the infinite 1D chain **6a** (m.p. 105 °C dec.) as white crystals which can be
stored and handled at room temperature and in the air (Scheme 1) (25). A similar
behaviour was shown by 1,4-diiodooctafluorobutane (**1b**), 1,6-diiodoperfluorohexane
(**1c**), and 1,8-diiodoperfluorooctane (**1d**) which afforded the corresponding non-
covalent co-polymers **6b-d** (Scheme 2) as stable and high-melting crystals. Elemental
analyses showed that diiodoperfluoroalkanes **1** and diamine **2** are present in a 1 : 1 ratio
in the solids **6** and the same 1 : 1 ratio between PFC and HC components was given by
^{1}H and ^{19}F NMR spectra of infinite chains **6a-d** in the presence of $(CF_3CH_2)_2O$. The
ability of the N$\cdots$I interaction to work as a general recognition motif for PFC-HC self-
assembly was confirmed by the formation of the crystalline 1D chains **7a-d**, **8a-d** and
9a-d (Scheme 2) generated by the already considered diiodides **1a-d** and tertiary or

Scheme 2. Infinite 1D networks **6b-d**, **7a-d**, **8a-d** and **9a-d** formed by α,ω diiodoperfluoroalkanes **1a-d** with TMEDA (**2**), DABCO (**3**), K.2.2.2. (**4**) and K.2.2. (**5**), respectively.

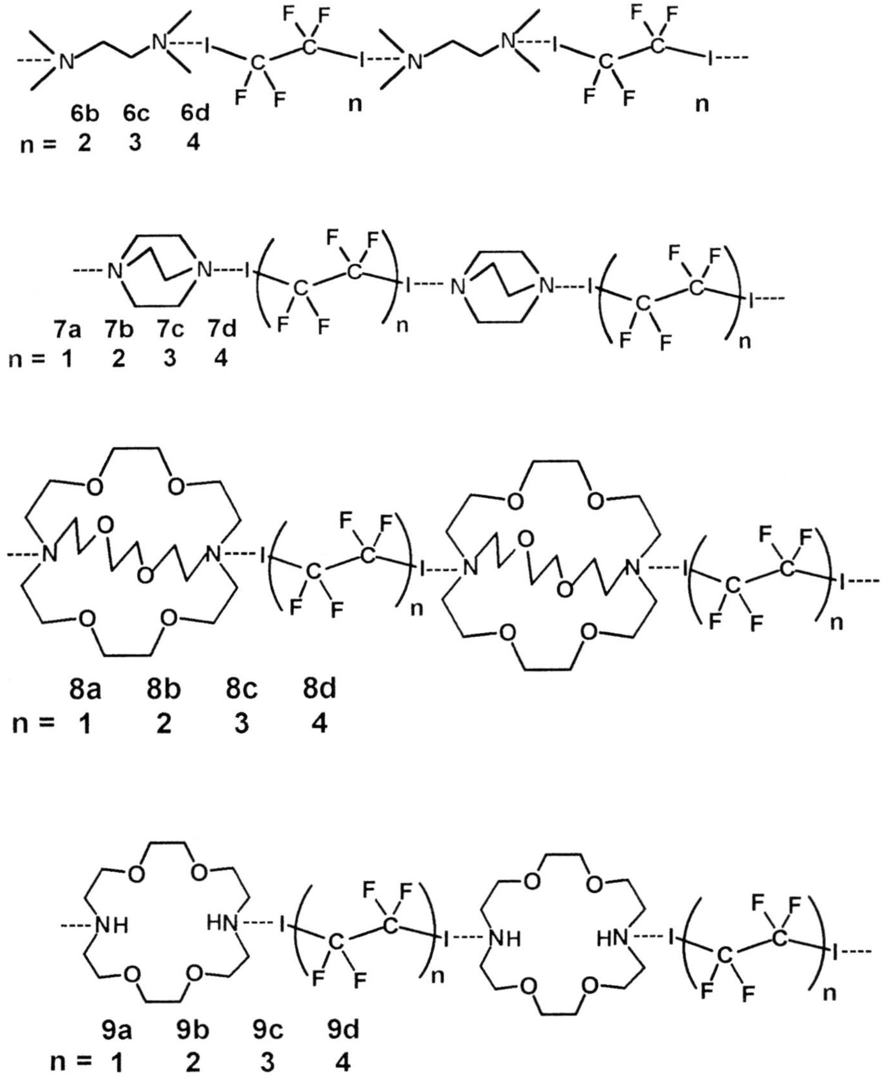

6b 6c 6d
n = 2 3 4

7a 7b 7c 7d
n = 1 2 3 4

8a 8b 8c 8d
n = 1 2 3 4

9a 9b 9c 9d
n = 1 2 3 4

secondary amines 1,4-diazabicyclooctane (DABCO) (**3**), Kryptofix.2.2.2. (**4**) and Kryptofix. 2.2. (**5**), respectively. In ^{19}F NMR spectra of concentrated solution of infinite chains **6-9** the -CF$_2$I group is shifted upfield compared to pure precursors **1**. On diluting these samples, the chemical shift of the -CF$_2$I groups moves back to the frequencies shown in pure iodides **1** and both ^{1}H and ^{19}F NMR spectra of highly diluted adducts **6-9** cannot be distinguished from those shown by pure precursors **1-5**. These observations prove that, on being dissolved, linear non-covalent co-polymers **6-9** give rise to the starting free components **1-5**. The interaction existing in the solid state between perfluorinated diiodides **1** and diamines **2-5** is reversible and similar in nature to the non-covalent binding described by Pullin (11-17) and Allred (22). In solution, perfluorinated diiodides **1** and diamines **2-5** enter a rapid association process (driven by the same n→σ* donation from nitrogen to iodine which allows the solid co-polymers **6-9** to be formed) and dimeric, trimeric, ... adducts are generated reversibly.

Figure 1. X-ray packing of the co-crystal **6a** formed by 1,2-diiodotetrafluoroethane (**1a**) and TMEDA (**2**) viewed down the *b*-axis. Molecules segregate into distinct layers joined by N····I — C interactions (dotted lines). The disorder of both species is modeled using rigid body restraints: the amine is adequately represented by two images of equal occupancy, the diiodide moiety is described by splitting the carbon and the fluorine atoms over four locations.

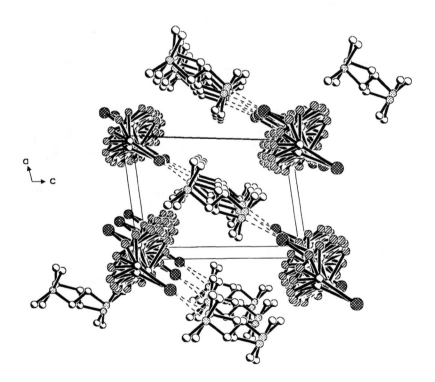

X-Rays. The structures of the infinite 1D coordination networks reported in Schemes 1 and 2 firmly rest on X-ray analyses (25). Crystal structures of **6a** (Figure 1) and **8a** (Figure 2) prove how the N····I interaction is largely responsible for keeping the perfluorinated diiodide in place in the infinite 1D chains. The value of the N····I distance

Figure 2. X-ray packing of the co-crystal **8a** formed by K.2.2.2. (**4**) and 1,2-diiodotetrafluoroethane (**1a**) viewed down the *c*-axis. The diiodide moiety is modeled using rigid body restraints: the disorder involving carbon atoms is described by splitting each atom over two locations with occupancy 0.5, while the disorder involving fluorine atoms is described by splitting each atom over four locations.

very close to 2.80 Å in both co-crystals, is longer than average covalent N—I bond length (2.07Å) (26) but less than 0.8 times the sum of van der Waals radii (1.98Å for iodine and 1.55Å for nitrogen). For both **6a** and **8a**, the N····I interaction is approximately on the extended C-I bond axis, the N····I—C angle being 166.7° (0.7) and 168.6° (0.7) respectively. In both co-crystals **6a** and **8a** the perfluorinated diiodoethane moiety of **1a** exhibits dramatic rotational disorder around the axis joining the two iodines, these two atoms being the only non-disordered atoms of **1a**. In the crystal of **6a** all the atoms of the diamine, including the nitrogens, also exhibit disorder over two positions corresponding to distinct conformers or to different molecular orientations. Both PFC and HC units **1a** and **2**, respectively, are liquid at room temperature as pure compounds and highly disordered in co-crystal **6a** so that the N····I interaction must be largely responsible for the **6a** crystal cohesion. Note that in **6a** the hydrocarbon amine and the perfluorinated diiodide segregate into layers (perpendicular to the *c* crystallographic axis) which are joined by N····I interactions running roughly along the the (111) direction. A qualitative measure of the importance of the interaction between **1a** and **2** results from the fact that the molar volume of the liquids **1a** and **2** is reduced by 21% in the co-crystallization process to give **6a**. These data seem particularly significant since in the co-crystals the PFC and the HC components segregate and volume reduction must results at the interface between the two components. The general reluctance of PFC and HC moieties to mix is apparent also from the structure of **8a** (Figure 2). Because of the different steric requirements of the two co-crystallizing species, the structure is not characterized by layers and the disordered perfluorinated diiodide **1a** gives rise to pairs of segregated and parallel columns linked to the diamine by the N····I interactions.

Figure 3. DSC of α,ω-diiodoperfluoroalkanes **1a-d** and K.2.2.2. (**4**).

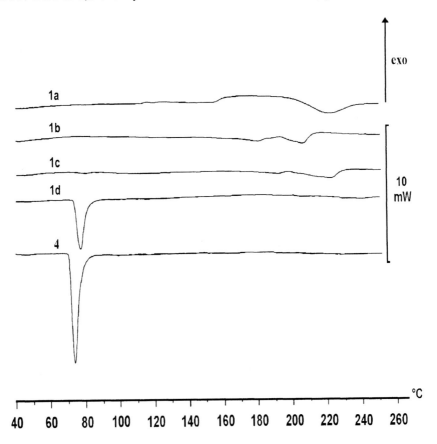

Differential Scanning Calorimetry (DSC). Further informations on the R_f-I····N interaction in the solid phase were obtained from DSC analyses. Heating curves of iodides **1a-d** and of K.2.2.2. (**4**) (pure compounds, temperature range 40-270 °C, heating rate 10 °C ·min⁻¹) are reported in Figure 3 and heating curves of infinite

Figure 4. DSC of adducts **8a-d** formed by K.2.2.2. (**4**) and α,ω-diiodoperfluoroalkanes **1a-d**.

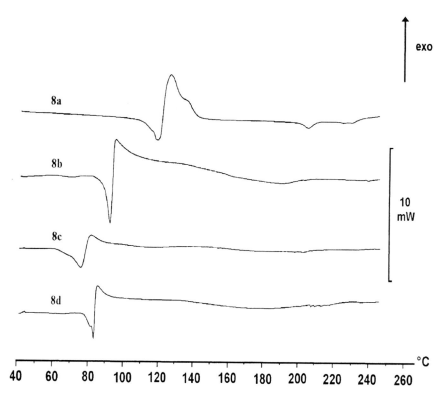

networks **8a-d** in the same temperature range and with the same heating rate are reported in Figure 4 (27). Interestingly, the sharp endotherm with a peak at 73 °C shown by the amine **4** and associated with its melting is not present in any network **8** and the same holds for the melting endotherm of diiodoperfluorooctane (**1d**) (peak at 77 °C) which is missing in the curve of **8d**. Infinite networks **8a-d** all melt at temperatures higher than starting amine **4**. For instance, when **4** (m.p. peak 73 °C) interacts in a 1:1 ratio with 1,2-diiodotetrafluoroethane (**1a**, b.p.112-113 °C), the formed non-covalent co-polymer **8a** melts with a peak at 118 °C. These thermal behaviours are consistent with the formation of well defined molecular aggregates on interaction of iodides **1** with amine **4**. They also confirm that the intermolecular interaction occurs through specific and relatively strong bindings and is not a non-specific association of solvent-solute type. The curves of infinite networks **8a-d** all show a similar trend: on heating the co-polymer, an endotherm is first observed and it is rapidly followed by an exotherm. The two processes are probably associated with the melting of the sample and the successive reaction/decomposition of the melted systems, respectively. Finally, it is worth noting that in **8a-c** the melting temperatures decrease with increased length of the fluorinated chain. In contrast, **8d** melts slightly higher than **8c**, but for this co-polymer it has to be remembered that **1d** is a solid at room temperature while **1c** is a liquid.

Figure 5. IR spectra of TMEDA (**2**) in the 2700-3000 cm^{-1} region, of 1,4-diiodoperfluorobutane (**1b**) in the 1000-1250 cm^{-1} region, and of their adduct **6b** in the same region.

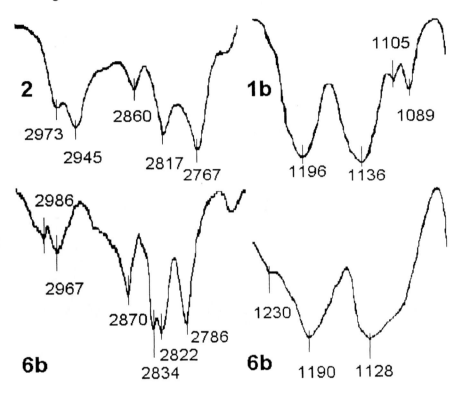

IR and Raman. Also IR and Raman spectroscopies afforded useful informations on the R_f-I····N interaction. Due to the fact that the central N····I interaction is weaker than covalent or ionic bonds, it is reasonable to discuss vibrational spectra of infinite networks **6-9** in terms of modified modes of starting iodides **1** and amines **2-5**. The validity of the adopted approach is supported by the fact that similar changes of the absorption bands of a given iodide, or amine, were observed on interaction with different amines, or iodides, respectively. As reported in Figure 5, on formation of the complex **6b** the C-H stretching mode of the donor amine **2** in the 2700-3000 cm^{-1} region shifts to higher frequencies. Similar upward shifts were observed for other networks **6 - 9** and have been described for other donor - acceptor adducts (28-30),

Figure 6. Raman spectrum of 1,4-diiodoperfluorobutane (**1b**) in the 200-1400 cm^{-1} region.

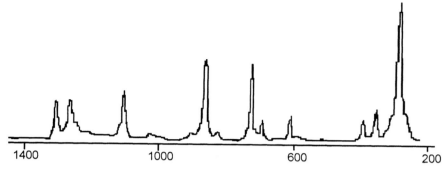

Figure 7. Raman spectrum of the adduct **6b** between 1,4-diiodoperfluorobutane (**1b**) and TMEDA (**2**) in the 200-1400 cm^{-1} region.

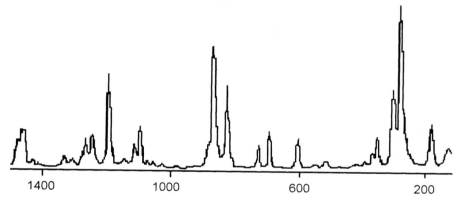

for instance the dimers given by trimethylamine with I$_2$, Br$_2$, and IBr (31, 32), or the trimers formed by 1,4-diazabicyclo[2.2.2]octane with boron trihalides (33). As to the acceptor iodides **1**, on complex formation the C-F stretching modes in the 1000-1250 cm^{-1} region move to lower frequencies (Figure 5b). Similarly, it has been described that the complex formation between iodotrifluoromethane and various trialkylamines caused a downward shift of the C-F band at ca.1170 both in the solid phase (80 K) and in solution (CCl$_4$). Raman spectra of co-crystals **6-9** were also diagnostic of the donor-acceptor interaction occurring between diamines **2-5** and diiodides **1a-d**. The spectra of co-polymers **6-9** were about the sum of the spectra of corresponding PFC and HC components, **1** and **2-5**, respectively, the N····I interaction resulting in slight shifts of the bands or change of their relative intensities. For instance, the typical absorptions shown by diamines **2-5** in the 2700-3000 region were present also in corresponding co-crystals **6-9**. The adduct formation strictly maintained the band shapes of pure diamines but induced a higher frequency shift, similar to that observed in IR spectra.

Band splitting and/or shift were seen in the C-F stretching region. As reported in Figures 6 and 7 for TMEDA (**2**) and its adduct **6b** with 1,4-diiodoperfluorobutane (**1b**) on adduct formation the relative intensities of the C-F stretching bands undergo a dramatic change. The absorption at 1168 cm^{-1} is probably activated by the amine-iodide self-assembly.

Study of the R_f-X····El Interaction in the Liquid Phase.

The ability of perfluoroalkyl halides and heteroatom substituted hydrocarbons to work as complementary motifs also in the liquid phase is quite general. Both iodo- and bromoperfluoroalkanes can work as acceptor motifs and not only nitrogen, as described above in the solid phase, but also sulphur and oxygen atoms can work as electron donors. The formation of the dimer between quinuclidine (**10**) and 1-iodo-heptafluoropropane (**1e**) has been used as a model system of the R_f-X····El interaction in solution. This system was preferred over those formed by the diiodides **1a-d** with the diamines **2-5** as multiple interactions leading to trimeric, tetrameric, ... species could not occur. On addition of equimolar amounts of **1e** to deuterochloroform solutions of **10**, the ^{1}H and ^{13}C NMR spectra of the electron donor **10** showed changes and observed chemical shift variations gave a first indication of a specific interaction between the two solutes (n → σ* donation) to give the complex **11**. Strong evidence of a specific

Scheme 3. Association equilibrium between quinuclidine (**10**) and 1-iodoheptafluoropropane (**1e**).

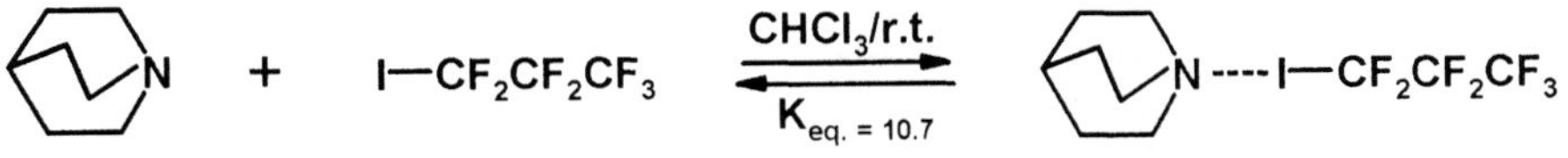

C-I····N interaction came from ^{19}F and ^{14}N NMR spectra. On complete complexation of **10** (**1e**/**10** ratio ≅ 150), a 7 ppm upfield shift for the ^{14}N chemical shift was observed and signal width passed from 960 Hz to 2160 Hz (34). The significant line broadening observed in the ^{14}N NMR spectrum is a clear indication of an increase in the quadrupolar relaxation due to a specific I····N intermolecular association. As for the iodoheptafluoropropane (**1e**), a several ppm shift to higher fields of the signal of the -CF$_2$I group in ^{19}F NMR spectra was observed upon forming complex **11**. The quinuclidine induced change of the chemical shifts of the other two signals in ^{19}F spectra decreased with increasing distance from the iodine atom. This is consistent with the generalisation that the specific I····N intermolecular interaction, firmly established in the solid phase (25), is the driving force for PFC-HC self-assembly in the liquid phase. On varying the iodide/amine ratio in a 1 M solution of the complex **11** in deuterochloroform, the chemical shift of the -CF$_2$I group changed from -61.13 ppm

Table 2. ^{19}F-NMR chemical shift differences $\Delta\delta_{CF2}{}^a$ of 1,2-diiodo- and 1,2-dibromotetrafluoroethane in different solvents.

Solvent	I(CF$_2$)$_2$I	Br(CF$_2$)$_2$Br	Solvent	I(CF$_2$)$_2$I	Br(CF$_2$)$_2$Br
	$\Delta\delta_{CF2I}{}^b$	$\Delta\delta_{CF2Br}{}^c$		$\Delta\delta_{CF2I}{}^b$	$\Delta\delta_{CF2Br}{}^c$
Piperidine	11.23	2.40	1,3-Propanediol	4.30	0.90
N-Methylpiperidine	8.89	2.07	1,3-Propanedithiol	3.45	0.58
Morpholine	10.59	1.95	Pyridine	7.32	1.12
N-Methylmorpholine	8.39	1.71	Furan	1.11	0.13
Thiomorpholine	9.94	1.62	Thiophene	1.41	0.26
Tetrahydrofuran	4.13	0.95	Acetone	3.63	0.90
Tetrahydrothiophene	5.38	0.76	DMSO	7.22	2.03
3,6-Dioxaoctane	3.36	0.96	HMPA	8.23	2.76
3,6-Dithiaoctane	5.50	0.77	Acetonitrile	2.74	0.45

[a] CFCl$_3$ was used as internal standard; $\Delta\delta = \delta_{n\text{-pentane}} - \delta_{used\ solvent}$.
[a] $\delta_{ICF2CF2I\ in\ n\text{-pentane}} = -52.42$ ppm. [b] $\delta_{BrCF2CF2Br\ in\ n\text{-pentane}} = -63.32$ ppm.

(δ_{free}) to -77.80 (δ_{max}) and by following this variation as a function of the **10**/**1e** ratio an equilibrium constant of 10.7 ± 0.4 l mol^{-1} has been derived for the complex formation at 25 °C. This association constant could be considered as an apparent binding constant since chloroform has been suggested to interact with quinuclidine (35). In order to prove that intermolecular interactions, similar to that described above between **1e** and **10**, occur in the solution when other donor and acceptor motifs are involved, the ^{19}F NMR spectra of 1,2-diiodo- and 1,2-dibromotetrafluorobutanes, **1a** and **1f**, respectively, have been recorded by using a wide set of structurally different solvents. In Tables 2 and 3 the chemical shift differences of the -CF$_2$I and -CF$_2$Br groups of **1a** and **1f** between values observed in *n*-pentane and in hydrocarbon solvents containing an heteroatom as electron donor motif are reported ($\Delta\delta_{used\ solvent} = \delta_{n\text{-pentane}} - \delta_{used\ solvent}$). Some trends revealed by the Tables are worth noting. As far as the acceptor motif R$_f$-X is concerned, the tendency to form strong complexes (I > Br) parallels the order of halogen atom polarisabilities, consistent with a key-role of halogen polarisation (and/or charge-transfer energies) in the interaction. The same order (I > Br) has been established in the solid state for the R$_H$ - X····El interaction that, while being definitively weaker, has been firmly recognised through searches of the Cambridge Structural Database (6, 7). The ranking of the electron donor ability of different heteroatom containing hydrocarbons obtained by using diiodide **1a** parallels perfectly that obtained with dibromide **1f**, thus implying that similar self-assembling processes are occurring with the two electron acceptors. As for the electron donor motif El, the order in which the R$_F$-X····El interaction becomes weaker is N > S ≥ O. n-Donors work better than π-donors (Table 2) and for a given heteroatom, stronger interaction is observed in higher hybridization states (Table 2). Both steric and electronic effects affect the strength of the R_F-X···El intermolecular interaction (Table 3). More crowded environments around the donor site make it less accessible at shorter distances for the incoming halogen atom and an electron withdrawing environment around the donor motif diminishes its donor ability. In both cases weaker interactions result. The presence of electron releasing residues on the donor molecule induce an opposite effect on the intermolecular interaction. Single $\Delta\delta_{CF2X}$ values reported in Tables 2 and 3 are highly consistent with each other and with the electronic and steric effects expected for an n → σ* electron donation in the R_F-X···El adducts. It thus follows that the $\Delta\delta_{CF2X}$ value is a simple, effective, and sensitive tool to rank the donor and acceptor motifs according to their ability to self-assemble, strong interactions resulting in large $\Delta\delta_{CF2X}$ values.

Table 3. Electronic and steric effects on $\Delta\delta_{CF2I}{}^a$ values shown by 1,2-diiodotetrafluoroethane.

Solvent	$I(CF_2)_2I$ $\Delta\delta_{CF2I}{}^b$	Solvent	$I(CF_2)_2I$ $\Delta\delta_{CF2I}{}^b$
Cis-2,6-dimethylpiperidine	8.29	Ethyl isonicotinate	5.68
2,2,6,6-Tetramethylpiperidine	6.92	Ethyl acetate	2.80
1,2,2,6,6-Pentamethylpiperidine	2.64	Ethyl monofluoroacetate	2.12
2-Methylpyridine	7.20	Ethyl difluoroacetate	1.30
2,6-Dimethylpyridine	5.86	Ethyl trifluoroacetate	0.72
4-Methylpyridine	7.62	Ethyl monochloroacetate	1.82
4-Isopropylpyridine	7.71	Ethyl dichloroacetate	1.54
4-Acetylpyridine	5.87	Ethyl monoiodoacetate	2.41

[a] CFCl$_3$ was used as internal standard; $\Delta\delta = \delta_{n\text{-pentane}} - \delta_{used\,solvent}$. [b] $\delta_{ICF2CF2I}$ in n-pentane = -52.42 ppm.

Resolution of 1,2-Dibromohexafluoropropane.

Differences in $\Delta\delta_{CF2I}$ and $\Delta\delta_{CF2Br}$ values reported in Table 2 clearly show that perfluoroalkylbromides are weaker electron acceptors than perfluoroalkyliodides. This is consistent with the fact that no solid co-polymer was obtained starting from racemic 1,2-dibromohexafluoropropane (**1g**) and several chiral and enantiopure diamines (e. g. (-)-1,2-diaminocyclohexane, (+)-Troger's base, (-)-sparteine). The desired co-crystal formation was obtained when better electron donors were available as was the case in the solid system **12** formed between (-) - sparteine hydrobromide (**13**) and (S)-1,2-

Scheme 4. System **12** formed by (-)-sparteine hydrobromide (**13**) and (*S*)-1,2-dibromohexafluoropropane (**1g**).

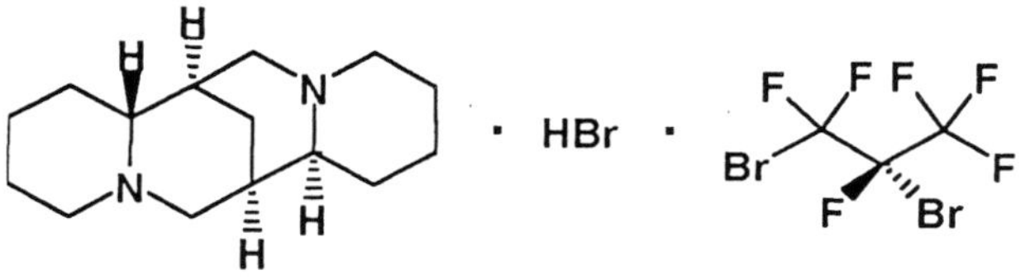

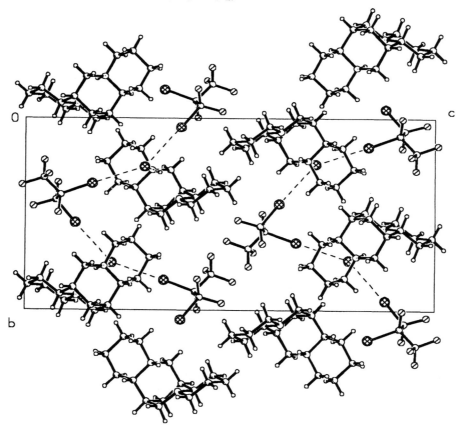

dibromohexafluoropropane (**1g**) (Scheme 4). In the co-crystal **12** the two nitrogen atoms of (-)-sparteine work as electron donors towards the proton and the bromide ion works as electron donor towards the two bromine atoms of **1g**. In fact, the absorption bands typical for a trialkylammonium cation were observed in the 2250-2700 cm^{-1} region of IR spectra of **12** and single crystal X-ray analysis showed that the bromide ion was in short contact with the primary bromine of a molecule of **1g** and the secondary bromine of another molecule of **1g** (Figure 8). The bromide ion is thus bridging

Figure 8. Crystal packing of the system **12** formed by (-)-sparteine hydrobromide (**13**) and (*S*)-1,2-dibromohexafluoropropane (**1g**).

between two molecules of dibromoperfluoropropane **1g** leading to the formation of an infinite 1D network. While the infinite 1D networks **6a** and **8a**, formed by diiodotetrafluoroethane (**1a**) with TMEDA (**2**) and K.2.2.2. (**4**), respectively, give rise to straight chains (Figures 1 and 2), the infinite 1D network of **12** is rolled in a chiral and enantiopure helix. The packing of both the HC and the PFC motif into the chiral helixes of the three-component system **12** allows the transfer of information between the two motifs to be maximixed, namely allows the enantiopure HC **13** to work as an effective resolving agent for the racemic PFC **1g**. Indeed, the co-crystal **12** containing exclusively the (*S*) enantiomer of dibromoperfluoropropane **1g** was formed when (-)-sparteine hydrobromide (**11**) and racemic 1,2-dibromoheptafluoropropane (**1g**) were crystallized from chloroform/carbon tetrachoride solutions.

References.

1. Smart, B. E. in: *Organofluorine Chemistry: Principles and Commercial Applications*, Banks, R. E.; Smart, B. E.; Tatlow, J. C. Eds., Plenum Press, New York, **1994**.
2. Dumas, J.-M.; Gomel, L.; Guerin, M. in: *The Chemistry of Functional Groups, Supplement D*, Patai, S.; Rappoport, Z. Eds., John Wiley and Sons, New York, **1983**, 985.
3. Bent, H. A. *Chem. Rev.* **1968**, *68*, 587.
4. Desiraju, G. R. *Angew. Chem. Int. Ed. Engl.* **1995**, *34*, 2311.
5. Lommerse, J. P. M.; Taylor, R. *J. Enzyme Inhybition* **1997**, *11*, 223.
6. Ramasubbu, N.; Parthasarathy, R.; Murray-Rust, P. *J. Am. Chem. Soc.* **1986**, *108*, 4308.
7. Lommerse, J. P. M.; Stone, A. J.; Taylor, R.; Allen, F. H. *J. Am. Chem. Soc.* **1996**, *118*, 3108.
8. Xu, K.; Ho, D. M.; Pascal, R. A.-Jr. *J. Am. Chem. Soc.* **1994**, *116*, 105.
9. Stevenson, D. P.; Coppinger, G. M. *J. Am. Chem. Soc.* **1962**, *84*, 149.
10. Bjorvatten, T. *Acta Chem. Scand.* **1968**, *22*, 410.
11. Cheetham, N. F.; Pullin, D. E. *Chem. Comm.* **1965**, *18*, 418.
12. Cheetham, N. F.; Pullin, D. E. *Chem. Comm.* **1967**, *20*, 233.
13. Cheetham, N. F.; McNaught, I. J.; Pullin, A. D. E. *Aust. J. Chem.* **1971**, *24*, 479.
14. Mishra, A.; Pullin, A. D. E. *Aust. J. Chem.* **1971**, *24*, 2493.
15. Cheetham, N. F.; McNaught, I. J.; Pullin, A. D. E. *Aust. J. Chem.* **1974**, *27*, 973.
16. Cheetham, N. F.; McNaught, I. J.; Pullin, A. D. E. *Aust. J. Chem.* **1974**, *27*, 987.
17. McNaught, I. J.; Pullin, A. D. E. *Aust. J. Chem.* **1974**, *27*, 1009.
18. Semin, G. K.; Babushkina, T. A.; Khrlakyan, S. P.; Pervova, E. Ya.; Shokina, V. V.; Knunyants, I. L. *Teor. Eksp. Khim.* **1968**, *4*, 275; *C. A.* **1968**, *69*, 72685z.
19. Chen, Q. Y.; Qiu, Z. M. *J. Fluorine Chem.* **1987**, *35*, 79.
20. Chen, Q.-Y.; Li, Z.-T.; Zhou, C. M. *J. Chem. Soc., Perkin Trans. 1* **1993**, 2457.
21. Legon, A. C.; Millen, J. D.; Rogers, S. C. *J. Chem. Soc., Chem. Comm.* **1975**, 580.
22. Larsen, D. W.; Allred, A. L. *J. Phys. Chem.* **1965**, *69*, 2400.

23. Bertran, J. F.; Rodriguez, M. *Org. Magnetic Res.* **1979**, *12*, 92.
24. Foster, R. *Organic Charge-Transfer Complexes*, Academic Press, London, **1969**, 100.
25. Amico, V.; Corradi, E.; Meille, S. V.; Messina, M. T.; Resnati, G. *J. Am. Chem. Soc.* **1998**, *120*, 8261.
26. Padmanabhan, K.; Paul, I. E.; Curtin, D. Y. *Acta Crystallogr., Sect. C*, **1990**, *46*, 88.
27. Lunghi, A.; Cardillo, P.; Panzeri, W.; Messina, M. T.; Metrangolo, P.; Resnati, G. *J. Fluorine Chem.* **1998**, *91*, 191.
28. Yokobayashi, K.; Watari, F.; Aida, K. *Spectochim. Acta A* **1968**, *24*, 1651.
29. Yada, H.; Tanaka, J.; Nagakura, S. *J. Mol. Spectrosc.* **1962**, *9*, 461.
30. VanPaasschen, J. M.; Geanangel, R. A. *Can. J. Chem.* **1975**, *53*, 723.
31. Gayles, J. N. *J. Chem. Phys.* **1968**, *49*, 1840.
32. Zingaro, R. A.; Tolberg, W. E. *J. Am. Chem. Soc.* **1959**, *81*, 1353.
33. McDivitt, J. R.; Humphrey, G. L. *Spectrochim. Acta A* **1974**, *30*, 1021.
34. Messina, M. T.; Metrangolo, P.; Panzeri, W.; Ragg, E.; Resnati, G. *Tetrahedron Lett.* **1998**, in press.
35. Maliniak, A.; Kowalewski, J.; Panas, I. *J. Phys. Chem.* **1984**, *88*, 5628.

Chapter 18

Catalytic Asymmetric Synthesis of Diastereomeric α- or β-CF₃ Liquid Crystalline Molecules

Conformational Probe for Anti-Ferroelectricity and Self-Assembly for Spontaneous Chiral Resolution of the Racemates

Koichi Mikami

Department of Chemical Technology, Tokyo Institute of Technology, Tokyo 152–8552, Japan (e-mail: kmikami@o.cc.titech.ac.jp)

Asymmetric synthesis of diastereomeric α- or β-CF₃ liquid crystalline (LC) molecules is established through the catalytic asymmetric carbonyl-ene reaction with fluoral by a chiral binaphthol (BINOL)-derived titanium catalyst (**1**). Diastereomeric α- or β-CF₃ LCs work as conformational probes for anti-ferroelectricity. The first example of spontaneous chiral resolution of racemates in fluid LC phases is further found with the diastereomeric LCs.

Current interest has been focused on a catalytic asymmetric synthesis of organofluorine compounds (1) through C-C bond forming reactions, which has been remained essentially unexplored thus far (2, 3). Special attention has to be paid to the unique characteristics of organofluorine substrates. 1) Fluorine greatly resists S$_N$2-type nucleophilic displacements due to the extremely high C-F bond strength (C-F: 485.7 kJ/mol; C-Cl: 329.6 kJ/mol; C-Br: 274.4 kJ/mol; C-H: 410.6 kJ/mol). 2) Fluorine easily departs from the carbon as fluoride ion in β-elimination fashion from β-F carbanions. 3) Fluorine facilitates the generation of β-carbanions due to its large electro-negative effect. 4) Fluorine destabilizes α-carbanions by the 2pπ-2pπ repulsion. 5) Fluorine stabilizes α-carbenium ions by the 2pπ-2pπ* conjugation. These electronic characters are the origins of the inductive and resonance effects of F and CF₃ substituents, such as the -I, +Iπ, and +R effects of F atom and the -I, -Iπ, and the intriguing -hyperconjugation (negative fluorine hyperconjugation: HCJ) effect (4) of CF₃. These electronic effects are also operative in the large bond moments of C-F and C-CF₃. Since the addition reaction to carbonyl compounds plays a central role in organic synthesis, asymmetric catalysis even with fluoro-carbonyl compounds may provide a powerful methodology for the catalytic asymmetric synthesis of organofluorine compounds. Recently, the carbonyl-ene reaction catalyzed by a chiral *Lewis* acid complex has emerged as an efficient method for asymmetric synthesis (5). In particular, the asymmetric catalytic glyoxylate-ene reaction, that involves a highly reactive enophile, provides an efficient method for asymmetric C-C bond formation by a chiral binaphthol (BINOL)-derived titanium catalyst (**1**) (Scheme 1).

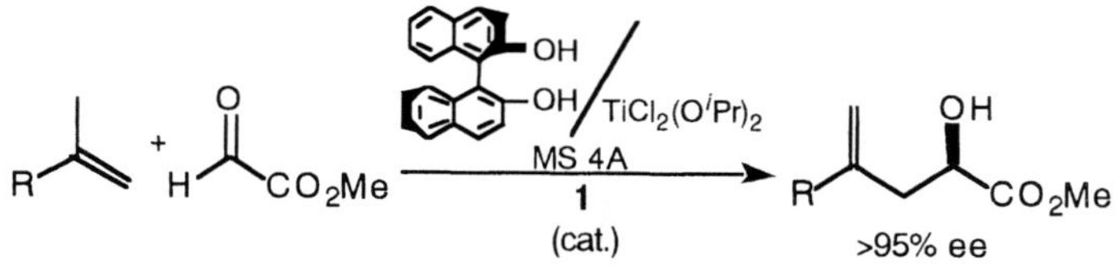

Scheme1 Glyoxylate-ene reaction catalyzed by chiral BINOL-Ti catalyst.

Enantioselective Catalysis of Carbonyl-Ene Reaction with Fluoral.

The asymmetric catalysis of carbonyl-ene reaction is also feasible with fluoral (6, 7) (**2a**), as a reactive enophile (LUMO energy level: CF_3CHO/H^+ -8.64 eV, MeO_2CCHO/H^+ -7.90 eV, *vide infra*), by the BINOL-Ti catalyst (**1**) (Table 1). The reaction was carried out by simply adding an olefin and then freshly dehydrated and distilled fluoral (**2a**) at 0 °C to the solution of the chiral titanium complex (**1**) prepared from (*R*)- or (*S*)-BINOL and diisopropoxytitanium dihalide in the presence of molecular sieves MS 4A as described for the glyoxylate-ene reaction (5). The reaction was completed within 30 min. The ene-type product, namely homoallylic alcohol (**3**) was obtained along with the allylic alcohol (**4**) (entries 1-4). The enantiomeric purities of both products were determined to be more than 95% ee by ^1H NMR analysis after transformation to the (*S*)- and (*R*)-MTPA ester derivatives. Thus, the absolute configuration of the products was determined by the Mosher method (8). The sense of asymmetric induction is, therefore, exactly the same as observed for the glyoxylate-ene reaction; the (*R*)-catalyst provides the (*R*)-alcohol products (5).

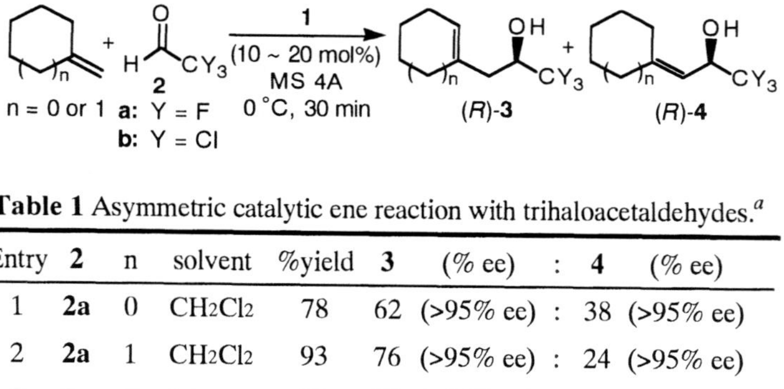

Table 1 Asymmetric catalytic ene reaction with trihaloacetaldehydes.[a]

Entry	2	n	solvent	%yield	3	(% ee)	:	4	(% ee)
1	**2a**	0	CH₂Cl₂	78	62	(>95% ee)	:	38	(>95% ee)
2	**2a**	1	CH₂Cl₂	93	76	(>95% ee)	:	24	(>95% ee)
3	**2a**	0	toluene	82	77	(>95% ee)	:	23	(>95% ee)
4	**2a**	1	toluene	82	79	(>95% ee)	:	21	(>95% ee)
5	**2b**	0	CH₂Cl₂	57	55	(26% ee)	:	45	(75% ee)
6	**2b**	1	CH₂Cl₂	49	52	(34% ee)	:	48	(66% ee)
7[b]	**2b**	1	toluene	35	63	(11% ee)	:	37	(66% ee)

[a] All reactions were carried out with 0.1 mmol (10 mol%) of **1**, 1.0 mmol of olefin, and *ca.* 2.0 mmol of **2a** in the presence of MS 4A (0.2 g), unless otherwise marked. [b] 0.2 mmol (20 mol%) of **1** was employed.

LUMO Energy Level *vs* Electron Density of F-Carbonyl Compounds (*vide supra*)

Interestingly, much lower ee's were observed and more allylic alcohols were formed when chloral (**2b**) was used as the enophile (entries 5-7). Thus, the reactivity of trihaloacetaldehydes including acetaldehyde (**2c**) has been analyzed in terms of the LUMO energy level and the electron density on the carbonyl carbon (C_1) on the basis of the MNDO, PM3, and 6-31G** calculations (Table 2). We carried out molecular orbital calculations on the protonated aldehydes as a model for **2a-c**/Lewis acid complexes. In the transition states, one of the halogen substituents on the α-carbon of the aldehydes is most likely to occupy the antiperiplanar region relative to the attacking reagent (9, 10). The structures of the model compounds were thus optimized with the torsional angle about F/Cl-C-C-O being fixed to 90°. Starting from the geometries optimized by MM2 calculations, MNDO and PM3 and *ab-initio* (RHF/6-31G**) geometry optimizations were carried out without any additional constraints. MNDO and PM3 atomic charges were evaluated from the positive core charges and the numbers of valence electrons. The natural bond orbital (NBO) analysis, instead of conventional Mulliken population analysis, was used for evaluating atomic charges in the *ab-initio* calculations (11). The refined results were obtained using the split-valence basis set with polarization functions (6-31G**).

Table 2 Computational analysis of CX_3CHO/H^+ complexes.[a]

		fluoral (**2a**)	chloral (**2b**)	acetaldehyde (**2c**)
ab initio (RHF/6-31G**)	LUMO (eV)	-5.40	-4.88	-4.09
	C_1 charge	+0.61	+0.64	+0.70
PM3	LUMO (eV)	-8.64	-7.83	-7.42
	C_1 charge	+0.36	+0.39	+0.44
MNDO	LUMO (eV)	-8.55	-8.14	-7.37
	C_1 charge	+0.36	+0.42	+0.42

[a] MO calculation was run on the aldehyde **2a-c** / H^+ complexes as a model of **2a-c** / Lewis acid complexes.

Inspection of Table 2 leads to MO analysis of the ene *vs.* cationic reactivity of trihaloacetaldehydes **2**. The frontier orbital interaction between the HOMO of the ene components and the LUMO of the carbonyl enophile is the primary interaction in ene reactions. Fluoral (**2a**) complex with the lower LUMO energy level is thus the more reactive enophile component to give mainly the homoallylic alcohols **3**. By contrast, the chloral (**2b**) complex bears the greater positive charge at the carbonyl carbon (C_1) and hence is the more reactive carbonyl compound in terms of the cationic reaction leading eventually to the allylic alcohols **4**. The reduced carbonyl C-O bond polarity of fluoral may be rationalized by electron-withdrawal from oxygen by fluorine substituents (10b). Thus, the ene reactivity of aldehydes including chloral is determined in terms of the balance of LUMO energy level *vs.* electron density on the carbonyl carbons (C_1).

Diastereoselective Catalysis of Fluoral-Ene Reactions.

The diastereo- and enantioselective catalysis of fluoral-ene reaction was established   in the same manner described above   except for the use of trisubstituted olefins **5** (Table 3). Significantly, all the fluoral-ene reactions provide preferentially the homoallylic alcohols **6** presumably because of the enhanced ene-reactivity of the trisubstituted olefins by the introduction of the electron-donating methyl group (HOMO energy level: 2-methyl-2-butene -9.39 eV, 2-butene -9.80 eV) along with remarkably high level of *syn*-diastereoselectivity. Thus, the catalytic fluoral-ene reaction provides an efficient route to the *syn*-diastereoselective and enantioselective synthesis of CF$_3$-containing compounds.

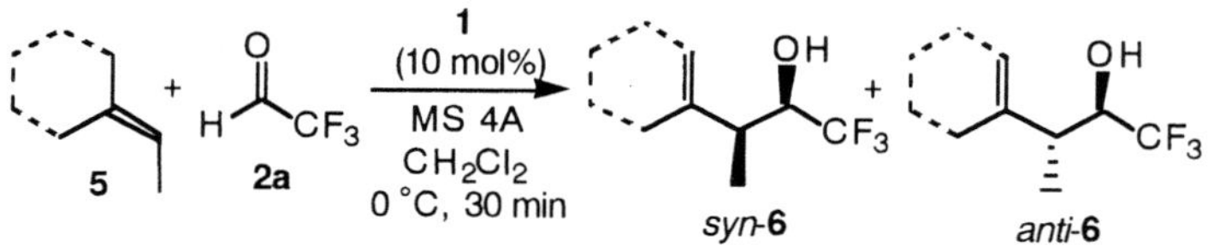

Table 3 Asymmetric catalytic fluoral-ene reaction.[a]

Entry	Olefin	%yield	*syn*-**6**			:	*anti*-**6**
1	a	94	98	(96% ee)	:		2
2[b]		46	96		:		4
3	b	76	94	(95% ee)	:		6
4	c	66	91	(78% ee)	:		9

[a] All reactions were carried out with 0.1 mmol (10 mol%) of **1**, 1.0 mmol of **5**, and *ca.* 2.0 mmol of **2a** in the presence of MS 4A (0.2 g), unless otherwise marked. [b] An equimolar amount of Me$_2$AlCl was used.

The *syn*-diastereomer **6c** of the fluoral-ene product with 2-methyl-2-butene (**5c**) is assigned by ^{13}C NMR analysis through ozonolysis to the aldol-type α-methyl-β-hydroxy ketone (**7c**) (eq. 1). The α-methyl carbon is in the range of *syn*-diastereomers (10.4 ppm) (12). The *syn*-diastereoselectivity is further confirmed after stereospecific transformation to dienes **8a** by *anti*-elimination (13) of **6a** (eq. 2). The most definitive feature is the ^{13}C NMR of the olefinic CH$_3$ carbon of diene **8a** obtained by the *anti*-elimination of the fluoral-ene product (**6a**). Thus, resultant diene **8a**, which shows the CH$_3$ carbon signal at higher field (14.6 ppm) than that (27.1 ppm) of photo-isomerized product **8a**, can be assigned to (*E*)-diene **8a**. Thus, the major diastereomer of the fluoral-ene product is assigned to be the *syn*-isomer.

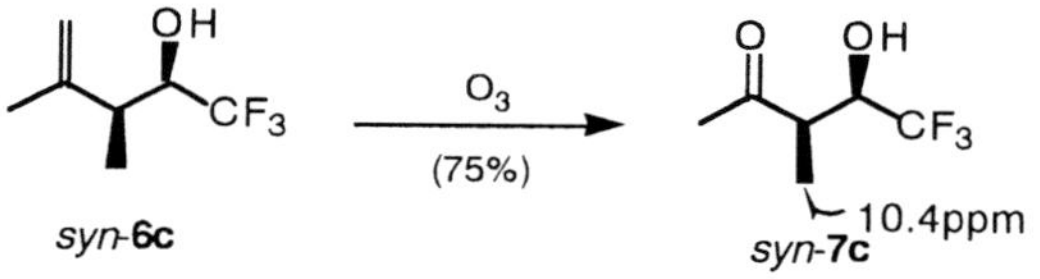

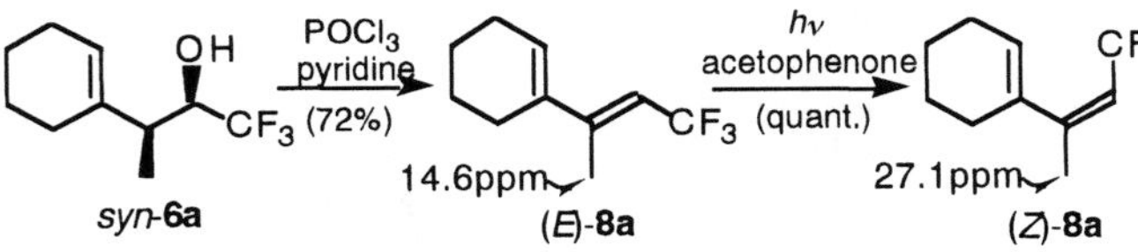

$$\text{(eq. 2)}$$

syn-**6a**   (E)-**8a**   (Z)-**8a**

6-Membered Ene Transition State (14): F and Metal Coordination.

Fluorine and metal coordination can be estimated by comparing the difference (D) between F-metal bond length determined by X-ray analysis (15) and the simple summation of F and metal *van der Waals* radii ($D < 0$) (Table 4). The Lewis basicity and "hard" nucleophlic character of fluorine are due to its extreme electro-negativity and three lone pairs.

Table 4 Fluorine and metal coordination.

Met.	Li	Na	K	Cs	Al	Ti
M-F	2.23-2.29	2.15-2.91	2.67-3.39	2.87-3.61	1.77-1.81	2.03-2.10
sum.	2.92	3.26	3.70	4.07	2.78	2.82
D	-0.66	-0.73	-0.67	-0.83	-0.99	-0.76
	Zr	Hf	Sn	Zn	Pd	Ag
	1.94-2.18	2.31-2.38	2.05-2.46	2.80	3.13-3.16	2.64-2.72
	2.95	2.94	2.93	2.72	2.72	2.79
	-0.89	-0.60	-0.68	+0.08	+0.43	-0.11

Syn-diastereoselectivity of the fluoral-ene reaction implies coordination between fluoral and the BINOL-derived titanium complex (**1**) (Figure 1). Interestingly, the *syn*-diastereoselectivity is analogous to that of the alkylaluminum triflate-promoted glyoxylate-ene reaction with *trans*- and *cis*-2-butene rather than the *anti*-selectivity with tin tetrachloride (14). This suggests that the present fluoral-ene reaction also proceeds through monodentate coordination rather than bidentate coordination (16) presumably because of the bulkiness of the BINOL-derived titanium complex (**1**). Thus, the fluoral-ene reaction would proceed through the equatorial transition state (**A**) involving a shift of hydrogen *cis* to the methyl group due to the "*cis* effect" (17). Because the axial transition state (**B**) should be disfavored by the 1,3-diaxial repulsion in the 6-membered ene transition state (14).

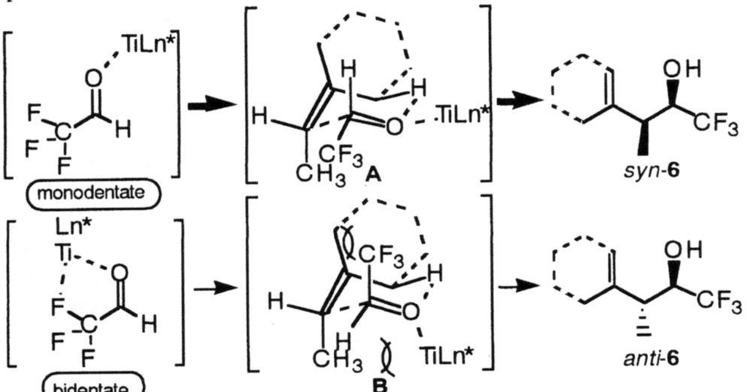

Figure 1 Transition state for *syn*-**6**.

Diastereomeric LC Molecules with CF₃-Group for Large Ps.

A number of molecules that exhibit a ferroelectric (F) phase or an anti-ferroelectric (AF) phase (Figure 2) such as methyl-substituted MHPOBC (18) and the trifluoromethyl analogue, TFMHPOBC (19) with greater spontaneous polarization (Ps) (Figure 3), have been prepared and their physical properties have been investigated because of potential application in electro-optic devices for liquid-crystalline displays (LCDs) (20). The AFLC molecules show quite useful characteristics such as a tri-stable switching, sharp DC threshold, and double-loop-hysteresis.

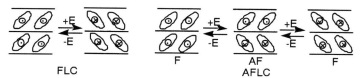

Figure 2 Bi-stable and Tri-stable switching.

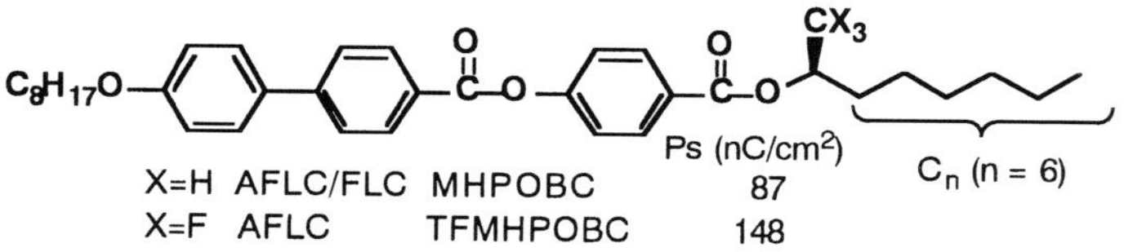

$C_8H_{17}O$— ... —$\overset{O}{\overset{\|}{C}}$—O— ... —$\overset{O}{\overset{\|}{C}}$—O—*$\overset{CX_3}{\diagup}$...

		Ps (nC/cm²)	C_n (n = 6)
X=H	AFLC/FLC MHPOBC	87	
X=F	AFLC TFMHPOBC	148	

Figure 3 MHPOBC and TFMHPOBC.

A number of AFLC molecules showing an AF (SmCA*) phase have been synthesized and an attempt to correlate the molecular structure of the AFLC molecules to the appearance of the SmCA* phase has been carried out. However, only a limited number of AFLC molecules have been reported, containing a chiral alkyl terminus with a *single stereogenic center* such as MHPOBC and the trifluoromethyl analog TFMHPOBC, wherein the odd-even effect is often observed (21). The characteristic phase of AFLC, SmCA*, appears in the molecules with an even numbered (n) alkyl chain. The SmC* phase of FLC appears in molecules with an odd n.

Conformational Probe for Anti-Ferroelectricity.

What kinds of inter-smectic layer interaction are responsible for SmCA* as against SmC* ? One of the most important interactions has been reported to be the transverse pairing of the dipole moments located in the chiral alkyl terminus (Figure 2) (22). There are, however, two twofold axes in the adjacent layers. One at the layer boundary is parallel to the X axis in the plane and the other in the center of the layer is located perpendicularly. Therefore, the "Px model" has been also proposed quite recently (23): spontaneous polarization exists in the vicinity of layer boundary, parallel to the tilt plane (the X axis). For the anti-ferroelectric pairing, either a bent or extended conformer (Figure 4) is proposed at the chiral alkyl terminus. The extended conformer is necessarily proposed for a main chain polyester LC (25). Furthermore, scanning tunnelling microscope (STM) image exhibited the extended conformation of MOBIPC (26), however, on a molybdenum disulfide

(MoS$_2$) rather than a polyimide alignment layer. By contrast, X-ray crystallographic studies showed the bent conformation of MHPOBC in its crystalline phase (24). However , it is very difficult to confirm the SmC$_A$* structure by X-ray diffraction (XRD), because of the liquid-like order rather than the crystalline order in smectic layers. Bent or extended ? That is the question !

Bent   Extended

Figure 4 Bent *vs.* Extended conformation for AFLC.

In order to investigate the relationship between the conformation and AFLC properties, we designed diastereomeric molecules with *double stereogenic centers* (27), β-methyl-substituted TFMHPOBC analogues with even and odd numbered (5, and 6) chiral alkyl chains (n). We found the remarkable phenomena that the transition temperature and the phase sequence depend critically on the diastereomeric excess (% de) of the *unlike* (28) ($R_\alpha S_\beta$)-diastereomers (29). Diastereomeric LC molecules are synthesized diastereo- and enantioselectively through the chiral titanium complex-catalyzed fluoral-ene reaction with ethylidenecycloalkanes **5a** (Scheme 2) (2). Thus, the *u-(syn)-($R_\alpha S_\beta$)*-alcohols (**6**) were obtained in almost quantitative yield in more than 95% ee and 95% de. Protection of the alcohol, ozonolysis, reduction, and deprotection sequence lead to the ($R_\alpha S_\beta$)-diastereomer (**11**) of the chiral portion in LC molecules in essentially stereo-pure form (>95% ee, >95% de). Inversion of stereochemistry of the α-hydroxy group through the triflate derivative (30) afforded stereochemically pure *like-($S_\alpha S_\beta$)*-diastereomer **11** (>95% ee, >95% de). Standard esterification of **11** afforded the diastereomeric LC molecules **12**.

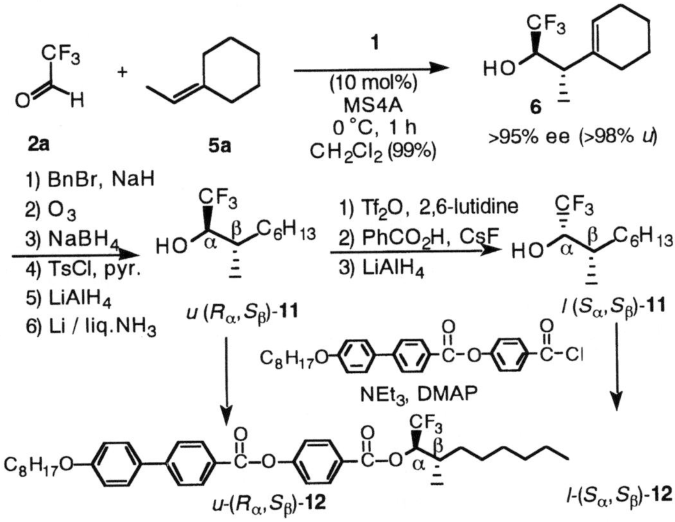

Scheme 2 Ene-based synthesis of β-Me-TFMHPOBC.

The phase transition temperatures of **12** were determined by varying the ratio of the diastereomeric mixture (Figure 5). These diastereomeric LC molecules show a direct transition from SmA to SmCA* in the region of high % de for (R,S)-diastereomer, while SmC* is injected in the (S,S)-diastereomer region. Furthermore, (R,S)-diastereomers have a wider SmCA* temperature range of more than 10 degrees than (S,S)-diastereomers. Thus, the SmCA* phase tends to be stabilized in (R,S)-diastereomers. Furthermore, the electro-optic response of (R,S)-diastereomers **12** showed AF double-loop-hysteresis. By contrast, (S,S)-diastereomers **12** exhibited single-loop-hysteresis, which is characteristic to FLC. These results suggest that the anti-ferroelectric property of (R,S)-diastereomers is due to the bent conformation (**B**), preferentially localized by *ab initio* (RHF/6-31G*) calculations on (R,S)-$PhCO_2CH(CF_3)CH(CH_3)C_5H_{11}$ (Figure 5) (31). We thus proposed the bent (L-shape) conformation as the key for AFLCs.

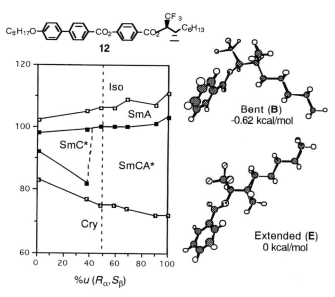

Figure 5 Bent conformation for AFLC.

On the basis of our paradigm of bent conformation for AFLC molecules, we designed the (S^*,S^*)-diastereomer of the β-CF_3-MHPOBC analogues (Figure 6). Indeed, the (S,S)-diastereomer of the $PhCO_2CH(CH_3)CH(CF_3)C_5H_{11}$ is preferentially localized in the bent conformation by *ab initio* (RHF/6-31G*) calculations; (S,S)-$PhCO_2CH(CH_3)CH(CF_3)C_5H_{11}$ is stabilized by 1.96 kcal/mol over the extended conformation. As expected, (S,S)-diastereomers have a much wider SmCA* temperature range (0 to 101 °C) than the (R,S)-diastereomer of β-CH_3-TFMHPOBC analogues (72 ~ 103 °C). Thus, the SmCA* phase is stabilized by the favored bent (S,S)-diastereomers of β-CF_3-MHPOBC analogues. These results further strengthen our paradigm for bent conformation in AFLCs (32).

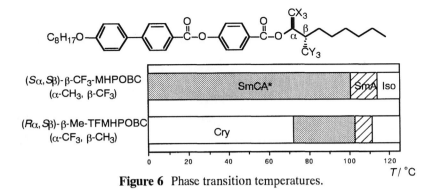

Figure 6 Phase transition temperatures.

Self-Assembly for Spontaneous Optical Resolution of Racemates.

The phase sequences and transition temperatures for β-CH₃-TFMHPOBC analogues **12** were taken with varying enantiomeric excesses for the *u*-(*R,S*)- and *l*-(*S,S*)-diastereomers, respectively. We found the intriguing feature that their phase sequence critically depends on the enantiomeric excesses (% ee) not only for (*R,S*)- but also for (*S,S*)-diastereomers (Figure 7) (33). Surprisingly, we also found a direct transition from SmA to a yet unknown (SmX) phase in the racemic region for either diastereomer.

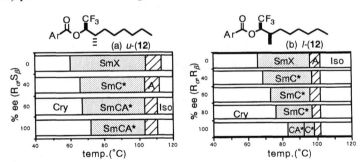

Figure 7 %Ee dependence of phase sequences.

Figure 8(a) shows the electrooptic response of the *u*-(*R*,S**)-racemate observed under the crossed polarizers so that one of the uniform domains becomes dark. The micrographs under the application of ±3V DC field are also shown in Figures 8(b) and 8(c), respectively. Generally, in the SmC phase of achiral or racemic compounds, molecules respond to a field only via dielectric interaction at high field, but no ferroelectric switching occurs. As shown in Figures 8(b) and 8(c), however, uniform domains partially grow as they corrode the others. The brightness of these domains switched when reversing the field. This corrosion corresponds to ferroelectric switching. With further increases in the field, the molecules rotated due to the dielectric anisotropy, changing the extinction direction normal to the smectic layer. This result clearly indicates that partial chiral resolution occurs and chiral (*R,S*)- and (*S,R*)-molecules form their chiral domains. We also detected a slight spontaneous polarization.

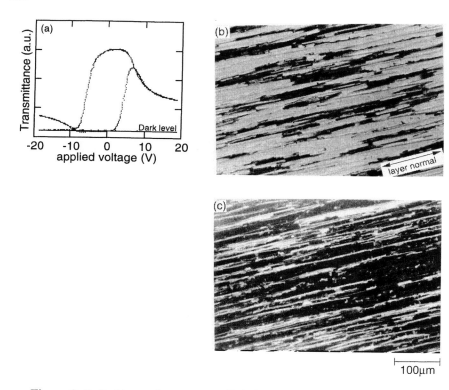

Figure 8 Optical transmittance *vs.* applied triangular wave electric field (90 °C).

A number of molecules show the SmC phase, where molecules tilt with respect to the layer normal. When the system is chiral, it has C_2 symmetry with the two-fold axis normal to the molecular tilting plane, so that a spontaneous polarization appears along this direction (34). In this ferroelectric phase, molecules can undergo switching due to ferroelectric interaction with an applied electric field. Even in an antiferroelectric phase where molecules in adjacent layers tilt in opposite directions and no net polarization exists, electrooptic switching occurs associated with the field-induced antiferroelectric to ferroelectric phase transition. When the system is racemic, *R* and *S* enantiomers are completely mixed in each smectic layer. Therefore, the racemate has, in principle, no net spontaneous polarization in a layer and shows no significant electrooptic response in either SmC or SmC_A. As expected, the racemic (TF)MHPOBC with single stereogenic center showed no significant electrooptic response. In sharp contrast, the more conformationally rigid diastereomeric β-methyl-TFMHPOBC analogues were found to exhibit distinct electrooptic switching even in their racemic composition (33), indicating spontaneous chiral resolution.

Finally, let's consider why this chiral resolution which no one has ever observed is detectable in the present racemate. The important factor may be the difference in their recognition energy. Moreover, it is assumed that the conformation of the double stereogenic part may be more firmly fixed in the diastereomer molecule (29c), compared with that of other chiral molecules with a single stereogenic center such as (TF)MHPOBC. Thus, the larger recognition energy containing more stable conformation bring about the spontaneous chiral resolution in the present diastereomer case. Actually, the racemic LC for (S*,S*)-diastereomer also exhibits electrooptic response (33), indicating chiral resolution (Figure 7(b)).

Just 150 years ago, Pasteur observed for the first time using optical microscopy that crystals of a sodium ammonium salt of "acide racemique" (NaNH₄tartrate·4H₂O) were spontaneously resolved into two enantiomorphous forms (35). This finding initiated the development of modern stereochemistry and the concept of molecular dissymmetry in nature (36). Ever since, molecular or supra-molecular dissymmetry have attracted chemists' interest to observe spontaneous chiral resolution from a racemate in a fluid state distinct from a static crystalline state. However, only a few examples have so far been reported. Echhardt et al. (37) deposited a monolayer film on a mica plate and observed the film by atomic force microscopy (AFM) to show chiral resolution. A similar two-dimensional (2-D) conglomerate has been observed from racemic liquid crystalline molecules on a crystalline graphite surface by STM (38). These results suggest a 2-D analog of Pasteur's system. However, it is believed that chiral resolution could not occur or could not be detected in fluid systems such as liquid crystals due to thermal fluctuation and/or molecular diffusion. Quite recently, Link et al. (39) reported the formation of a chiral layer(s) due to the tilt of achiral bent-core molecules (40). In their case, the formation of chiral domain originates from a uniform tilt of _achiral_ bent molecules within each smectic layer due to packing entropy (excluded volume) effect but not from accurate enantiomer recognition of the racemate. We have found, however, spontaneous chiral resolution of racemic CF₃-containing diastereomeric LCs into 3-D conglomerate in fluid liquid crystalline phase: the racemic LC for (S*,R*)-diastereomer spontaneously self-assembles into the enantiomeric domains to exhibit the electrooptic switching and striped textures indicating right- and left-handed enantiomers. It is important to note that this spontaneous optical resolution is not caused by tilting geometry as in the case of bent-core molecules but by molecular chiral discrimination. Thus, this is the first example of spontaneous chiral resolution in fluid liquid crystalline phase (41).

Conclusion

The enantio- and diastereo-selective catalysis of the fluoral-ene reaction provides a powerful methodology for catalytic asymmetric synthesis of fluorine containing compounds not only of biological importance but also of importance for other applications. Abiological applications of ene products are thus exemplified in the design and synthesis of diastereomeric liquid crystalline molecules.

Acknowledgment

The author is grateful to Drs. M. Maruta and A. Ishii of Central Glass Co., Mrs. Y. Suzuki and I. Kobayashi of Showa Shell Sekiyu K. K., Prof. H.

266

Takezoe and Drs. Y. Takanishi and S. Kawauchi of Tokyo Institute of Technology, and Drs. M. Terada and T. Yajima of my research group in Tokyo Institute of Technology.

Literatures cited

1. Reviews: (a) Ojima, I.; McCarthy, J. R.; Welch, J. T. Eds. *Biomedical Frontiers of Fluorine Chemistry*; ACS Books, ACS: Washington, D. C., **1996**. (b) Resnati, G.; Soloshonok, V. A. Eds. *Tetrahedron: Symposia-in-Print* **1996**, *52*, No. 1. (c) Kukhar, V. P.; Soloshonok, V. A. Eds. *Fluorine-containing Amino Acids: Synthesis and Properties*; John Wiley & Sons: Chichester, **1995**. (d) Hudlicky, M.; Pavlath, A. E. Eds.*Chemistry of Organic Flurine Compounds*; ACS Monograph 187, ACS: Washington, D. C., **1995**. (e) Banks, R. E.; Smart, B. E.; Tatlow, J. C. Eds. *Organofluorine Chemistry: Principles and Commercial Applications*; Plenum Press: New York, NY, **1994**. (f) Filler, R.; Kobayashi, Y.; Yagupolskii, L. M. Eds. *Organofluorine Compounds in Medicinal Chemistry and Biomedical Applications*; Elsevier: Amsterdam, **1993**. (g) Resnati, G. *Tetrahedron* **1993**, *49*, 9385.
2. For our asymmetric catalytic fluoral-ene reactions: (a) Mikami, K.; Yajima. T.; Terada, M. and Uchimaru, T. *Tetrahedron Lett.* **1993**, *34*, 7591. (b) Mikami, K.; Yajima, T.; Terada, M.; Kato, E. and Maruta, M. *Tetrahedron: Asymm.* **1994**, *5*, 1087. (c) Mikami, K.; Yajima, T.; Takasaki, T.; Matsukawa, S.; Terada, M.; Uchimaru, T. and Maruta, M. *Tetrahedron* **1996**, *52*, 85. Asymmetric catalytic aldol reactions with fluoral and difluoroacetaldehyde: (d) Mikami, K.; Takasaki, T.; Matsukawa, S. and Maruta, M. *Synlett* **1995**, 1057.
3. Asymmetric catalytic aldol reactions with fluorinated aromatic aldehydes and ketones: (a) Soloshonok, V. A. and Hayashi, T. *Tetrahedron Lett.* **1994**, *35*, 2713. (b) Soloshonok, V. A. and Hayashi, T. *Tetrahedron: Asymm.* **1994**, *5*, 1091. An asymmetric catalytic ene reaction with pentafluorobenzaldehyde: (c) Maruoka, K.; Hoshino, Y.; Shirasaka, T. and Yamamoto, H. *Tetrahedron Lett.* **1988**, *29*, 3967. Asymmetric catalytic addition reactions of dialkylzincs to fluorinated aromatic aldehydes: (d) Soai, K.; Hirose, Y. and Niwa, S. *J. Fluor. Chem.* **1992**. *59*, 5.
4. For leading references on negative HCJ, see: (a) Scheleyer, P. v. R.; Kos, A. J.; *Tetrahedron* **1983**, *39*, 1141. (b) Streitwieser, A. Jr.; Berke, C. M.; Schriver, G. W.; Grier, D.; Collins, J. B. *Tetrahedron, Suppl.* **1981**, *37*, (1)345. (c) Farnham, W. B.; Smart, B. E.; Middleton, W. J.; Calabrese, J. C.; Dixon, D. A. *J. Am. Chem. Soc.* **1985**, *107*, 4565. (d) Smart, B. E.; Middleton, W. J.; Farnham, W. B. *J. Am. Chem. Soc.* **1986**, *108*, 4905. (e) Farnham, W. B.; Dixon, D. A. ; Calabrese, J. C. *J. Am. Chem. Soc.* **1988**, *110*, 2607.
5. Reviews: (a) Mikami, K. *Pure Appl. Chem.* **1996**, *68*, 639. (b) Mikami, K. In *Advances in Asymmetric Synthesis*; Hassner, A., Ed.; JAI Press: Greenwich, Connecticut, **1995**; Vol. 1, p 1. (c) Mikami, K.; Terada, M. and Nakai, T. In *Advances in Catalytic Processes*; Doyle, M. P. , Ed.; JAI Press: Greenwich, Connecticut, **1995**; Vol. 1, pp 123. (d) Mikami, K.; Terada, M.; Narisawa, S. and Nakai, T. *Synlett* **1992**, 255. (e) Mikami, K. and Shimizu, M. *Chem. Rev.* **1992**, *92*, 1021.
6. Review on the ene-type addition reaction of CF3-containing compounds with achiral *Lewis* acid promoters: Nagai, T. and Kumadaki, I. *J. Synth. Org. Chem. Jpn.* **1991**, *49*, 624.

7. Diastereoselective aldol reactions with fluorinated carbonyl compounds: (a) Iseki, K.; Oishi, S. and Kobayashi, Y. *Chem. Lett.* **1994**, 1135. (b) Iseki, K.; Oishi, S.; Taguchi, T. and Kobayashi, Y. *Tetrahedron Lett.* **1992**, *33*, 8147.

8. Dale, J. A. and Mosher, H. S. *J. Am. Chem. Soc.* **1973**, *95*, 512.

9. Houk, K. N. *et al. Science* **1986**, *231*, 1108. Mikami, K. and Shimizu, M. 'Stereoelectronic Rules in Addition Reactions: "Cram's Rule" in Olefinic Systems' in *Advances in Detailed Reaction Mechanisms*; Coxon, J. M., Ed.; JAI Press: Connecticut, U. S. A. Vol. 3, **1993**.

10. (a) An *ab-initio* MO study on α-fluoro propanal in support of the *Felkin-Anh* model: Wong, S. S. and Paddonrow, M. N. *Chem. Commun.* **1990**, 456. (b) A semi-empirical and *ab-initio* MO study on fluoroketones: Linderman, R. J. and Jamois, E. A. *J. Fluor. Chem.* **1991**, *53*, 79.

11. Reed, A. E.; Weinstock, R. B. and Weinhold, F. *J. Chem. Phys.* **1985**, *83*, 735.

12. (a) Heathcock, C. H. In *Asymmetric Synthesis*; Morrison, J. D., Ed.; Academic Press: New York, **1984**, Vol. 3; Chapter 2. (b) Evans, D. A.; Nelson, J. V. and Taber, T. R. *Top. Stereochem.* **1982**, *13*, 1. (c) Heathcock, C. H.; Pirrung, M. C. and Sohn, J. E. *J. Org. Chem.* **1979**, *44*, 4294.

13. Nagai, T.; Hama, M.; Yoshioka, M.; Yuda, M.; Yoshida, N.; Ando, A.; Koyama, M.; Mikai, T. and Kumadaki, I. *Chem. Pharm. Bull.* **1989**, *37*, 177. Also see ref. 6.

14. Mikami, K.; Loh, T.-P. and Nakai, T. *Tetrahedron Lett.* **1988**, *29*, 6305.

15. Review: Plenio, H. *Chem. Rev.* **1997**, *97*, 3363.

16. *Ab initio* MO calculations reveal that bidentate transition structures in the reactions of 2-fluoropropanal or monofluoroacetaldehyde with lithium hydride are stabilized by the electrostatic attraction between Li and F: Wong, S. S. and Paddon-Row, M. N. *J. Chem. Soc. Chem. Commun.* **1991**, 327.

17. For general discussion on the "*cis* effect", see: (a) Houk, K. N.; Williams, J. C., Jr.; Mitchell, P. A.; Yamaguchi, K. *J. Am. Chem. Soc.* **1981**, *103*, 949. (b) Shulte-Elte, K. H.; Rautenstrauch, V. *J. Am. Chem. Soc.* **1980**, *102*, 1738. (c) Orfanpoulos, M.; Gardina, M. B.; Stephenson, L. M. *J. Am. Chem. Soc.* **1979**, *101*, 275.

18. (a) Furukawa, K.; Terashima, K.; Ichihashi, M.; Saitoh, S.; Miyazawa, K. and Inukai, T. *Ferroelectrics* **1988**, *85*, 451. (b) Hiji, N.; Chandani, A. D. L.; Nishiyama, S.; Ouchi, Y.; Takezoe, H.; Fukuda, A. *Ferroelectrics*, **1988**, *85*, 99. (c) Chandani, A. D. L.; Hagiwara, T.; Suzuki, Y.; Ouchi, Y.; Takezoe, H. and Fukuda, A. *Jpn. J. Appl. Phys.* **1988**, *27*, L729. (d) Chandani, A. D. L.; Gorecka, E.; Ouchi, Y.; Takezoe, H.; Fukuda, A. *Jpn. J. Appl. Phys.*, **1989**, *28*, L1265. (e) Fukuda, A.; Takanishi, Y.; Isozaki, T.; Ishikawa, K. and Takezoe, H. *J. Mater. Chem.* **1994**, 4, 997.

19. (a) Suzuki, Y.; Nonaka, O.; Koide, Y.; Okabe, N.; Hagiwara, T.; Kawamura, I.; Yamamoto, N.; Yamada, Y. and Kitazume., T. *Ferroelectrics* **1993**, *147*, 109. (b) Suzuki, Y.; Hagiwara, T.; Kawamura, I.; Okamura, N.; Kitazume, T.; Kakimoto, M.; Imai, Y.; Ouchi, Y.; Takezoe, H. and Fukuda, A. *Liq. Cryst.* **1989**, *6*, 167.

20. (a) Miyachi, et al., *Handbook of Liquid Crystals*, Vol. 2B, Wiley-VCH, p. 664 (1998). (b) Goodby, J. W. *Ferroelectric Liquid Crystals*, Gordon and Breach, Philadelphia, 1991. Nohira, H. *J. Synth. Org. Chem., Jpn.* **1991**, *49*, 467. (c) Skarp, K.; Handschy, M. A. *Mol. Cryst. Liq. Cryst.* **1988**, *165*, 439. (d) Clark, N. A. and Lagerwall, S. T. *Appl. Phys. Lett.* **1980**, *36*, 899. (e) Meyer, R. B. *Mol. Cryst. Liq. Cryst.* **1977**, *40*, 33.

21. Takezoe, H.; Fukuda, A.; Ikeda, A.; Takanishi, Y.; Umemoto, T.; Watanabe, J.; Iwane, H.; Hara, M. and Itoh, K. *Ferroelectrics* **1991**, *122*, 167.

22. Takanishi, Y.; Hiraoka, K.; Agrawal, V. K.; Takezoe, H.; Fukuda, A. and Matsushita, M. *Jpn. J. Appl. Phys.* **1991**. *30*. 2023.

23. Miyachi, K.; Matsushima, J.; Takanishi, Y.; Ishikawa, K.; Takezoe, H. and Fukuda, A. *Phys. Rev. E (Rapid Communication)* **1995**, *52*. R2153.

24. X-ray crystal analysis of bent conformations of AFLC molecules: (a) Hori, K. and Endo, K. *Bull. Chem. Soc. Jpn.* **1993**, 66, 46. (b) Hori, K.; Kawahara, S. and Ito, K. *Ferroelectrics* **1994**, *147*, 91. (c) Okuyama, K.; Kawano, N.; Uehori, S.; Noguchi, K.; Okabe, N.; Suzuki, Y. and Kawamura, I. *Mol. Cryst. Liq. Cryst.* **1996**, *276*. 193.

25. (a) Watanabe, J. and Hayashi, M. *Macromolecules* **1989**, *22*, 4083. (b) Watanabe, J. and Kinoshita, S. *J. Phys. II (France)*, **1992**, 2, 1237. (c) Watanabe, J.; Komura, H. and Niori, T. *Liq. Cryst.* **1993**, *13*, 455.

26. Takezoe, H.; Fukuda, A; Ikeda, A; Takanishi, Y.; Umemoto, T.; Watanabe, J.; Iwane, H.; Hara, M. and Itoh, K. *Ferroelectrics* **1991**, *122*, 167.

27. For vicinally stereogenic FLC molecules, see: β-Cl- or -Br-γ-Me: (a) Sakurai, T.; Mikami, N.; Ozaki, M. and Yoshino, K. *J. Chem. Phys.* **1986**, *85*, 585. α-Cl-β-Me: (b) Sakurai, T.; Mikami, N.; Higuchi, R.; Honma, M.; Ozaki, M. and Yoshino, K. *J. Chem. Soc., Chem. Commun.* **1986**, 978. (c) Yoshino, K.; Ozaki, M.; Kishio, S.; Sakurai, T.; Mikami, N.; Higuchi, R. and Honma, M. *Mol. Cryst. Liq. Cryst.* **1987**, *144*, 87. α-Me-β-Me: (d) Mikami, N.; Higuchi, R. and Sakurai, T. *J. Chem. Soc., Chem. Commun.* **1990**, 1561. β-OMe-γ-Me: (e) Mikami, N.; Higuchi, R. and Sakurai, T. *J. Chem. Soc. Chem. Commun.* **1992**, 643. α-CN-β-Me: (f) Kusumoto, T.; Nakayama, A.; Sato, K.; Hiyama, T.; Takehara, S.; Osawa, M. and Nakamura, K. *J. Mater. Chem*. **1991**, *1*, 707.

28. For the definition of u and l nomenclature, see: Prelog, V. and Helmchen, G. *Angew. Chem. Int. Ed. Engl.* **1982**, *21*, 567. Seebach, D. and Prelog, V. *Angew. Chem. Int. Ed. Engl.* **1982**, *21*, 654

29. (a) Mikami, K.; Yajima. T.; Siree, N.; Terada, M.; Suzuki, Y. and Kobayashi, I. Abstr., 3H218 in the 69th Annual Meeting of the Chemical Society of Japan, Tokyo, March, 1995. (b) Mikami, K.; Yajima. T.; Siree, N.; Terada, M.; Suzuki, Y. and Kobayashi, I. *Synlett* **1996**, 837. (c) Mikami, K.; Yajima. T.; Terada, M.; Kawauchi, S.; Suzuki, Y. and Kobayashi, I. *Chem. Lett.* **1996**, 861. (d) Kobayashi, I.; Suzuki, Y.; Yajima. T.; Kawauchi, S.; Terada, M. and Mikami, K. *Mol. Cryst. Liq. Cryst.* **1997**, *303*, 165.

30. Hagiwara, T.; Tanaka, K. and Fuchigami, T. *Tetrahedron Lett.* **1996**. *37*, 8187.

31. Quite recently, Prof. Fukuda has determined that the chiral alkyl chain makes a bent angle $>54.7°$ (the magic angle) with the core axis and that the carbonyl group near the chiral center lies on the tilt plane in antiferroelectric SmCA* on the basis of the polarized infrared spectroscopy. Jin, B.; Ling, Z.; Takanishi, Y.; Ishikawa, K.; Takezoe, H.; Fukuda, A.; Kakimoto, M.; Kitazume, T. *Phys. Rev.*, **1996**, E*53*, R4295.

32. Mikami, K.; Yajima. T.; Terada, M.; Kawauchi, S.; Suzuki, Y.; Kobayashi, I.; Takanishi, Y.; Takezoe, H. *Mol. Cryst. Liq. Cryst*. in press.

33. Mikami, K.; Yajima. T.; Terada, M.; Suzuki, Y. and Kobayashi, I. *J. Chem. Soc.,Chem. Commun* **1997**, 57.

34. Meyer, R. B.; Lieber, L.; Strzelcki, L.; Keller, P. *J. de Phys.* **1975**,*36*, L69.

35. Pasteur, L. *Ann. Chim. Phys.* **1848**, *24*, 442.
36. Pasteur, L. *"Researches on the Molecular Asymmetry (sic) of Natural Organic Products"* ("Recherches sur la Dissymmetric Moleculaire des Produits Organiques Naturels") Alembic Club Reprint No. 14, W. F. Clay, Edinburgh, UK, p. 43, (1860).
37. Echardt, C. J.; Peachy, N. M.; Swanson, D. R.; Takacs, J. M.; Khan, M. A.; Gong, X.; Kim, J.-H.; Wang, J.; Uphaus, R. A. *Nature* **1993**, *362*, 614.
38. Stevens, F.; Dyer, D. J.; Walba, D. M. *Angew. Chem. Int. Ed. Engl.* **1996,***35*, 900.
39. Link, D. R.; Natale, G.; Shao, R.; Maclennan, J. E.; Clark, N. A.; Korblova, E.; Walba, D. M. *Science* **1997**, *278*, 1924.
40. Niori, T.; Sekine, T.; Watanabe, J.; Furukawa, T.; Takezoe, H. *J. Mater. Chem.* **1996**, *6*, 1231.
41. Takanishi, Y.; Takezoe, H.; Suzuki, Y.; Kobayashi, I.; Yajima, T.; Terada, M.; Mikami, K. *Angew. Chem. Int. Ed. Engl.* in press.

ASYMMETRIC FLUOROORGANIC CHEMISTRY IN AGROCHEMISTRY AND PHARMACY

Chapter 19

Chiral 3-Aryl-4-halo-5-(trifluoromethyl)pyrazoles

Synthesis and Herbicidal Activity of Enantiomeric Lactate Derivatives of Aryl-Pyrazole Herbicides

Bruce C. Hamper, Kindrick L. Leschinsky, Deborah A. Mischke, and S. Douglas Prosch

Monsanto Life Science Company, AG Sector, 800 North Lindbergh Boulevard, St. Louis, MO 63167

The 3-aryl-4-halo-5-(trifluoromethyl)pyrazoles are a structurally unique class of herbicides and have resulted in the identification of commercial candidate JV 485. These compounds are potent inhibitors of protoporphyrinogen oxidase and result in rapid necrosis of plant tissue in pre- and postemergent applications. The herbicidal activity of the enantiomers of chiral phenoxylactate derivatives **1** and lactate esters **2**, obtained either by synthesis from enantiomerically pure lactate derivatives or by resolution using preparative HPLC, were compared in side by side preemergent and postemergent tests. The enantiomeric purity of **1** and **2** has been determined by HPLC using a chiral stationary phase and ^{1}H NMR employing a chiral solvating agent.

In the early 1980's, we became interested in exploring the chemistry of fluorinated heterocycles as potential intermediates for new biologically active materials (*1*). Introduction of the trifluoromethyl group into aromatic or heterocyclic rings (*2*) has resulted in the discovery of numerous agrochemical and medicinal products including acifluorfen, fluazifop, trifluralin, dithiopyr and, most recently, the COX-2 inhibitor celecoxib (*3*). We have made extensive use of trifluoromethyl-acetylenes and ketones to prepare herbicidal pyrazole phenyl ethers and isoxazoles (*4,5*). Both the pyrazole phenyl ethers and isoxazoles were found to inhibit protoporphrin IX oxidase (PROTOX), a well established mode of action for diphenyl ether and N-phenylimide herbicides (*6,7*). Extension of our understanding of the structural requirements for PROTOX inhibitors led to the discovery of 3-aryl-4-halo-5-(trifluoromethyl)pyrazoles as highly active herbicides (*8*) and the identification of JV 485 (Figure 1) as a commercial candidate for broad spectrum pre-emergent weed control in winter wheat (*9*).

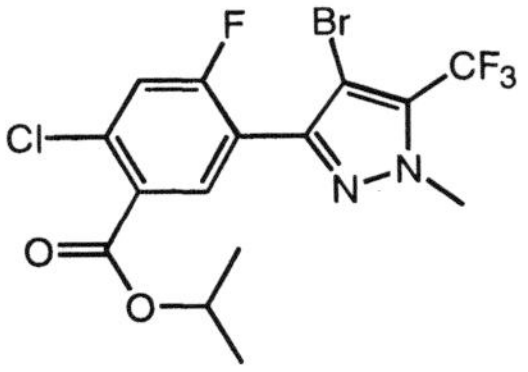

JV 485
Pre-emergent Herbicide
Broad Spectrum Control in Winter Wheat
Jointly Developed by
Monsanto Company and Bayer AG

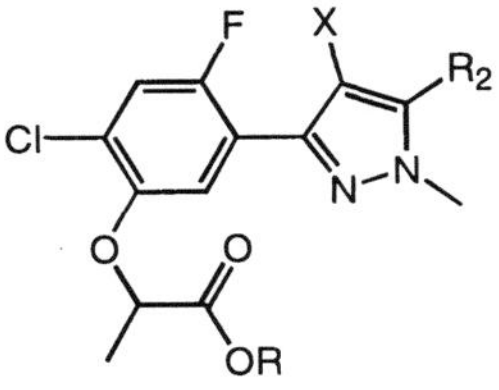

1

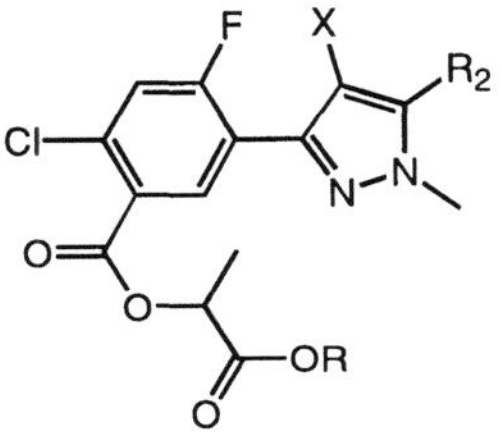

2

Figure 1. Chiral Derivatives of 3-Arylpyrazole Herbicides

In the course of investigating the structure-activity relationships of the arylpyrazoles, a number of chiral derivatives were prepared including phenoxylactates **1** and lactate esters **2**. Differences in the herbicidal activity of the enantiomers of chiral PROTOX inhibitors has been previously established for derivatives of pyrazole phenyl ethers and N-phenylimide herbicides (*10*). As has been observed in the pharmaceutical industry, there is great potential for the commercialization of enantiomerically pure versions of both new and existing chiral agricultural products. Recently, enantiomerically pure or enriched 2-aryloxypropionate herbicides fluazifop P-butyl and quizalofop P-ethyl (Fusilade 2000® and ASSURE II®) have been introduced as new versions (sometimes termed a 'racemic switch') replacing the previous racemic versions of these products (*11*). The advantages of improved unit activity resulting from the use of a single enantiomer can result in significant cost savings and in some instances can alleviate undesirable effects of the inactive or less active enantiomer. These potential advantages led to our investigation of the synthesis, resolution and comparative herbicidal activity of chiral lactate derivatives **1** and **2**.

Preparation of the lactate ester and ether derivatives required acetophenones **4** and **10** with the proper 2,4,5 substitution pattern (Figure 2). The 5'-carbon derivatives were obtained from 2-fluoro-4-chloro-5-methylacetophenone **4** which can be prepared by direct acylation of the toluene **3** (*12*). Preparation of the acetophenone **10** required for the 5'-oxygen derivatives was not as straightforward (*13*). Acylation of the anisole **6** led to a mixture of products arising from initial cleavage-acetylation of the methyl ether followed by Fries rearrangement to give **7** and partial acetylation of the phenol to give **8**. Cleavage of the ether can be avoided by Friedel-Crafts alkylation with dichloromethyl methyl ether to provide aldehyde **9**. The aldehyde was converted to

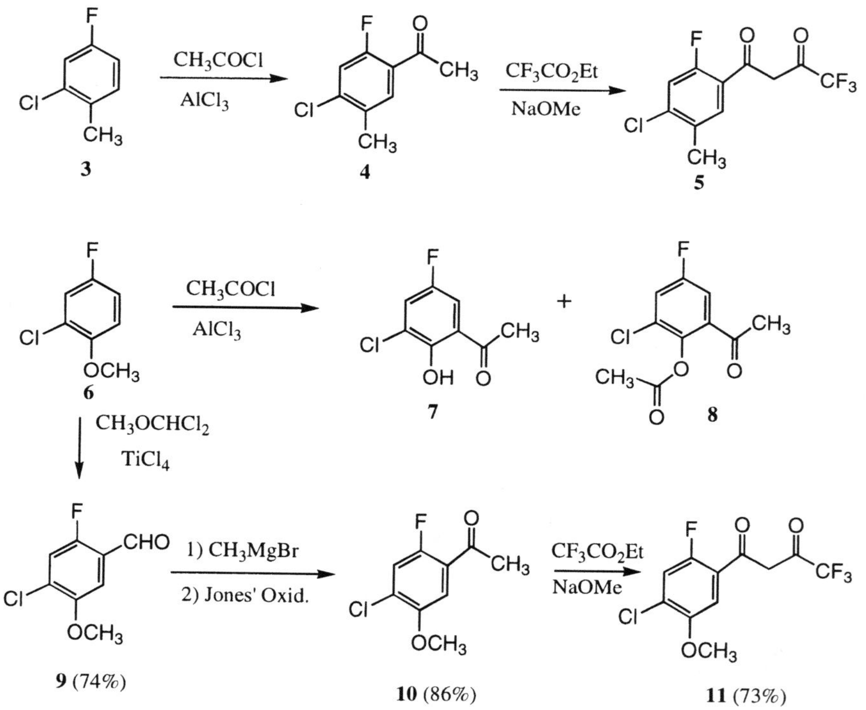

Figure 2. Preparation of 2,4,5-Substituted Acetophenones

acetophenone **10** by addition of methyl Grignard followed by Jones' oxidation. Both acetophenones **4** and **10** were treated with trifluoroacetate to afford diketones **5** and **11**, respectively.

The diketones **12** were converted to pyrazoles either by cyclocondensation with methyl hydrazine or by a two step process with hydrazine followed by methylation to give regioisomers **14** and **15** (Figure 3). Previous studies had shown that compounds derived from the 3-arylpyrazole **15** were significantly more active than those prepared from the 5-arylpyrazole **14** (*14*). Therefore, an effort was made to find regioselective routes towards the 3-arylpyrazole isomer. Direct cyclocondensation provided the 5-aryl **14** as the major product. Treatment with hydrazine gave the N-H pyrazole **13** which can be alkylated under basic conditions to give the 3-aryl isomer **15** as the major product. In the absence of base, pyrazole **13** was treated with methyl sulfate in toluene at reflux to afford 3-arylpyrazole **15** with less than 5% of the undesired 5-aryl isomer (*15*).

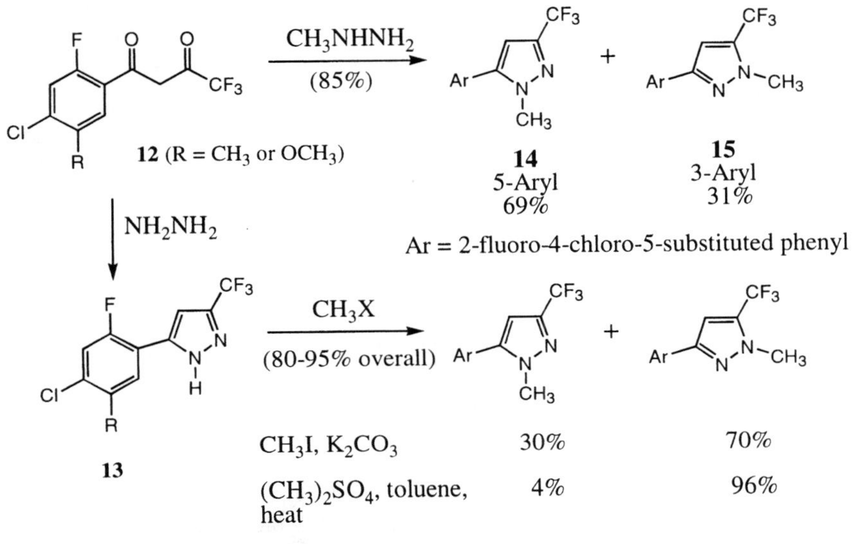

CH₃I, K₂CO₃	30%	70%
(CH₃)₂SO₄, toluene, heat	4%	96%

Figure 3. Regioselective Pyrazole Synthesis

Phenoxylactate derivatives 1 were prepared from 5'methoxyphenylpyrazole 16 by cleavage of the methylether and alkylation with suitable chiral lactate building blocks (Figure 4). Treatment of 16 with chlorine followed by hydrolysis of the methyl ether with BBr₃ provided the phenolic precursor 18. Enantiomerically enriched lactate derivatives of diphenyl ether herbicides have been previously prepared from the phenols by S_N2 displacement of either bromo- or tosylpropionates (16). Either of these substrates can give partial racemization at the stereogenic center due to competition of non-selective SN1 displacement processes. The sulfonates, however, typically afford greater enantioselectivity than the bromides and we felt that by careful control of the reaction conditions that these intermediates would give the best results. A crystalline (S)-(-)-tosylate was obtained from (S)-ethyl lactate that can be stored without loss of enantiomeric purity. The corresponding (R)-lactate was only available as an isobutyl ester and the resulting (R)-tosylate was obtained as an oil. This oil was purified chromatography and used immediately. Tosylate derivatives of lactate esters were found to provide the highest enantioselectivity for the S_N2 displacement using acetonitrile as a solvent. The CF₃ pyrazole (R)-20 was obtained in 99.2% ee as determined by HPLC using a chiral stationary phase (vida infra). In a similar manner, the enantiomer (S)-21 was obtained by displacement of the (R)-isobutyl tosylate in 98.8% ee. For comparison of the relative herbicidal activity of the ethyl esters, the isobutyl ester 21 was converted by transesterification to (S)-22.

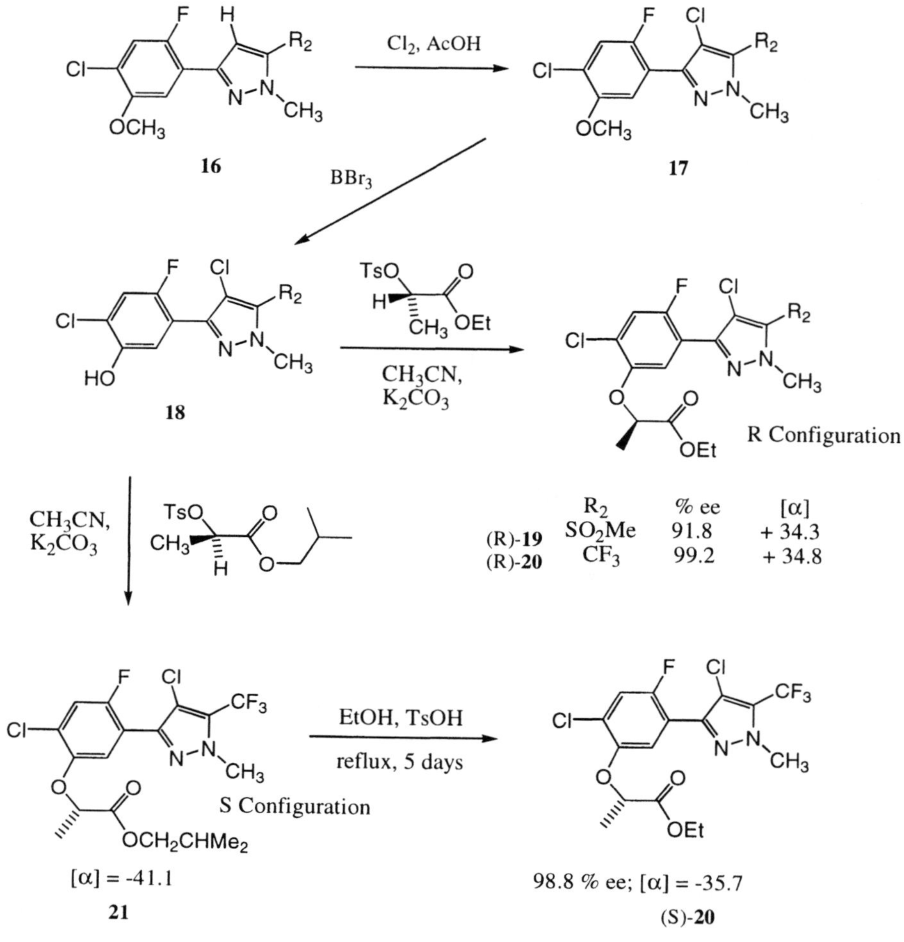

Figure 4. Synthesis of Phenoxylactate Derivatives

Lactate esters of the 5'-carbon phenylpyrazoles **2** and JV485 were prepared from common intermediate **23** (Figure 5). Conversion of the 5'-methyl group of phenylpyrazole **23** to carboxylic acid **24** was achieved by 'Mid-Century oxidation' with molecular oxygen in the presence of cobalt and manganese catalysts (*17*). Halogenation of **24** with chlorine gas or bromine in acetic acid gave **25** which can be esterified to provide either JV 485 or other derivatives such as chiral lactate ester **26**.

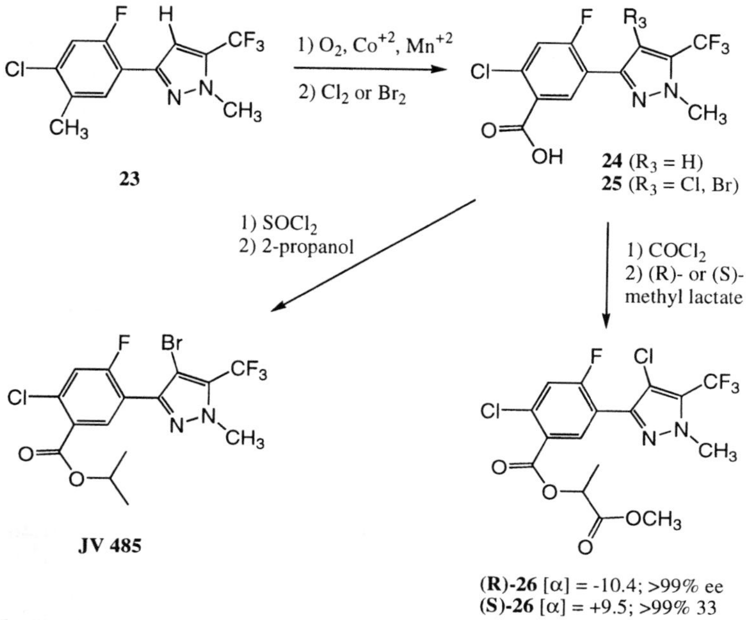

Figure 5. Synthesis of Lactate Ester Derivatives

Enantiomeric purity of **1** and **2**, as well as many of the chiral intermediates, was determined by HPLC using a chiral stationary phase (Table 1). The commercially available covalent phenylglycine (CSP 1) and Whelk-O (CSP 2) were found to provide resolution of these analytes using normal phase conditions. For most of the compounds, CSP 2 provided better resolution as determined by the chromatographic separation factor (α). Baseline resolution was obtained for all the listed compounds except for ethyl lactate enantiomers **20** and **22** (X = O, R_1 = CH$_3$, R_2 = OEt). In this case, better resolution was obtained using CSP 1 which gave a separation factor of 1.13. In general, CSP 2 provided better resolution of the amide derivatives than the esters. The 5'-diester **26** was also well resolved by CSP 2. A preparative scale column (10 mm i.d. x 25 cm) containing CSP 1 was used for the resolution of a racemic mixture of **20** (Table 2). Direct chromatographic resolution is particularly attractive for compounds in which there are not readily available chiral precursors and has been used for resolution of related phenoxybutyrate compounds (*18*). Using an automated injection and collection system, the eluent from the first peak of a total of 54 runs (16 mgs per run) was collected and combined. Although the first few runs appeared to provide baseline resolution, the combined fractions for the first eluted peak were a 80:20 mixture of enantiomers. By reducing the amount per injection to 2.5 mg and increasing the total runs to 246, we were able to collect 0.27 g of (R)-**20**. The combined second fractions were enriched in the S enantiomer, but were of lower enantiomeric purity.

Table 1. HPLC Resolution of Chiral Phenylpyrazoles.

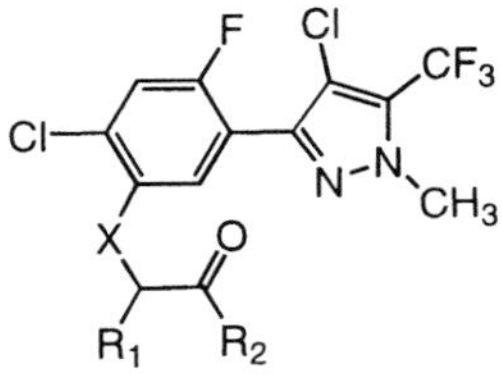

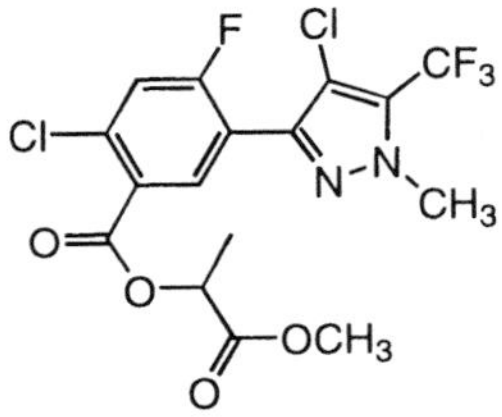

26

X	R_1	R_2	CSP 1 α	CSP2 α
O	CH_3	NH_2	NS	1.25
O	CH_3	$NHCH_3$	1.05	1.38
O	CH_3	$NHCH_2CH_2CO_2CH_3$	1.05	1.46
O	Et	$NHCH_3$	1.04	1.38
NH	CH_3	OEt	NS	1.41
O	CH_3	OEt	1.13	1.05
O	Et	OCH_3	1.13	1.14
5'-diester			1.08	1.38

CSP 1 - covalent phenylglycine; CSP 2 - Whelk-O 1

Table 2. Preparative Chromatographic Resolution by HPLC on a Chiral Stationary Phase.

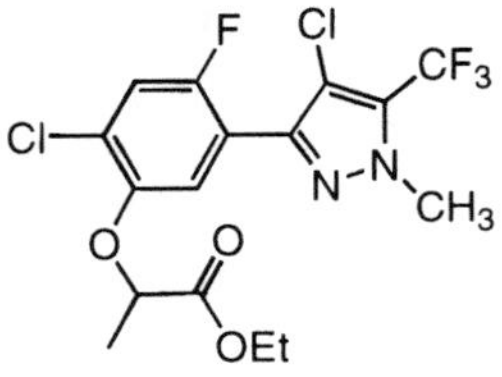

20

amount per injection	total runs	recovered wt.	enantiomeric purity	configuration
16 mg	54	-	60 % ee	enriched in R
2.5 mg	246	0.27 g	97.3 % ee	R
			43.7 % ee	enriched in S

The ^{1}H NMR spectra of lactate derivatives **1** were investigated in the presence of a chiral solvating agent (CSA), (R)-(+)-2,2,2-trifluoro-1-(9-anthryl)ethanol (Table 3). Addition of the CSA to mixtures enriched in one enantiomer gave non-equivalent NMR resonances for almost all of the observable proton nuclei of **20**. The proton signals for the alkyl ester and the methyl group attached to the stereogenic center of the R enantiomer appear upfield of the S enantiomer for enriched samples of **19** and **20**. Non-equivalence was greatest for the methylene protons of the ester (d), the lactate methyl group (c) and the aromatic proton ortho to the lactate functionality (b). NMR non-equivalence was used as a means of assigning absolute configuration to related chiral phenoxy derivatives such as butyrates.

Table 3. Proton NMR Non-equivalence of the Enantiomers of Phenoxylactate Derivatives in the Presence of a CSA.

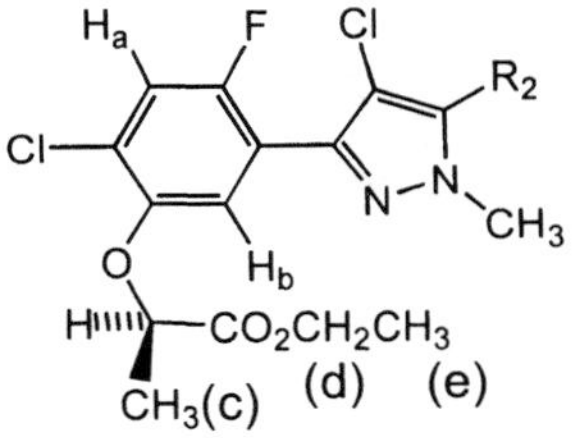

20 ($R_2 = CF_3$)

19 ($R_2 = SO_2Me$)

R_2	Proton Assignment	chemical shift (δ, major peak)	$\Delta\delta$	(R); sence of non-equivalence
CF3	H_a	7.05	0.030	upfield
	H_b	6.87	0.039	upfield
	CH_2 (d)	4.51	0.062	upfield
	CH_3 (c)	1.50	0.037	upfield
	CH_3 (e)	1.12	0.009	upfield
SO2Me	CH_3 (c)	1.28	0.018	upfield
	CH_3 (e)	0.69	0.004	upfield

Herbicidal activity was determined for the enantiomerically pure compounds in both preemergent and postemergent screens (Figure 6). The activity is expressed as a GR_{80} value which indicates the amount of material, in grams per hectare, required to inhibit 80% of plant growth relative to an untreated control. Values were averaged across a representative set of narrow leaf (NL) and broadleaf (BL) species. In the case of the phenoxylactates **20**, the R enantiomer is significantly more active averaging an 8X improvement depending on the particular weed species. The lactate esters **26**, however, are nearly identical in activity. Enantiomeric purity is critical for the evaluation of the relative activity, since even a 5% impurity will limit the difference in the comparative tests to 20X. The enantiomeric purity of the samples of **20** and **26** used in these tests was at least 99%, which would allow for the detection of an 100X difference in activity in the case of one totally inactive enantiomer. Clearly, there is a significant difference in the sensitivity of plants to the enantiomers of lactate ethers (derivatives **1**) and the lactate esters (derivatives **2**). Although the mechanism accounting for the difference is unknown, the esters **26** could undergo non-selective ester hydrolysis to the achiral carboxylic acid resulting in equivalent activity for both enantiomers. If so, the conversion would need to be quite fast, since these are very fast acting herbicides. An *in vitro* test of the PROTOX activity of the enantiomers of **26** could determine whether the R and S forms have the same activity or are both converted to a non-enantiomeric derivative *in vivo*. In a related study, the enantiomeric pairs of isoxazole PROTOX inhibitors were found to have identical activity at the enzyme level, but different activities *in vivo*. (*6*)

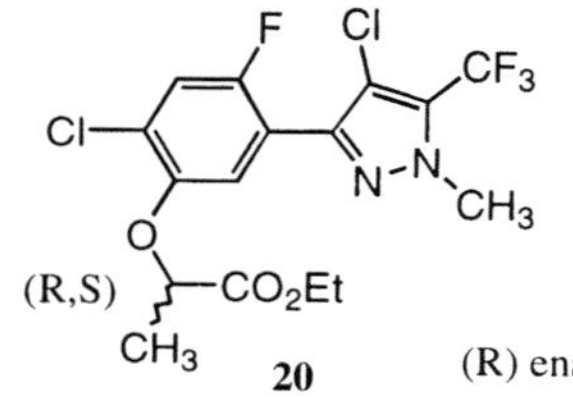

		Preemergent GR80 (g/ha)		Postemergent GR80 (g/ha)	
		NL	BL	NL	BL
20	(R)	800	50	200	10
	(S)	>5000	250	>5000	120

(R) enantiomer is 5 to 20 X more active

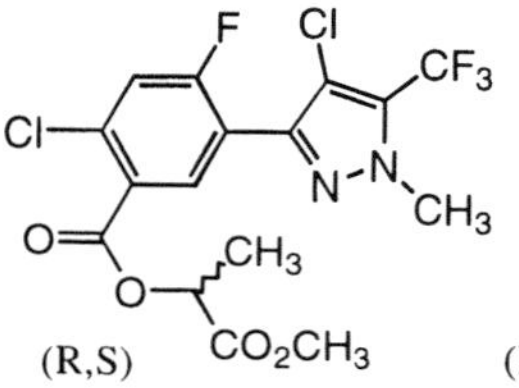

		Preemergent GR80 (g/ha)		Postemergent GR80 (g/ha)	
		NL	BL	NL	BL
26	(R)	175	70	60	8
	(S)	250	50	>60	8

(R) and (S) are nearly identical

Figure 6. Relative Herbicidal Activity of Chiral Phenylpyrazoles

The herbicidal activity of the enantiomers of a number of chiral 3-aryl-4-halo-5-(trifluoromethyl)pyrazoles has been investigated. It is advantageous that one obtain enantiomerically pure materials (in excess of 98% ee) in order to obtain meaningful biological data, since even minor amounts of an 'active' enantiomer impurity can give rise to biological activity for the 'inactive' enantiomer. Compounds were obtained by direct synthesis and by preparative HPLC using a chiral stationary phase. Assignments of enantiomeric purity were obtained by synthesis from known precursors or by a combination of ^{1}H NMR and HPLC determination using CSAs and elution order from CSPs, respectively. We have already seen the impact of enantiomeric agricultural products in the marketplace and will need to evaluate the potential activity of enantiomers of new candidates in the future. Currently, we are investigating enantiomers of biologically active compounds in our screens as a standard practice whenever possible.

Literature Cited

1. Welch, J. T. *Tetrahedron* **1987**, *43*, 3123-3197.
2. Elguero, J.; Fruchier, A.; Jagerovic, N.; Werner, A. *Organic Preparations and Procedures Int.* **1995**, *27*, 33-74.
3. Penning, T. D.; Talley, J. J.; Bertenshaw, S. R.; Carter, J. S.; Collins, P. W.; Docter, S.; Graneto, M. J.; Lee, L. F.; Malecha, J. W.; Miyashiro, J. M.; Rogers, R. S.; Rogier, D. J.; Yu, S. S.; Anderson, G. D.; Burton, E. G.; Cogburn, J. N.; Gregory, S. A.; Koboldt, C. M.; Perkins, W. E.; Seibert, K.; Veenhuizen, A. W.; Zhang, Y. Y.; Isakson, P. C. *J. Med. Chem.* **1997**, *40*, 1347-1365.
4. Lee, L. F.; Moedritzer, K.; Rogers, M. D. *U. S. Patent* 4,855,442, August 8, 1989.
5. Hamper, B. C.; Leschinsky, K. L.; Massey, S. S.; Bell, C. L.; Brannigan, L. H.; Prosch, S. D. *J. Agric. Food Chem.* **1995**, *43*, 219-228.
6. Dayan, F. E.; Duke, S. O.; Reddy, K. N.; Hamper, B. C.; Leschinsky, K. L. *J. Agric. Food Chem.* **1997**, *45*, 967-975.
7. Dayan, F. E.; Duke, S. O. *Herbicide Activity: Toxicology, Biochemistry and Molecular Biology;* R. M. Roe, et.al. (Eds.) IOS Press, 1997, pp. 11-35.
8. Woodard, S. S.; Hamper, B. C.; Moedritzer, K.; Rogers, M. D.; Mischke, D. A.; Dutra, G. A. *U. S. Patent* 5,281,571, 1994.
9. Prosch, S. D.; Ciha, A. J.; Grogna, R.; Hamper, B. C.; Feucht, D.; Dreist, M. *Proc. Br. Crop Prot. Conf.-Weeds* **1997**, Issue 1, 45-50.
10. Nandihalli, U. B.; Duke, M. V.; Ashmore, J. W.; Musco, V. A.; Clark, R. D.; Duke, S. O. *Pestic. Sci.* **1994**, *40*, 265-277; Okamoto, M.; Sato, R.; Nagano, E.; Nakazawa, H. *Agric. Biol. Chem.* **1991**, *55*, 3151-3153.
11. Zeiss, H. J.; Mildenberger, H. Eur. Pat. Appl. EP 492629, July 1, 1992; Dicks, J. W.; Slater, J. W.; Bewick, D. W. *Proc. Br. Crop Prot. Conf.-Weeds* **1985**, Issue 1, 271-80.
12. Hamper, B. C.; Leschinsky, K. L. *U. S. Patent* 5,532,416, July 2, 1996.
13. Hamper, B. C.; Leschinsky, K. L. *U. S. Patent* 5,600,008, February 4, 1997.
14. Hamper, B. C.; Mischke, D. A.; Leschinsky, K. L.; McDermott, L. L.; Prosch, S. D. *Eighth International Congress of Pesticide Chemistry* Conference Proceedings Series, American Chemical Society, Washington, D. C., 1995, pp 42-48.
15. Hamper, B. C.; Mao, M. K.; Phillips, W. G. *U. S. Patent* 5,698,708, Dec. 16, 1997.
16. Bauer, K.; Bieringer, H.; Hacker, E.; Koch, V.; Willms, L. *J. Agri. Food Chem.* **1990**, *38*, 1071-1073.
17. Chupp, J. P.; Hamper, B. C.; Wettach, R. H. *U. S. Patent* 5,587,485, Dec. 24, 1996.
18. Hamper, B. C.; Dukesherer, D. R.; Moedritzer, K. *J. Chromatogr.* **1994**, *666*, 479-484.

Chapter 20

Chiral Fluorinated Anesthetics

Keith Ramig

Department of Natural Sciences, Baruch College, City University of New York, Box A–0506, 17 Lexington Avenue, New York, NY 10010

The syntheses of fluorinated anesthetic enantiomers are summarized. All strategies start with preparation and resolution of a chiral halogenated carboxylic acid. Then, the carboxylic acid group is converted to either a fluoroalkyl group or a hydrogen atom. In the latter case where the carboxylic acid group functions as a latent hydrogen atom, the carboxyl group may be attached directly to the chiral center. In the subsequent decarboxylation step, a very high degree of stereoselectivity is seen.

Four of the currently marketed inhaled anesthetics, halothane, enflurane, isoflurane, and desflurane (Scheme 1), contain a chiral carbon atom but are administered as racemic mixtures. In the late 1980's, an ambitious project was begun in the discovery group of Anaquest (1) in Murray Hill, NJ: the synthesis and preliminary pharmacological testing of individual enantiomers of the fluoroether anesthetics isoflurane and desflurane, and some analogues. There had been previous efforts (2) at obtaining individual enantiomers of halothane, but the product was of relatively low enantiomeric enrichment. This made the initial disclosure (3) of the pharmacology of the halothane enantiomers suspect. Even with the exciting developments in the field of asymmetric synthesis and chiral resolution, these molecules had resisted preparation in highly enantioenriched form because they are so small and simple, lacking the useful functionality needed.

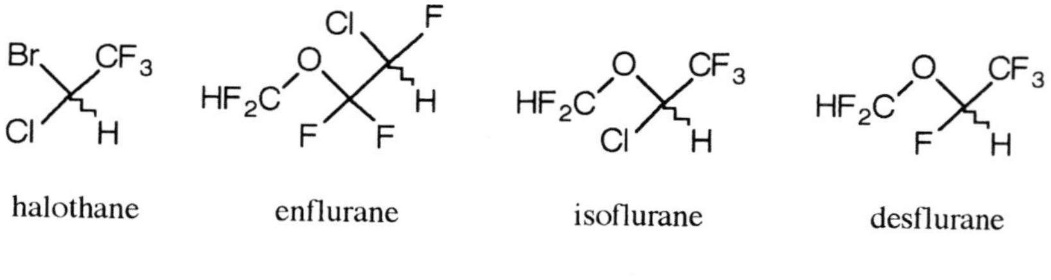

Scheme 1. Chiral Racemic Inhaled Anesthetics

There are two good reasons for the synthesis of anesthetic enantiomers. From a purely practical standpoint, it is possible that the desirable anesthetic properties may reside in only one of the enantiomers. If this is the case, then a new marketable pharmaceutical may emerge. From a scientific standpoint, chiral non-racemic bio-active molecules are always needed for the study of the physiological mechanism of action of a drug. This is especially the case for anesthetics, as not much is known about their mode of action, although a receptor-based mechanism is strongly suggested by the finding that the potency of isoflurane is stereospecific (4).

Meinwald and Pearson were the first to disclose preparation of highly pure anesthetic enantiomers (5). Using an improvement of the procedure of Edamura and Larsen (2), they were able to prepare both enantiomers of halothane and enflurane with high enantiomeric purity (e.p., defined as relative amount of major enantiomer vs. minor enantiomer x 100%) (Scheme 2). First, an appropriate chiral halogenated acid derivative was resolved by fractional crystallization of the α–methylbenzylamine salt. After obtaining either enantiomer with high enantiomeric purity, the absolute configuration of each acid was determined. For the case of chlorofluoroacetic acid, this information was already known (6), and the absolute configuration of bromochloroacetic acid was determined by an X-ray crystal structure. Then, the carboxyl group was converted to either a trifluoromethyl group or a fluoroether group under conditions that did not epimerize the chiral center. By this method, each enantiomer was obtained in high enantiomeric purity, and the absolute configurations of halothane and enflurane were proven. At that time and even to this day, the absolute configurations of many chiral molecules which are highly halogenated at the chiral center are not known. This type of information about these small molecules should be of great theoretical interest for researchers probing the nature of chirality.

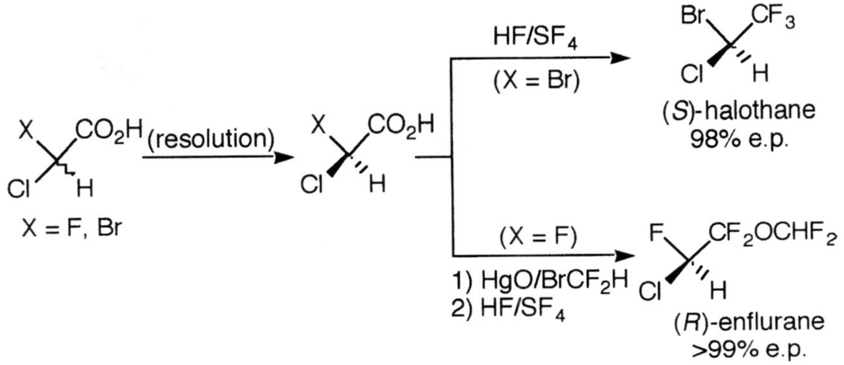

Scheme 2. Meinwald and Pearson's Synthesis of Halothane and Enflurane Enantiomers

Meinwald and Pearson in collaboration with König made another advancement in the field of chiral anesthetics when they discovered that these molecules could be resolved on an analytical scale using capillary gas chromatography (GC) on derivatized cyclodextrin stationary phases (7). Before this method became available, the only method of determining the enantiomeric purity of the anesthetics was based on optical rotation. Because the optical rotations of halothane and enflurane are not large, uncertainty and inaccuracy were sure to have plagued these results. This new GC method allows quick and accurate determination of the enantiomeric purity of an anesthetic. Subsequently, we (8) and others (9) have reported the analytical, and in some cases preparative, separation of anesthetic enantiomers and derivatives using GC with various cyclodextrin-based stationary phases.

Halothane, enflurane, and isoflurane enantiomers have also been separated preparatively by enantioselective desorption from clathrate complexes, albeit with relatively low enantioenrichment. Various researchers have used molecules such as brucine (5,10), α–cyclodextrin (11), and tri-o-thymotide (12) as enantioselective complexing agents. The highest degree of success of which we are aware is isolation of isoflurane with an enantiomeric purity of 96%, after five successive cycles using brucine as the resolving agent (Halpern, D.; unpublished results).

The first syntheses of highly pure enantiomers of isoflurane (13) and desflurane (14) were reported by Huang and Rozov (Scheme 3). The strategy here was somewhat different from that used to synthesize halothane and enflurane. First, a chiral halogenated carboxylic acid was prepared and resolved by fractional crystallization of the dehydroabietylamine salt. Each enantiomer was obtained with very high enantiomeric purity. To produce the anesthetic enantiomer, the carboxyl group had to be replaced by a hydrogen atom. This was accomplished by pyrolysis of the potassium salt of the acid in a protic media. Anionic decarboxylation of highly fluorinated acid derivatives is a well-known reaction (15), and this particular version delivered either enantiomer of isoflurane with no epimerization of the chiral center, despite the harsh and highly caustic conditions. In collaboration with Anaquest researchers, Polavarapu then determined the absolute configuration of isoflurane using a method based on circular dichroism spectra (16).

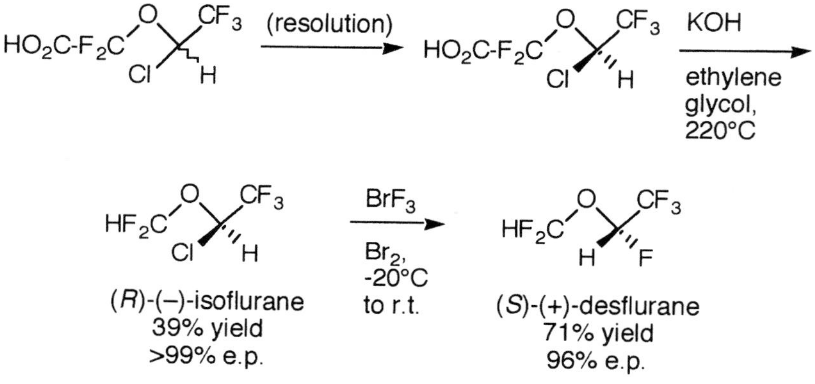

Scheme 3. Synthesis of Isoflurane and Desflurane Enantiomers

The first synthesis of desflurane enantiomers (14) was accomplished by an apparently simple substitution reaction (Scheme 3). The best conditions produced desflurane from isoflurane with 96% stereoselectivity. However, these best conditions were the result of much trial and error. Use of the fluorination system of HF/SbCl$_5$ in liquid HF gave a low yield of racemic desflurane. Also, when the unconverted isoflurane was isolated and analyzed, it was found to be partially racemized itself. This pointed to intervention of an S$_N$1-type of mechanism, which was undoubtedly favored by the highly polar solvent. The switch to the fluoride source BrF$_3$ in the non-polar bromine as solvent apparently caused the reaction to proceed by an S$_N$2-type mechanism, as evidenced by the almost complete inversion of configuration.

After these initial successes, there was a need for improvements in the methodology. For example, the resolution process in the isoflurane synthesis was laborious because the chiral center in the halogenated acid is two atoms removed from the carboxyl group. For ease of chiral resolution, the carboxyl group should be directly attached to the chiral center so that the resolving agent is as close as possible to the substrate's chiral center when the diastereomeric salts are formed. However, that would necessitate a stereoselective decarboxylation reaction as part of the synthesis, if the carboxyl group is to be used as a latent hydrogen atom (Scheme 4). A few examples of such reactions are known (*17*), but only in certain cases was the stereoselectivity at a synthetically useful level. Most pertinent to the current investigation was the example of bromochlorofluoroacetic acid, which was partially resolved and decarboxylated to give bromochlorofluoromethane with a useful level of stereoselectivity (*18*). At the time, the enantiomeric purity of the product was not known with great certainty because it was based on optical rotation values. König was the first to report the analytical-scale enantiomeric separation of bromochlorofluoromethane using capillary GC on a cyclodextrin-derived stationary phase (*19*). Using this method of product analysis, subsequent studies have shown that the reaction which produces this very simple chiral molecule proceeds with 89% retention of configuration (*20*).

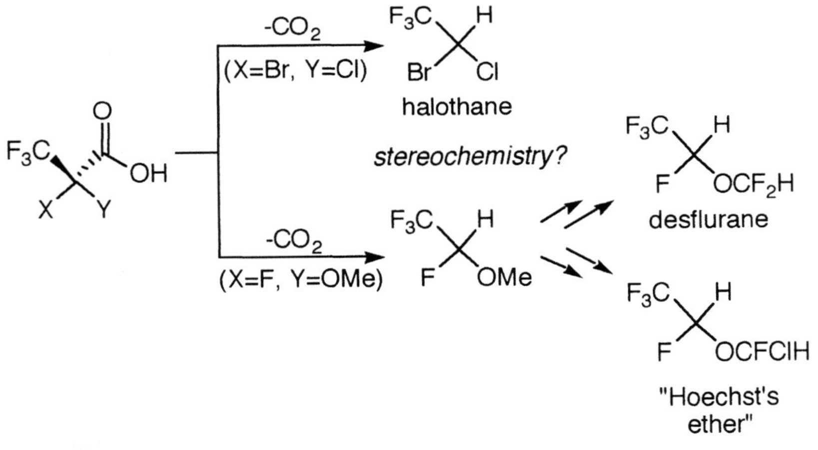

Scheme 4. Enantiomer Synthesis Using Stereoselective Decarboxylation

This encouraging precedent led us to develop new asymmetric syntheses of halothane (*21*) and desflurane plus the desflurane analogue "Hoechst's ether" (*22*). The cornerstone of the second-generation syntheses is stereoselective decarboxylation. In both cases, a chiral halogenated acid derivative was prepared which had the carboxyl group attached directly to the chiral center. Rather than use salt formation and repeated crystallization for resolution of the acids, we chose to form secondary amides and use chromatography for separation of the resulting diastereomers. The diastereomerically enriched amides would then be converted to enantiomerically enriched acids to be used as decarboxylation substrates.

The acid halide used in the halothane synthesis (*21*) was prepared in racemic form by classical organofluorine chemistry (Scheme 5) (*23*). Treatment of the acid chloride with (S)-α-methylbenzylamine gave two diastereomeric amides which were separated by preparative HPLC. Up to 4g/run of mixture could be separated, giving a faster-eluting secondary amide with 98.5% isomeric purity. The slower-eluting secondary amide had an isomeric purity of 93%. We then converted the faster-eluting amide to the required acid by the two-step process of debenzylation with sulfuric acid and hydrolysis of the resulting primary amide. At this point, we were unable to determine directly the enantiomeric purity of the acid, or its absolute configuration. However, there was no reason to believe that any scrambling had occurred during the conversion of the secondary amide to the acid.

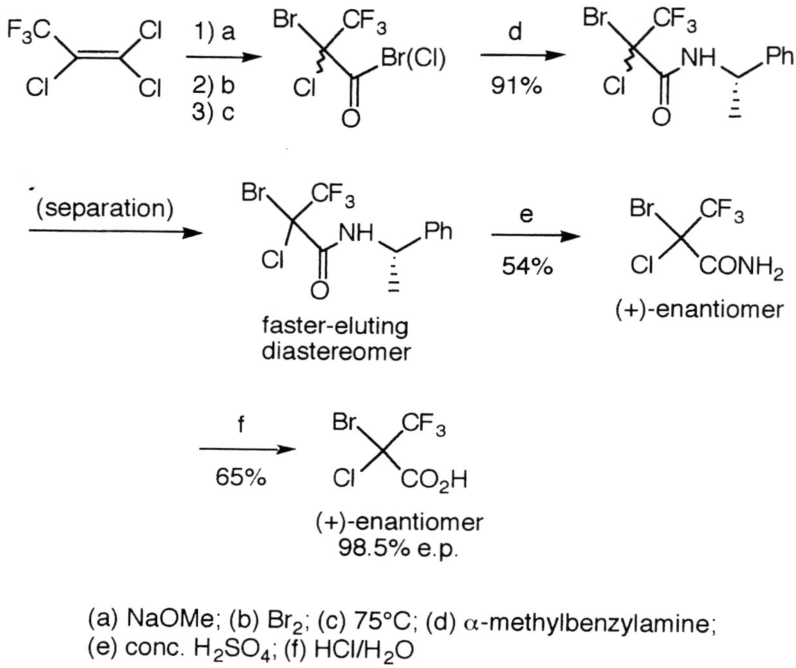

(a) NaOMe; (b) Br$_2$; (c) 75°C; (d) α-methylbenzylamine; (e) conc. H$_2$SO$_4$; (f) HCl/H$_2$O

Scheme 5. Synthesis of Enantioenriched Halothane Precursor Acid

Then we attempted the crucial decarboxylation reaction (Scheme 6). The first conditions tried were pyrolysis of the potassium salt of the acid in a protic solvent, the same conditions that were used in the isoflurane synthesis (13). These conditions gave halothane, but with an enantiomeric purity of only 20%. When amine salts of the acid were made and pyrolyzed, the outcome was much better. In this case, the reaction proceeded with 99% stereoselectivity (defined as e.p. of product/e.p. of starting material x 100%). (–)-Strychnine gave the best chemical yield, but triethylamine could also be used with no loss of enantiomeric purity. As far as we are aware, this is the highest degree of stereoselectivity ever seen in a carbon-carbon bond-breaking process. Because the halothane we isolated had the S configuration, and because the few stereoselective decarboxylations in the literature all yielded retention of configuration when the outcome was clean, we have tentatively assigned the S configuration to the acid. We are currently trying to confirm this by X-ray crystallography.

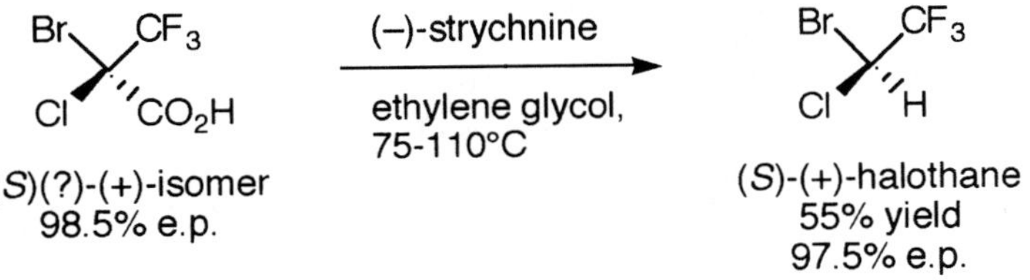

(S)(?)-(+)-isomer
98.5% e.p.

(−)-strychnine

ethylene glycol,
75-110°C

(S)-(+)-halothane
55% yield
97.5% e.p.

Scheme 6. Stereoselective Decarboxylation Yielding Halothane

Using a similar strategy, we also envisioned the asymmetric synthesis of another anesthetic compound, "Hoechst's ether" (see Scheme 4). This molecule was first synthesized in racemic form by chemists at Hoechst (24). The mixture of four diastereomers was patented as an anesthetic, but its properties were not desirable enough for it to be brought to market. The situation here is more complicated because there are two chiral centers to deal with in an asymmetric synthesis. To keep things manageable, we decided to control the stereochemistry at only one of the chiral centers. The product in this case would be a pair of diastereomers. If we were successful, we hoped that one or both of the diastereomeric pairs would have improved anesthetic properties vs. the racemate. The first big hurdle was a synthesis of the enantioenriched acid, which turned out to be relatively simple (Scheme 7).

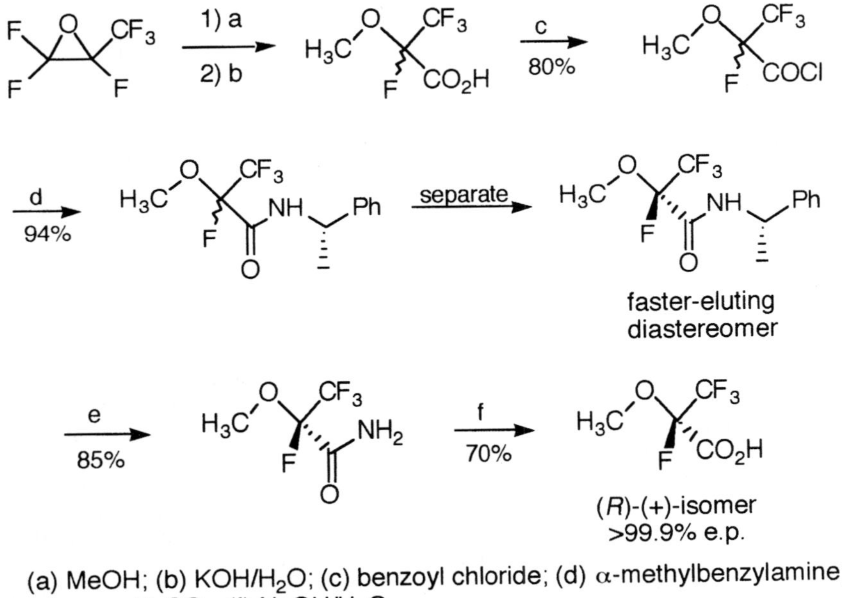

faster-eluting
diastereomer

(R)-(+)-isomer
>99.9% e.p.

(a) MeOH; (b) KOH/H$_2$O; (c) benzoyl chloride; (d) α-methylbenzylamine; (e) conc. H$_2$SO$_4$; (f) NaOH/H$_2$O

Scheme 7. Synthesis of Enantiopure Decarboxylation Substrate

The racemate of the required acid has been known for over 30 years (25), and a similar enantiomeric resolution has been performed on an analogue of the acid by Kawa (26). Preparation of the acid consisted of bubbling commercially available hexafluoropropylene oxide through methanol. Isolated in good yield was the methyl ester of the desired acid, which was saponified with aqueous sodium hydroxide (25). The acid chloride was prepared, followed by amide formation with (S)-α–methylbenzylamine. As in the case of our halothane synthesis, we used preparative chromatography to separate the diastereomers. In this case, however, the separation was so good that we were able to use flash chromatography on regular silica gel. At 100 g of mixture per run, we soon had half a kilogram of each diastereomer. The faster-eluting isomer had almost perfect isomeric purity, while the slower-eluting one had an isomeric purity of 99.5%. The amides were converted back to the acids by the usual two-step process. At that point, X-ray crystallography showed that the acid derived from the faster-eluting diastereomer had the R configuration (22).

Now that we knew which enantiomer of the acid we had, we attempted the decarboxylation (Scheme 8). The decarboxylation reaction had already been performed successfully on the racemate (27), so we used similar conditions. The acid was converted to its potassium salt and pyrolyzed in a mixture of triethylene glycol (TEG) and 1,3-dimethyl-3,4,5,6-tetrahydro-2(1H)-pyrimidinone (DMPU). Despite the harsh conditions, the product was isolated in high yield and required no additional purification. The product was levorotatory, and analysis by GC indicated that almost no stereochemical scrambling had taken place. To determine the absolute configuration of the product, we converted it to (–)-desflurane. At the time, Polavarapu had published that, based on studies of circular dichroism, the levorotatory enantiomer of desflurane had the S configuration (28), so it appeared that our reaction had yielded nearly complete inversion of configuration. However, a crystal structure of desflurane was recently reported that showed that (–)-desflurane is of the R configuration (29). It appears that Polavarapu's spectroscopic method is sound, but the sample vials containing the enantiomers of desflurane he used had been accidentally switched before analysis (30).

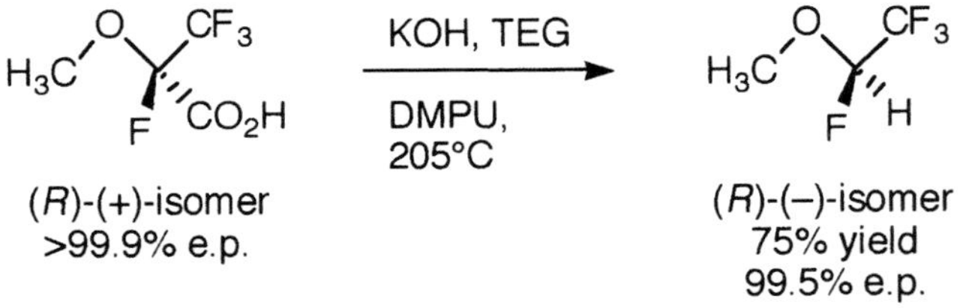

H_3C—O—CF_3, F, CO_2H KOH, TEG → DMPU, 205°C H_3C—O—CF_3, F, H

(R)-(+)-isomer
>99.9% e.p.

(R)-(–)-isomer
75% yield
99.5% e.p.

Scheme 8. Stereoselective Decarboxylation Yielding Desflurane Analog

We have also subjected Mosher's acid to the same conditions, and found that racemization is the predominant outcome, with a slight preference for *inversion* of configuration (Rozov, L.; Gall, M.; Kudzma, L.; unpublished results) (Scheme 9). This reversed outcome is obviously due to the phenyl group, and the reasons for it are unclear. It could be that the presumed anionic intermediate is more planar than when the fluorine substituent is present, and is therefore more prone to inversion. It appears that as long as the substituents of the acid stabilize a negative charge by induction, the outcome of a decarboxylation can be very clean and this type of reaction should be applicable to a wide range of chiral halogenated substrates.

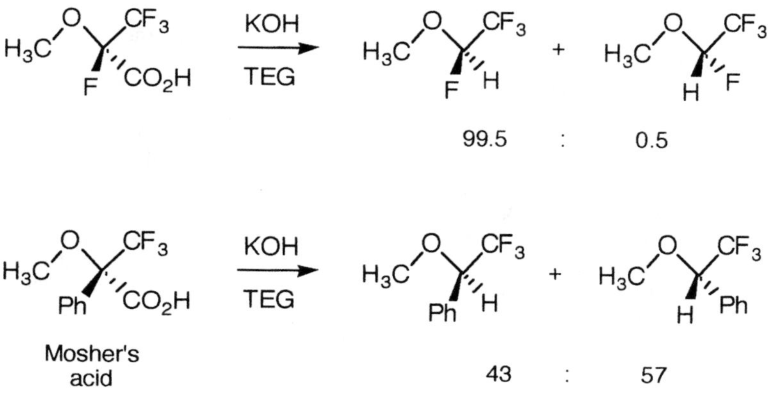

Scheme 9. Comparison of the Fluorine Atom and Phenyl Group as α-Substituents

Completion of the synthesis of diastereomeric pairs of Hoechst's ether required conversion of a methoxy group to a chlorofluoromethoxy group (Scheme 10). We accomplished this using a sequence of trichlorination, monofluorination, and selective photoreduction using 2-propanol as the reductant (*31*). The fluorination reaction was only slightly selective for monofluorination vs. difluorination using the standard fluorination reagent of SbF$_3$/Br$_2$. It is likely that a new selective fluorination method recently reported could be used on this substrate (*32*). Noteworthy is the photoreduction, which was completely selective for monoreduction. Chiral GC analysis of the product indicated negligible scrambling had occurred at the oxygen-bearing chiral center, and that the product was a 1:1 ratio of diastereomers as expected. We also took the other enantiomer of the acid through the entire sequence to the complimentary diastereomeric pair of Hoechst's ether.

Hoechst's Ether
(two diastereomers)

(a) Cl$_2$, hv; (b) SbF$_3$, Br$_2$; (c) 2-propanol, hv

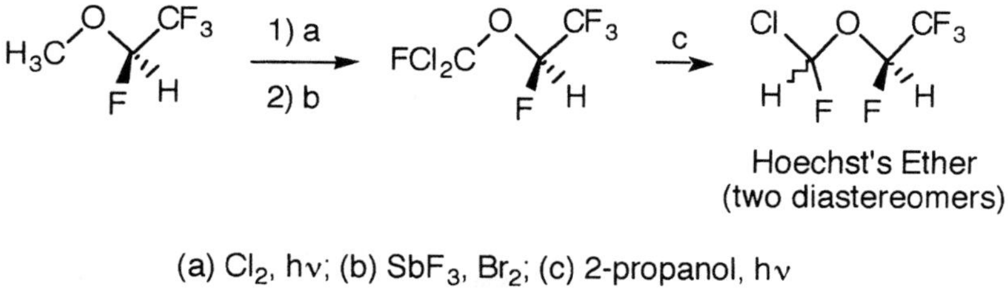

Scheme 10. Completion of the Synthesis of a Diastereomeric Pair of Hoechst's Ether

The groundbreaking synthetic efforts of the groups at Cornell University and at the company now known as Baxter Pharmaceutical Products have resulted in the availability of significant quantities of highly pure enantiomers of all the chiral fluorinated anesthetics. The isoflurane enantiomers have been used in studies designed to validate a protein receptor-based mechanism of anesthesia (*4*). Also, studies from several research groups indicate that an enantiomer of isoflurane may be a promising candidate for commercialization. In rodents, the *S* isomer is more potent than the racemate (*4*), which means in theory less of the anesthetic would need to be administered if the same trend is seen in humans. For the other anesthetics, the case is not as clear cut. Desflurane enantiomers show no difference in potency, but one of them has the advantage of causing a quicker recovery time vs. the racemate (*14*). The diastereomeric pairs of Hoechst's ether show no difference in potency or recovery time (Del Vecchio, R.; Casto, R.; unpublished results). Preliminary studies on isolated receptors from a nematode indicate that the enantiomers of halothane show differing potency (*33*).

Perhaps more important than potency differences are toxicity differences. Halothane would have the most potential for an improved anesthetic if it can be shown that one enantiomer is less toxic than the racemate. Outside the U.S., halothane is still used but its liver toxicity is so great that a long time period must pass between surgeries that use halothane for anesthesia (*34*). It has recently been found that both halothane (*35*) and enflurane (*36*) show stereoselective hepatic metabolism, a possible indication that one enantiomer of the anesthetics may be less toxic than the other.

Much remains to be done. Large amounts of the enantiomers will be needed for human trials, which will ultimately determine whether any possible improvements realized by use of a single enantiomer outweigh the expected extra cost associated with production of it. The current syntheses are a good start, but all use a chiral resolution step. This step will have to be eliminated if the synthesis is to be practical, as half of the product is lost in the form of an unwanted enantiomer. If one anesthetic enantiomer does emerge as a new drug candidate, then a practical industrial-scale synthesis of it will require that the chirality be introduced (not just maintained, as it is in the stereoselective decarboxylations described earlier) in an asymmetric step. The development of asymmetric methods for synthesis of these "exotic" highly halogenated compounds has lagged behind that of other types of compounds. Some halting first steps have been taken by Radford and Meinwald, who have attempted a fully asymmetric synthesis of halothane (37). This field appears to be ready for some new chiral chemistry. It is likely that the unusual nature of the fluorine atom will make the discovery of this new chemistry vexing as well as rewarding.

Acknowledgments. The important contributions of all Anaquest/Ohmeda/Baxter collaborators listed in the reference section are gratefully acknowledged. Dr. Leonid Rozov and Ms. Linda Brockunier are thanked for editing and proofreading the manuscript.

Literature Cited

1. All the chiral fluoroether anesthetics were discovered at a company called Ohio Medical Anesthetics. This was renamed Anaquest in the early 80's. This name survived until 1994, when the name was changed again to Ohmeda Pharmaceutical Products Inc. The company has since been purchased by Baxter International. It is now known as Baxter Pharmaceutical Products Inc.

2. Edamura, F.; Larsen, E.; Peters, H. *Abstracts of Papers* 159th American Chemical Society Meeting, **1970**, organic abstract 84.

3. Kendig, J. J.; Trudell, J. R.; Cohen, E. N. *Anesthesiology* **1973**, *39*, 518.

4. Harris, B. D.; Moody, E. J.; Basile, A. S.; Skolnick, P. *Eur. J. Pharmacol.* **1994**, *267*, 269. Lysko, G.; Robinson, J.; Casto, R.; Ferrone, R. *Eur. J. Pharmacol.* **1994**, *263*, 25. Eger, E. I.; Koblin, D. D.; Laster, M. J.; Schurig, V.; Juza, M.; Ionescu, P.; Gong, D. *Anesth. Analg.* **1997**, *85*, 188. Reviews:

Moody, E. J.; Harris, B. D.; Skolnick, P. *Trends in Pharmacol. Sci.* **1994**, *15*, 387. Franks, N. P.; Lieb, W. R. *Nature* **1994**, *367*, 607.

5. Pearson, D. L. Ph.D. Dissertation, Cornell University, **1990** (University Microfilm Int., Dissertation Information Service, 300 N. Zeeb Rd., Ann Arbor, Michigan 48106, USA).

6. Bellucci, G.; Berti, G.; Bettoni, C.; Macchia, F. *J. Chem. Soc. Perk. Trans. II* **1973**, 292.

7. Meinwald, J.; Thompson, W. R.; Pearson, D. L.; König, W. A.; Runge, T.; Francke, W. *Science* **1991**, *251*, 560.

8. Ramig, K.; Krishnaswami, A.; Rozov, L. *Tetrahedron* **1996**, *52*, 319.

9. Juza, M.; Braun, E.; Schurig, V. *J. Chromatogr. A* **1997**, *769*, 119 and references therein. Staerk, D. U.; Shitangkoon, A.; Vigh, G. *J. Chromatogr. A* **1994**, *677*, 133 and references therein.

10. Wilen, S. H.; Bunding, K. A.; Kascheres, C. M.; Wieder, M. J. *J. Am. Chem. Soc.* **1985**, *107*, 6997.

11. Knabe, J.; Agarwal, N. S. *Dtsch. Apoth. Ztg.* **1973**, *113*, 1449.

12. Gnaim, J. M.; Schurig, V.; Grosenick, H.; Green, B. S. *Tetrahedron: Asymmetry* **1995**, *6*, 1499.

13. Huang, C. G.; Rozov, L. A.; Halpern, D. F.; Vernice, G. G. *J. Org. Chem.* **1993**, *58*, 7382.

14. Rozov, L. A.; Huang, C. G.; Halpern, D. F.; Vernice, G. G. US Patent 5 283 372, **1994**. Rozov, L. A.; Huang, C. G.; Halpern, D. F.; Vernice, G. G.; Ramig, K. *Tetrahedron: Asymmetry* **1997**, *8*, 3023.

15. LaZerte, J. D.; Hals, L. J.; Reid, T. S.; Smith, G. H. *J. Am. Chem. Soc.* **1953**, *75*, 4525. Hudlicky, T.; Fan, R.; Reed, J. W.; Carver, D. R.; Hudlicky, M. *J. Fluorine Chem.* **1992**, *59*, 9.

16. Polavarapu, P. L.; Cholli, A. L.; Vernice, G. G. *J. Am. Chem. Soc.* **1992**, *114*, 10953.

17. Cram, D. J.; Wingrove, A. S. *J. Am. Chem. Soc.* **1963**, *85*, 1100. Cram, D. J.; Haberfield, P. *J. Am. Chem. Soc.* **1961**, *83*, 2363. Cram, D. J.; Haberfield, P. *J. Am. Chem. Soc.* **1961**, *83*, 2354.

18. Doyle, T. R.; Vogl, O. *J. Am. Chem. Soc.* **1989**, *111*, 8510.

19. König, W. A. *Gas Chromatographic Enantiomer Separation with Modified Cyclodextrins* Hüthig: Heidelberg, **1992**, p. 126. Schurig, who was apparently unaware of König's result, has subsequently reported the separation of the enantiomers: Grosenick, H.; Schurig, V.; Costante, J.; Collet, A. *Tetrahedron: Asymmetry* **1995**, *6*, 87.

20. Costante, J.; Hecht, L.; Polavarapu, P. L.; Collet, A.; Barron, L. D. *Angew. Chem., Int. Ed. Engl.* **1997**, *36*, 885.

21. Rozov, L. A.; Ramig, K. *Chirality*, **1996**, *8*, 3.

22. Rozov, L. A.; Ramig, K. *Tetrahedron Lett.* **1994**, *35*, 4501. Rozov, L. A.; Rafalko, P. W.; Evans, S. M.; Brockunier, L.; Ramig, K. *J. Org. Chem.* **1995**, *60*, 1319. Ramig, K.; Brockunier, L.; Rafalko, P. W.; Rozov, L. A. *Angew. Chem., Int. Ed. Engl.* **1995**, *34*, 222.

23. Henne, A. L.; Whaley, A. M.; Stevenson, J. K. *J. Am. Chem. Soc.* **1941**, *63*, 3478. Park, J. D.; Sweeney; W. M., Lacher; J. R. *J. Org. Chem.* **1956**, *21*, 1035. Park, J. D.; Sweeney; W. M., Lacher; J. R. *J. Org. Chem.* **1956**, *21*, 220.

24. Siegemund, G.; Muschaweck, R. Ger. Patent 2361058, **1975** (Hoechst).

25. Sianesi, D.; Pasetti, A.; Tarli, F. *J. Org. Chem.* **1966**, *31*, 2312.

26. Kawa, H.; Yamaguchi, F.; Ishikawa, N. *Chem. Lett.* **1982**, 745.

27. Rozov, L. A.; Huang, C. G.; Vernice, G. G. US Patent 5205914, **1993** (Anaquest).

28. Polavarapu, P. L.; Cholli, A. L.; Vernice, G. G. *J. Pharm. Sci.* **1993**, *82*, 791.

29. Schurig, V.; Juza, M.; Green, B. S.; Horakh, J.; Simon, A. *Angew. Chem., Int. Ed. Engl.* **1996**, *35*, 1680.

30. Polavarapu, P. L.; Cholli, A. L.; Vernice, G. G. *J. Pharm. Sci.* **1997**, *86*, 267.

31. Paleta, O.; Dadak, V.; Dedek, V. *J. Fluorine Chem.* **1988**, *39*, 397. Liska, F.; Fikar, J.; Kuzmic, P. *Collect. Czech. Chem. Commun.* **1993**, *58*, 565. Rozov, L. A.; Quiroz, F.; Vernice, G. G. US Patent 5416244, **1995**. Ramig, K.; Quiroz, F.; Vernice, G. G.; Rozov, L. A. *J. Fluorine Chem.* **1996**, *80*, 101.

32. Rozov, L. A.; Lessor, R. A.; Kudzma, L. V.; Ramig, K. *J. Fluorine Chem.* **1998**, *88*, 51.

33. Cascorbi, H. F.; Morgan, P. G.; Sedensky, M. M. *Abstracts of Scientific Papers*, 1994 Meeting of the American Society of Anesthesiologists, abstract A440 (see: *Anesthesiology* **1994**, *81* (supplement)).

34. For leading references, see: Martin, J. L; Dubbink, D. A.; Plevak, D. J.; Peronne, A.; Taswell, H. F.; Hay, E. J.; Pumford, N. R.; Pohl, L. R. *Anesth. Analg.* **1992**, *74*, 605. Ellis, F. R. *Brit. J. Anaest.* **1992**, *69*, 468.

35. Martin, J. L.; Meinwald, J.; Radford, P.; Liu, Z.; Graf, M. L. M.; Pohl, L. R. *Drug Metab. Rev.* **1995**, *27*, 179.

36. Garton, K. J.; Yuen, P.; Meinwald, J.; Thumel, K. E.; Kharasch, E. D. *Drug Metab. Disp.* **1995**, *23*, 1426.

37. Radford, P. M. Ph.D. Dissertation, Cornell University, **1997** (University Microfilm Int., Dissertation Information Service, 300 N. Zeeb Rd., Ann Arbor, Michigan 48106, USA).

INDEXES

Author Index

Abouabdellah, Ahmed, 84
Bégué, Jean-Pierre, 84
Bernacki, Ralph J., 158
Bonnet-Delpon, Danièle, 84
Bravo, Pierfrancesco, 98, 127
Brown, Herbert C., 22
Bruché, Luca, 98
Chakravarty, Subrata, 158
Cresteil, Thierry, 158
Crucianelli, Marcello, 98
Daly, J. W., 194
Filler, Robert, 1
Gyenes, Ferenc, 182
Hamper, Bruce C., 272
Harper, D. B., 210
Haufe, G., 194
Herbert, B., 194
Hiraoka, Shiuchi, 142
Hiyama, Tamejiro, 226
Hoffman, Robert V., 52
Inoue, Tadashi, 158
Iseki, Katsuhiko, 38
Ishii, Akihiro, 60
Jayachandran, B., 194
Kirk, K. L., 194
Kitazume, Tomoya, 142
Kornilov, Andrei, 84, 182
Kuduk, Scott D., 158
Kusumoto, Tetsuo, 226
Laue, K. W., 194

Leschinsky, Kindrick L., 272
Lin, Songnian, 158
Lu, S.-F., 194
Matsutani, Hiroshi, 226
Messina, Maria Teresa, 239
Metrangolo, Pierangelo, 239
Mikami, Koichi, 60, 255
Mischke, Deborah A., 272
Monsarrat, Bernard, 158
Nga, Truong Thi Thanh, 84
O'Hagan, D., 210
Ojima, Iwao, 158
Olufunke, O., 194
Padgett, W. L., 194
Pera, Paula, 158
Prosch, S. Douglas, 272
Ram Reddy, M. Venkat, 117
Ramachandran, P. V., 22, 117
Ramig, Keith, 282
Resnati, Giuseppe, 239
Rodrigues, Isabelle, 84
Rudd, Michael T., 117
Slater, John C., 158
Soloshonok, Vadim A., 74
Tao, Junhua, 52
Viani, Fiorenza, 98
Volonterio, Alessandro, 127
Walsh, John J., 158
Welch, John T., 182
Yamazaki, Takashi, 142
Zanda, Matteo, 98, 127

Subject Index

A

Acetophenones
    asymmetric reduction, 23
    reduction of ring-substituted, 29–30
α-Acetylenic trifluoromethyl ketones, Morita–Baylis–Hillman reaction, 121
Acyl-CoA cholesterol O-transferase (ACAT), inhibitors, 14, 17f
Adriamycin (doxorubicin), anticancer drug, 11
Agrochemical products. See Aryl-pyrazole herbicides
Aldehydes, fluorinated. See Allylboration, asymmetric
Aldol reactions
    enantioselective aldol condensations to fluorinated threo-3,4-dihydroxyphenlserine, 204, 206
    Evans, 136–138
    highly diastereoselective asymmetric reactions on chiral Ni(II)-complex of glycine with CF$_3$COR, 4f
    See also Evans aldol reaction; Ketene silyl acetals, fluorine-substituted; Nitroaldol reaction of 2,2-difluoroaldehydes
Aldols, fluorinated. See Sugars containing fluorinated methyl groups
1-Alkoxy(polyfluoro)alkyl sulfonates
    alkylation with aluminum reagents, 233t
    comparison of organometallic reagents for ethyl substitution, 233t
    ethylation with aluminum reagents, 232t
    nucleophilic substitution, 231–234
    nucleophilic substitution of hemiacetal sulfonates, 231–234
    optical resolution, 231
    use of aluminate reagents derived from 1-alkenes, 234, 235t
    See also Liquid crystals (LCs), fluorine-containing
Alkylation
    hemiacetal tosylate with aluminum reagents, 233–234
    using aluminates from 1-alkenes, 234, 235t

Alkylfluoroalkyl ketones, reduction, 25–26
α-Alkynyl α'-fluoroalkyl ketones, reduction, 26–27
Allyl alcohols, functionalized fluorinated. See Morita–Baylis–Hillman (MBH) reaction; Vinylmetalation
Allylboration, asymmetric
    B-allyldiisopinocampheylborane (Ipc$_2$BAll) and B-allyldi-2-isocaranylborane (2-Icr$_2$BAll) syntheses, 30
    ring-fluorinated benzaldehydes with Ipc$_2$BAll and 2-Icr$_2$BAll, 31f
    synthesis and reaction of B-allyldiisopinocampheylborane, 31f
    synthesis of 4-perfluoroalkyl-g-butyrolactones, 31, 32f
    See also Fluoroorganic compounds via chiral organoboranes
B-Allyldi-2-isocaranylborane (2-Icr$_2$BAll), synthesis, 30–31
B-Allyldiisopinocampheylborane (Ipc$_2$BAll), synthesis, 30–31
Amination. See Reductive amination of fluoro-carbonyl compounds
Amino acids. See Fluoroamino acids
Amino alcohols
    important targets in human disorder treatment, 84
    interest in β-amino β-fluoroalkyl alcohols, 89–90
    See also Fluoroalkyl peptidomimetic units
Aminohydroxylation, Sharpless asymmetric, step in synthesis of fluorinated norepinephrines, 199
α-Amino ketones, fluorination. See Monofluoroketomethylene peptide isosteres
2-Amino-4,4,4-trifluorobutanoic acid, synthesis via trifluoromethylation, 3, 5f
(2R,3S)2-Amino-3-trifluormethyl-3-hydroxy alkanoic acid derivatives, synthesis, 3, 4f
Anesthetics
    chiral racemic inhaled, 282
    comparison of fluorine and phenyl groups as α-substituents, 288–289
    completion of synthesis of diastereomeric pair of Hoechst's ether, 289

discovery of resolution by capillary gas chromatography, 283–284

enantiomer synthesis using stereoselective decarboxylation, 285

groundbreaking synthetic efforts affording quantities of highly pure enantiomers, 289–290

halothane, enflurane, isoflurane, and desflurane, 282

Meinwald and Pearson's synthesis of halothane and enflurane enantiomers, 283

preparation of highly pure enantiomers by Meinwald and Pearson, 283

quantities for human trials, 290

reasons for synthesis of anesthetic enantiomers, 283

separation by enantioselective desorption from clathrate complexes, 284

stereoselective decarboxylation yielding desflurane analog, 288

stereoselective decarboxylation yielding halothane, 286–287

syntheses of highly pure enantiomers of isoflurane and desflurane, 284

synthesis of enantio-enriched halothane precursor acid, 285–286

synthesis of enantiopure decarboxylation substrate, 287

toxicity differences of enantiomers, 290

See also Inhalation anesthetics

Anticancer agents, fluorine-containing chiral compounds, 11

Anti-ferroelectricity, conformational probe, 260–263

Antifungal agents, fluorine-containing chiral compounds, 14, 18

Aromatic compounds

asymmetric activation, 62–63

catalytic reactions with fluoral, 61*t*

Friedel–Crafts reactions, 60–63

transition state for ortho Friedel–Crafts product, 62

Aromatic trifluoromethyl ketones, Morita–Baylis–Hillman reaction, 120–121

Aryl fluoroalkyl ketones, reduction, 24–25

Aryl-pyrazole herbicides

chiral derivatives, 273*f*

conversion of diketones to pyrazoles by cyclocondensation or two step process with hydrazine, 274, 275*f*

determining enantiomeric purity, 277

herbicidal activity in preemergent and post-emergent screens, 279–280

herbicidal activity of enantiomers of chiral 3-aryl-4-halo-5-(trifluoromethyl)pyrazoles, 280–281

HPLC resolution of chiral phenylpyrazoles, 278*t*

inhibiting protoporphrin IX oxidase (PRO-TOX), 272

investigating structure-activity relationships, 273

preparation of 2,4,5-substituted acetophenones, 274*f*

preparation of lactate ester and ether derivatives, 273–274

preparative chromatographic resolution by HPLC on chiral stationary phase, 278*t*

proton NMR spectra of lactate derivatives, 279

regioselective pyrazole synthesis, 275

relative herbicidal activity of chiral phenylpyrazoles, 280*f*

synthesis of lactate ester derivatives, 276, 277*f*

synthesis of phenoxylactate derivatives, 275, 276*f*

Aryl trifluoromethyl ketones, Morita–Baylis–Hillman reaction, 120

Asymmetric allylboration. *See* Allylboration, asymmetric

Asymmetric enolboration-aldolization. *See* Enolboration-aldolization, asymmetric

Asymmetric reductions. *See* Reductions, asymmetric

Azetidinones

acidic methanolysis to methyl isoserinates, 93

alkylation, 94

cyclocondensation route to, 92

preparation of *cis*-fluoroalkyl, 90

preparation of protected, 90–91

preparing tri- and tetrapeptide isosteres, 91

reaction of benzyloxyketene with imine for high *cis/trans* stereoselectivity, 92

Aziridines, ring-opening of chiral, 202, 204

B

Baccatins. *See* Fluorine-containing taxoids

Benzaldehydes, diastereomeric, asymmetric Morita–Baylis–Hillman reaction, 123

Benzaldehydes, fluorinated

exploratory study of terpenyl alcohols for MBH reaction, 124

Morita–Baylis–Hillman (MBH) reaction, 118

Biclutamide (Casodex)

anticancer agent, 11

structure, 12

Binaphthol-derived titanium catalysts. *See* Friedel–Crafts reactions

Binding affinities, fluorinated norepinephrines, 207

Biomimetic transamination. *See* Reductive amination of fluoro-carbonyl compounds

Bisperfluoroalkyl aryl diketones, reduction, 27–28

Borane reagents. *See* Allylboration, asymmetric; Fluoroorganic compounds via chiral organoboranes; Reductions, asymmetric

Bromofluoroketene silyl acetals
catalytic aldol reaction by Lewis acids, 42
effect of reaction temperature on stereoselection during aldol reaction, 43–44
preparation, 39
See also Ketene silyl acetals, fluorine-substituted

C

Camphor sultam, asymmetric Morita–Baylis–Hillman reaction, 123
Carbonyl-ene reaction, enantioselective catalysis with fluoral, 256
Carbonyl reduction, asymmetric, step in synthesis of fluorinated norepinephrines, 199, 202
Casodex (biclutamide)
anticancer agent, 11
structure, 12
Catecholamines
naturally occurring, 194
synthetic approaches to enantiomers of fluorinated threo-3,4-dihydroxyphenylserine, 202, 204
See also Dopamine (DA); Epinephrine (EPI); Norepinephrine (NE)
Central nervous system (CNS) agents
chemoenzymatic route to *(R)*-fluoxetine, 15*f*
fluorine-containing chiral compounds, 11, 14
Chair-like transition state, model for asymmetric elaborations leading to γ-trifluoromethyl-GABOB, 108
Chiral anesthetics. *See* Anesthetics
Chiral bioactive fluoroorganic compounds, range of research studies, 1–3
Chiral compounds. *See* Fluorine-containing chiral compounds
Chiral herbicides. *See* Aryl-pyrazole herbicides
Chiral organoboranes. *See* Fluoroorganic compounds via chiral organoboranes
Chiral sulfoxides. *See* Fluoroalkyl amino compounds
2-Chloro-2,2-difluoroacetophenone, Morita–Baylis–Hillman reaction, 121
B-Chlorodiisopinocamphenylborane (DIP-Chloride)
reagent for asymmetric reductions, 23
See also Reductions, asymmetric
Cholesterol absorption inhibitors
acyl-CoA cholesterol O-transferase (ACAT) inhibitors, 17*f*
fluorine-containing chiral compounds, 14
Cram's cyclic model, stereoselective transformations involving pyruvaldehyde *N,S*-ketals, 107
Cyanohydrin acetates, lipase-catalyzed hydrolysis, 199
Cyanohydrin synthesis, asymmetric, step in synthesis of fluorinated norepinephrines, 202, 203
Cytochrome P450 enzymes. *See* P450 enzymes

D

DA-125, fluoro analog of anticancer drug doxorubicin (Adriamycin), 11, 13
Densely functionalized molecules, methods to introduce fluorine, 52–53
Desflurane
enantiomer synthesis using stereoselective decarboxylation, 285
first synthesis by simple substitution reaction, 284–285
groundbreaking synthetic efforts affording quantities of highly pure enantiomers, 289–290
inhaled anesthetic, 282
stereoselective decarboxylation yielding analog, 288
synthesis of enantiomers, 284
See also Anesthetics
(S)-(+)-Dexfenfluramine (Redux), CNS agent, 11
Diastereomeric α- or β-CF$_3$ liquid crystalline molecules
asymmetric catalytic ene reaction with trihaloacetaldehydes, 256*t*
asymmetric catalytic fluoral-ene reaction, 258*t*
bent conformation for antiferroelectric liquid crystal (AFLC), 262*f*
bent versus extended conformation for AFLC, 261*f*
bi-stable and tri-stable switching, 260*f*
computational analysis of CX$_3$CHO/H$^+$ complexes, 257*t*
conformational probe for anti-ferroelectricity, 260–263
diastereomeric LC molecules with CF$_3$-group for large spontaneous polarization (Ps), 260
diastereoselective catalysis of fluoral-ene reactions, 257–259
%ee dependence of phase sequences, 263*f*
enantioselective catalysis of carbonyl-ene reaction with fluoral, 256
ene-based synthesis of molecule with double stereogenic centers, 261*f*
fluorine and metal coordination, 259*t*
glyoxylate-ene reaction catalyzed by chiral binaphthol-titanium (BINOL-Ti) catalyst, 256

LUMO energy level versus electron density of F-carbonyl compounds, 257
optical transmittance versus applied triangular wave electric field, 263, 264f
phase transition temperatures, 262, 263f
self-assembly for spontaneous optical resolution of racemates, 263–265
six-membered ene transition state, 259
spontaneous resolution by Pasteur using optical microscopy, 265
transition state for *syn*-diastereomer, 259f
unique characteristics of organofluorine substrates, 255
Diflucan (fluconazole)
antifungal agent, 14, 18
structure, 17
2,2-Difluoroaldehydes. *See* Nitroaldol reaction of 2,2-difluoroaldehydes
Difluoroketene silyl acetals
catalytic asymmetric aldol reaction using Lewis acids, 41–42
effect of reaction temperature on stereoselection during aldol reaction, 42–43
molecular orbital calculation, 40–41
preparation, 39
See also Ketene silyl acetals, fluorine-substituted
threo-3,4-Dihydroxyphenylserine
enantioselective aldol condensations, 204, 206
ring-opening of chiral aziridines, 202, 204
scheme for *syn*- and *anti*-products, 205
synthetic approaches to enantiomers of fluorinated, 202, 204
See also Catecholamines
Diisopinocampheylboron triflate (Ipc₂BOTf), enolborating agent, 32
DIP-Chloride (*B*-chlorodiisopinocamphenylborane)
enolboration-aldolization of ring-fluorinated acetophenones and benzaldehydes, 32–33
reagent for asymmetric reductions, 23
See also Reductions, asymmetric
Docetaxel
acting as spindle poisons, 159
anticancer agents, 11
bioassay for new fluorine-containing taxoids, 161
conformational analysis, 171, 172f
cytotoxicities, 163t, 169t, 171t
fluorine-containing analog, 12f
primary sites of hydroxylation by P450 family of enzymes, 164f
structure, 159
studies on metabolism of, 159
synthesis and biological activity of 3′-difluoromethyl-docetaxel analogs, 169–171
synthesis and biological activity of 3′-trifluoromethyl-docetaxel analogs, 163, 166–169
treatment of advanced breast cancer, 158
See also Fluorine-containing taxoids
Dopamine (DA)
effects of fluorine substitution, 196
naturally occurring catecholamine, 194
ring fluorinated derivatives, 196, 197
structure, 195
Doxorubicin (Adriamycin), anticancer drug, 11

E

Electron donor-acceptor complexes. *See* Perfluorocarbon (PFC) compounds
Enamines. *See* α-Fluoroalkyl β-sulfinyl enamines and imines
Endothermic transition, liquid crystal compound, 237
Enflurane
inhaled anesthetic, 282
Meinwald and Pearson synthesis, 283
separation by enantioselective desorption from clathrate complexes, 284
See also Anesthetics
Enolboration-aldolization, asymmetric
applying DIP-Chloride for ring-fluorinated acetophenones and benzaldehydes, 32–33
diisopinocampheylboron triflate (Ipc₂BOTf) reagent, 32
effect of fluorine on tandem enolboration-aldolization-intermolecular reduction, 35f
effect of fluorine on tandem enolboration-aldolization-reduction, 34f
effect of fluorine substitution in enolboration, 33f
increase in enantioselectivity at higher temperatures for pentafluoroacetophenone-pentafluorobenzaldehyde, 33f
kinetic resolution, 34f
product of tandem enolboration-aldolization-intermolecular reduction, 35
stereochemistry, 32f
tandem enolboration-aldolization-intramolecular reduction, 33f
See also Fluoroorganic compounds via chiral organoboranes
Epinephrine (EPI)
effects of fluorine substitution, 196
functioning through interactions with membrane-bound adrenergic receptors, 194, 196
limitation of fine-tuning to enhance selectivity, 198
naturally occurring catecholamine, 194
ring fluorinated derivatives, 196, 197
structure, 195

Epoxy ethers
  oxirane ring opening as access to α-amino ketones, 85–86
  preparation of 1-trifluoromethyl, 85
  reaction with aluminum amide leading to anti amino alcohols, 89
  reaction with chiral amide, 87
  reaction with dimethylaluminum amide and reduction, 86
  reaction with *(R)* phenethyl amine, 89
  ring opening of homochiral, 88–89
  See also Fluoroalkyl peptidomimetic units
Ethylation
  comparison of organometallic reagents, 232, 233*t*
  hemiacetal sulfonates with aluminum reagents, 232*t*
Evans aldol reaction
  investigating reaction of chiral α-hydroxy acetic anion equivalent with *N*-Cbz imine of trifluoropyruvate, 136, 137
  See also β-Fluoroalkyl β-amino alcohol units

F

Felkin–Anh models
  model for asymmetric elaborations leading to γ-trifluoromethyl-GABOB, 108–109
  stereoselective transformations involving pyruvaldehyde *N,S*-ketals, 107–108
Fluconazole (Diflucan)
  antifungal agent, 14, 18
  structure, 17
Fluoral
  asymmetric catalytic fluoral-ene reaction, 258*t*
  diastereoselective catalysis of fluoral-ene reactions, 257–259
  enantioselective catalysis of carbonyl-ene reaction, 256
  polymerization, 118, 119
  syn-diastereoselectivity of fluoral-ene reaction, 259
  See also Friedel–Crafts reactions
Fluorinated aldehydes. *See* Allylboration, asymmetric
Fluorinated aldols. *See* Sugars containing fluorinated methyl groups
Fluorinated benzaldehydes
  Morita–Baylis–Hillman reaction, 118
  oxynitrilase-catalyzed addition of HCN to, 199
Fluorinated methyl groups. *See* Sugars containing fluorinated methyl groups
ω-Fluorinated lipids, natural product, 212
Fluorination of α-amino ketones. *See* Monofluoroketomethylene peptide isosteres

Fluorine
  altering reaction courses, 38
  electronic effect on neighboring groups, 38
  metal coordination, 259
  methods to introduce into densely functionalized molecule, 52–53
Fluorine-containing chiral compounds
  acyl-CoA cholesterol O-transferase (ACAT) inhibitors, 17*f*
  analogs of paclitaxel and docetaxel, 12*f*
  anticancer agents, 11
  antifungal agents, 14, 18
  asymmetric synthesis of β-(fluoroalkyl)-β-amino acids via transamination, 6, 8*f*
  azidization of α-fluoro-α′-sulfinyl alcohol under Mitsunobu conditions, 7
  central nervous system (CNS) agents, 11, 14
  chemo-enzymatic approach to chiral β-trifluoromethyl-β-amino acids, 6, 8*f*
  chemoenzymatic route to *(R)*-fluoxetine, 15*f*
  cholesterol absorption inhibitors, 14
  complete diastereoselective fluorination in synthesis of 2-fluoronucleosides, 12*f*
  enzymatically controlled reactions of organofluorine compounds, 18
  fluoroamino acids, 3–6
  fluoronucleosides, 6, 10–11
  fluoro sugars, 6, 9
  highly diastereoselective asymmetric aldol reactions of chiral Ni(II)-complex of glycine with CF₃COR, 4*f*
  HIV inhibitor, 18
  HIV reverse transcription inhibitor, 19*f*
  inhalation anesthetics, 14, 16
  new approaches in medicinal chemistry, 1
  non-carbohydrate synthesis of sugars, 9*f*
  one-pot transformation through Pummerer reaction, 7
  range of research studies, 1–3
  synthesis of (2R,3S)-2-amino-3-trifluoromethyl-3-hydroxy alkanoic acid derivatives, 4*f*
  synthesis of β-2-fluorodideoxy nucleosides, 9*f*
  synthesis of enantiomerically pure (2R)-N-Boc-2-amino-4,4,4-trifluorobutanoic acid via trifluoromethylation, 3, 5*f*
  synthesis of (R)-4,4,4,4′,4′,4′-hexafluorovaline, 6, 7*f*
Fluorine-containing taxoids
  baccatin preparation, 159–161
  baccatins by modified Chen's method, 161, 162
  bioassay for new, 161
  Chem3D representation of 3′-CF₃-taxoid, 176*f*

conformational analysis of paclitaxel and
docetaxel, 171, 172f
cytotoxicities, 163t
cytotoxicity of 3'-CF$_2$H-taxoids, 171t
cytotoxicity of 3'-CF$_3$-taxoids, 169t
determination of binding conformation of
taxoids to microtubules using fluorine
probes, 175, 177
effects of C-2 position modification on cytotoxicity, 161
energetically similar conformers of F$_2$-tax-
oid maintaining F–F distance in restrained molecular dynamics (RMD)
study, 177, 179f
estimation of F–F distances for microcrystalline and tubulin-bound forms of F$_2$-10-
Ac-docetaxel by radio frequency driven
dipolar recoupling (RFDR) technique,
177, 178f
F$_2$-paclitaxel and F-docetaxel for studying
solution structure and dynamic behavior,
173
fluorine probe approach, 171, 173–177
fluorine probe confirming hydrophobic clustering as driving force for conformational
stabilization, 175
metabolism by P450 enzymes, 165f
metabolism study of new, 163
Newman projections of three conformers
for F$_2$-paclitaxel, 174f
possible transition states for kinetic resolution of 4-CF$_3$-β-lactam, 166, 168f
primary sites of hydroxylation on paclitaxel
and docetaxel by P450 family of enzymes, 164f
RMD studies, 173
solution-phase structure and dynamics of
taxoids, 171, 173–177
synthesis, 161, 162
synthesis and biological activity of 3'-diflu-
oromethyl-docetaxel analogs, 169–171
synthesis and biological activity of 3'-triflu-
oromethyl-docetaxel analogs, 163,
166–169
synthesis of 3'-CF$_2$H-docetaxel analogs,
169, 171, 172
synthesis of 3'-CF$_3$-taxoids through kinetic
resolution, 166t
synthesis of difluoro-βb-lactam, 169, 170
synthesis of series of 3'-CF$_3$-taxoids, 163,
166, 167
use of ^{19}F NMR for variable temperature
study, 171, 173
Fluorine probe approach
Chem3D representation of 3'-CF$_3$-taxoid,
176f
confirming hydrophobic clustering, 175
determination of binding conformation of
taxoids to microtubules, 175, 177
energetically similar conformers of F$_2$-tax-

oid maintaining F–F distance of restrained molecular dynamics (RMD)
study, 177, 179f
estimation of F–F distances for microcrystalline and tubulin-bound forms of F$_2$-10-
Ac-docetaxel by radio frequency driven
dipolar recoupling (RFDR) technique,
177, 178f
Newman projections of three conformers
for F$_2$-paclitaxel probe, 174f
probes for study, 173
RMD studies, 173
solution-phase structure and dynamics of
taxoids, 171, 173–177
See also Fluorine-containing taxoids
Fluoroacetaldehyde
oxidation to fluoroacetate by NAD dependent aldehyde dehydrogenase, 223f
precursor to fluorometabolites, 221–222
See also Fluorometabolite biosynthesis by
Streptomyces cattleya
Fluoroacetate
biosynthesis by Streptomyces cattleya,
214–221
natural product, 211–212
oxidation from fluoroacetaldehyde by NAD
dependent aldehyde dehydrogenase, 223f
See also Fluorometabolite biosynthesis by
Streptomyces cattleya
Fluoroacetone, natural product, 212–213
(R)-β-Fluoroalaninols, stereoselective synthesis, 131
(S)-β-Fluoroalaninols and alanines, stereoselective synthesis, 132
Fluoro-aldehydes and ketones
vinylalumination, 121–122
vinylcupration, 122
α-Fluoroalkyl β-sulfinyl enamines and imines
β-amino nitro derivatives as precursors of
α-fluoroalkyl β-diamino compounds,
103–104
general synthesis strategy, 100
model for highly stereocontrolled reduction
of β-naphthylsulfinyl imines, 103
oxidative cleavage of p-methoxyphenyl
(PMP) group, 102–103
preparation and use, 99–104
preparing N-aryl, N-alkoxycarbonyl, and N-
unsubstituted polyfluoroalkyl enamines
and imines, 100
stereoselective reduction of C=N bond of
β-sulfinyl imines, 102
stereoselective reduction with hydride reducing agents, 101–102
synthesizing fluorinated β-sulfinyl imines,
99, 100
synthetic potentialities of fluorinated β-sulfinyl imines as chiral building blocks,
104
See also Fluoroalkyl amino compounds

Fluoroalkyl amino compounds

1,3-dipolar cycloadditions involving fluorinated sulfinyl dipolarophiles, 112–113

approach to enantiopure quaternary fluoro amino compounds via sulfinyl epoxide chemistry, 111–112

β-amino nitro derivatives potential precursors of α-fluoroalkyl β-diamino compounds, 103–104

building blocks in pharmaceutical field, 98–99

diastereoselective elaboration of pyruvaldehyde N,S-ketals into fluorinated ephedraalkaloids, 107

fluoropyruvaldehyde N,S-ketals from α-(fluoroalkyl)-β-sulfinylenamines, 105–106

fluoropyruvaldehyde N,S-ketals: nonracemic synthetic equivalents of α-fluoroalkyl α-amino aldehydes, 105–109

general strategy to prepare nonracemic compounds via chiral sulfoxide chemistry, 99, 100

intermediates to obtain sulfur-free γ-tri- and γ-difluorinated β-amino acids, 107–108

man-made analogs α-difluoromethyl (Dfm) α-amino acids, 103

models for asymmetric elaborations leading to γ-Tfm-GABOB, 108–109

models for stereoselective transformations involving pyruvaldehyde N,S-ketals and derivatives, 107

oxidative cleavage of p-methoxyphenyl (PMP) group, 102

preparation and use of α-fluoroalkyl β-sulfinyl enamines and imines, 99–104

preparing N-aryl, N-alkoxycarbonyl, and N-unsubstituted polyfluoroalkyl enamines and imines, 100–101

retrosynthetic strategy to β-fluoroalkyl β-amino alcohols from pyruvaldehyde N,S-ketals, 105

stereocontrol in reduction of naphthyl derivatives, 102–103

stereoselective reduction of α-(fluoroalkyl) β-sulfinyl enamines and imines, 101

stereoselective reduction of C=N bond of β-sulfinyl imines, 102

stereoselective synthesis of 3-fluoro-D-alanine, 111

stereoselective synthesis of deuterium labeled fluoro amino compounds, 101, 102

sulfinimines of trifluoropyruvate: toward library of nonracemic α-trifluoromethyl (Tfm) α-amino acids, 109–111

syntheses of monofluoro-amino compounds, 111–113

synthesis and elaboration of 6-fluoromethyl furo[3,4-c]isoxazolidines, 113

synthesis of library of nonracemic α-Tfm-amino acids from sulfinimines of trifluoropyruvate, 109, 110

synthesis of monofluorinated nucleosides, 112

synthesis of nonracemic γ-Tfm-GABOB from pyruvaldehyde N,S-ketals, 108

synthesis of (R)-α-Tfm-phenylalanine with regeneration and recycle of sulfinyl chiral auxiliary, 109, 110

synthesizing fluorinated β-sulfinyl imines, 99, 100

synthetic approaches to β-sulfinyl imines and tautomerism, 100

synthetic potentialities of fluorinated β-sulfinyl imines as chiral building blocks, 104

β-Fluoroalkyl-β-amino acids, asymmetric synthesis via transamination, 6, 8f

β-Fluoroalkyl β-amino alcohol units

additions of lithium alkyl aryl sulfoxides to N-PMP (p-methoxyphenyl) fluoroalkyl/aryl imines, 129–136

dominant Zimmerman–Traxler (aldol) chair-like transition state, 132

general stereoselective approaches, 129

investigating Evans aldol reaction of chiral α-hydroxy acetic anion, 136

lithium alkyl aryl sulfoxides in synthesis of, 135–136

mechanism of non-oxidative Pummerer (NOP) reaction, 130–131

preparation of biological and pharmaceutical substrates, 127

preparation of series of α-fluoroalkyl and α-(fluoro)aryl glycinols, 131

retrosynthetic approach to 2-Tfm-sphingolipids, 136–137

stereoselective application of NOP reaction, 132–133

stereoselective synthesis of, having quaternary amine center via Evans aldol reaction with imines of trifluorpyruvate, 136–138

stereoselective synthesis of γ-Tfm-GABOB, 133–134

stereoselective synthesis of (S)-β-fluoroalaninols and alanines, 132

stereoselective synthesis of trifluoronorephedrine, 133

stereoselective tandem imino-aldol reaction, 137

synthesis of both enantiomers of α-Tfm-phenylalanine, -alanine, and -serine (trifluoromethyl = Tfm), 134, 135

synthesis of enantiopure α-Tfm-β-hydroxy aspartic units and derivatives, 136, 138

synthesis of enantiopure α-Tfm-butyrine, -threoninate, and -allo-threoninate, 134–135

synthetic approaches to racemic and nonracemic, 128

use of lithiated sulfoxides in stereoselective synthesis, 129
Fluoroalkyl peptidomimetic units
acidic methanolysis of azetidinones for methyl isoserinates, 93
alkylation of azetidinone, 94
β-amino β-fluoroalkyl alcohols as peptidomimetic units, 89–90
chiral epoxidation of enol ethers, 88
configuration in formation of oxazolidinone, 86, 87
cyclocondensation route to azetidinone, 92
dimethylaluminum amide preparation, 86
fluoroalkyl β-amino alcohols precursors to fluoroalkyl peptidyl ketones, 85
homochiral *syn* methyl 3-trifluoromethyl isoserinates, 93
non racemic trifluoromethyl azetidinones, 92
oxirane ring opening of epoxy ethers to α-amino ketones, 85
preparation of 1-trifluoromethyl epoxy ethers, 85
preparation of anti methyl 3-trifluoromethyl isoserinates, 93
preparation of *cis*-fluoroalkyl azetidinones, 90
preparation of methyl isoserines and protected azetidinones, 90–91
reaction of benzyloxyketene with imine for azetidinones, 92
reaction of chiral epoxy ethers with aluminum amide, 89
reaction of epoxy ethers with chiral amide, 87
reaction of epoxy ethers with dimethylaluminum amide and reduction, 86
reaction of epoxy ethers with secondary amines and reduction, 85–86
ring opening of fluoroalkyl β-lactams, 90
ring opening of homochiral epoxy ethers, 88–89
searching for chiral approach to syn and anti trifluoromethyl amino alcohols, 87–89
Staudinger reaction of ketenes to aldimines to *cis*-β-lactams, 90
stereochemistry of reduction of amino ketones, 87f
synthesis of α-fluoroalkyl β-amino alcohols, 85
target of β-amino β-fluoroalkyl alcohols due to trifluoromethyl isoserine, 89–90
trifluoromethyl isoserine derivatives, 91
Wittig olefination of ethyl trifluoroacetate, 85
Fluoro-amines. *See* Reductive amination of fluoro-carbonyl compounds

Fluoroamino acids
asymmetric synthesis of β-(fluoroalkyl)-β-amino acids via transamination, 8f
chemo-enzymatic approach to chiral β-trifluoromethyl-β-amino acids, 8f
chiral compounds, 3–6
highly diastereoselective asymmetric aldol reactions on chiral Ni(II)-complex of glycine with CF₃COR, 4f
synthesis of (2R, 3S)-2-amino-3-trifluoromethyl-3-hydroxy alkanoic acid derivatives, 4f
synthesis of enantiomerically pure (2R)-N-Boc-2-amino-4,4,4-trifluorbutanoic acid via trifluoromethylation, 5f
synthesis of *(R)*-4,4,4,4′,4′,4′-hexafluorovaline, 7f
See also Reductive amination of fluoro-carbonyl compounds
(S)-Fluoroaryl glycinols, stereoselective synthesis, 131
Fluorobenzaldehydes, exploratory study of terpenyl alcohols for Morita–Baylis–Hillman (MBH) reaction, 124
Fluoro-carbonyl compounds
asymmetric vinylalumination, 125
Morita–Baylis–Hillman reaction, 118–121
vinylmetalation, 121–122
See also Reductive amination of fluoro-carbonyl compounds
Fluorocitrate, natural product, 212
β-2-Fluorodideoxy nucleosides, synthesis, 6, 9f
Fluorometabolite biosynthesis by *Streptomyces cattleya*
¹⁹F {¹H} NMR spectrum of fluoroacetate and 4-fluorothreonine after incubation in 30% D₂O, 216f
¹⁹F {¹H} NMR spectrum of fluoroacetate and 4-fluorothreonine after incubation in [1,2-¹³C₂]-glycine, 217f
¹⁹F {¹H} NMR spectrum of fluoroacetate and 4-fluorothreonine after incubation in [2-¹³C]-glycine, 217f
¹⁹F {¹H} NMR spectrum of fluoroacetate and 4-fluorothreonine after incubation in [²H₄]-succinate, 220f
¹⁹F {¹H} NMR spectrum of fluoroacetate and 4-fluorothreonine after incubation in [3-¹³C]-pyruvate, 219f
¹⁹F {¹H} NMR spectrum of fluoroacetate and 4-fluorothreonine after incubation in [3-¹³C]-serine, 219f
fluoride consumption, cell growth, and fluorometabolite production versus time relationships, 216f
incorporation of glycerol, 218, 221

incorporation of glycine, serine, and pyruvate, 215, 218
incorporation of succinate, 218
labeling rationale from *(R)*-[1-²H₂]-glycerol, 221*f*
overview of metabolic relationships in *S. cattleya*, 223*f*
oxidation of fluoroacetaldehyde to fluoroacetate by NAD dependent aldehyde dehydrogenase, 223*f*
rational for incorporation of single deuterium atom from [²H₄]-succinate into fluorometabolites, 220*f*
role of fluoroacetaldehyde, 221–222
Fluoronucleosides
chiral compounds, 6, 10–11
complete diastereoselective fluorination in synthesis of 2-fluoronucleosides, 12*f*
synthesis of *β*-2-fluorodideoxy nucleosides, 6, 9*f*
2-Fluoronucleosides, complete diastereoselective fluorination in synthesis of, 12*f*
Fluoroorganic compounds, yet to develop full potential, 22
Fluoroorganic compounds via chiral organoboranes
α-alkynyl *α'*-fluoroalkyl ketones, 26–27
allylboration of fluorinated aldehydes, 30–32
asymmetric reduction, 22–30
bisperfluoroalkyl aryl diketones, 27–28
enolboration-aldolization of fluoro-ketones and aldehydes, 32–35
reduction of alkyl fluoroalkyl ketones, 25–26
reduction of aryl fluoroalkyl ketones, 24–25
ring-substituted acetophenones, 29–30
synthesis and ring-cleavage reactions of trifluoromethyloxirane, 28–29
See also Allylboration, asymmetric; Enolboration-aldolization, asymmetric; Reductions, asymmetric
Fluoropyruvaldehyde *N,S*-ketals
addition of Grignard reagents, 107
adducts as means to sulfur-free *γ*-tri- and *γ*-difluorinated *β*-amino alcohols, 107–108
diastereoselective elaboration into fluorinated ephedra-alkaloids, 107
enantioselectivity with each transfer of stereogenic center, 106
models for asymmetric elaborations leading to *γ*-Tfm-GABOB, 108–109
models for stereoselective transformations, 107
nonracemic synthetic equivalents of *α*-fluoroalkyl *α*-amino aldehydes, 105–109
preparation from *α*-(fluoroalkyl)-*β*-sulfinylenamines, 105–106

retrosynthetic strategy to prepare *β*-fluoroalkyl *β*-amino alcohols, 105
synthesis of nonracemic *γ*-Tfm-GABOB, 108
Fluorostyrene, Wittig olefination of aldehyde, 199, 200
Fluoro sugars
chiral compounds, 6, 9
non-carbohydrate synthesis, 9*f*
4-Fluorothreonine
asymmetric synthesis, 214
biosynthesis by *Streptomyces cattleya*, 214–221
natural product, 213–214
See also Fluorometabolite biosynthesis by *Streptomyces cattleya*
5-Fluorouracil, development, 1
Fluoxetine (Prozac)
chemoenzymatic route to *(R)*-fluoxetine, 15*f*
depression treatment, 11, 14
ω-Fluorinated lipids, organic fluorine in waxy seed kernel, 213*f*
Friedel–Crafts reactions
aromatic compounds, 60–63
asymmetric activation, 62–63
diastereoselective oxidation reaction of vinyl ether product, 66
¹H NMR spectra of silyl enol ethers, 69*f*
¹H NMR spectra of vinyl ether products and *β*-methylstyrene, 65*f*
possible mechanism of Mukaiyama aldol reactions, 66–67
reactions of aromatic compounds with fluoral catalyzed by chiral binaphthol-derived titanium catalysts (BINOL-Ti), 61*t*
reactions of aromatic compounds with fluoral catalyzed by chiral BINOL-Ti through asymmetric activation, 63*t*
reactions of methyl vinyl ethers with fluoral catalyzed by BINOL-Ti catalysts, 64*t*
reactions of silyl enol ethers with fluoral catalyzed by BINOL-Ti, 68*t*
sequential diastereoselective reactions of resultant silyl enol ethers, 69
sequential diastereoselective reactions of resultant vinyl ethers, 65–66
silyl enol ethers, 66–69
transition state for *ortho*-Friedel–Crafts product, 62
transition state for product of silyl enol ethers, 69
transition state of diastereoselective oxidation reaction of vinyl ether product, 66*f*
vinyl ethers, 63–66
Functionalized fluorinated allyl alcohols. *See* Morita–Baylis–Hillman (MBH) reaction; Vinylmetalation

G

Glycerol
  incorporation in fluorometabolites, 218, 221
  See also Fluorometabolite biosynthesis by
    Streptomyces cattleya
Glycine
  incorporation in fluorometabolites, 215,
    217*f*
  See also Fluorometabolite biosynthesis by
    Streptomyces cattleya

H

Halothane
  development, 1
  enantiomer synthesis using stereoselective
    decarboxylation, 285
  inhaled anesthetic, 282
  Meinwald and Pearson synthesis, 283
  separation by enantioselective desorption
    from clathrate complexes, 284
  stereoselective decarboxylation yielding,
    286–287
  synthesis of enantio-enriched precursor
    acid, 285–286
  See also Anesthetics
Hemiacetal
  asymmetric synthesis of 1-alkoxy-2,2,2-tri-
    fluoroethanol derivatives, 228–230
  determination of absolute configurations of
    $CF_3CH(OCH_2Ph)OCOPh$, 229–230
  effect of alkoxy ligand, 230
  optimization of synthesis, 228
  sulfonates, nucleophilic substitution,
    231–234
  See also Liquid crystals (LCs), fluorine-con-
    taining
Henry reaction. *See* Nitroaldol reaction of
  2,2-difluoroaldehydes
Herbicides. *See* Aryl-pyrazole herbicides
Heteroaryl trifluoromethyl ketones, 1,4-addi-
  tion, 121
Hexafluoroacetone, Morita–Baylis–Hillman
  reaction, 120
(*R*)-4,4,4,4′,4′,4-Hexafluorovaline, synthesis,
  7*f*
Hoechst's ether
  asymmetric synthesis, 287
  enantiomer synthesis using stereoselective
    decarboxylation, 285
  See also Anesthetics
Human immunodeficiency virus (HIV)
  HIV inhibitor, 18
  HIV reverse transcription inhibitor, 19*f*
Hunsdiecker oxidative decarboxylation
  nonselectivity, 188
  step in synthesis of tribactams, 187
Hydrogenation reactions
  difluoromethylenated materials, 146*f*

exo-monofluoromethylenated materials,
  148*f*
mechanism, 152–153
possible mechanism, 153*f*
preparation of CF_3-containing substances,
  149–150
substituent effect on hydrogenation stereoselectivity, 151
See also Sugars containing fluorinated
  methyl groups
Hydrolysis, lipase-catalyzed, cyanohydrin acetates, 199

I

Imines. *See* α-Fluoroalkyl β-sulfinyl enamines
  and imines
Imino-aldol reaction, stereoselective tandem,
  136, 137
Inhalation anesthetics, fluorine-containing chiral compounds, 14, 16
Intermolecular recognition. *See* Perfluorocarbon (PFC) compounds
Isoflurane
  groundbreaking synthetic efforts affording
    quantities of highly pure enantiomers,
    289–290
  inhaled anesthetic, 282
  separation by enantioselective desorption
    from clathrate complexes, 284
  synthesis of enantiomers, 284
  See also Anesthetics
B-Isopinocampheyl-9-
  borabicyclo[3.3.1]nonane (Alpine-Borane)
  reducing agent, 24
  See also Reductions, asymmetric
Isoserines
  preparation of methyl, 90–91
  trifluoromethyl derivatives, 91
  See also Fluoroalkyl peptidomimetic units

K

Ketene silyl acetals, fluorine-substituted
  boron complexes, Masamune's catalyst and
    Kiyooka's catalyst, 41
  catalytic aldol reaction of bromofluoroket-
    ene silyl acetal by Lewis acids, 42
  catalytic asymmetric aldol reaction, 39–46
  catalytic asymmetric aldol reaction of bro-
    mofluoroketene acetal, 42*t*
  catalytic asymmetric aldol reaction of diflu-
    oroketene acetal, 42*t*
  catalyzed asymmetric aldol reaction by chi-
    ral Lewis acids, 41–42
  closed chair-like transition states, 46
  effects of reaction temperature on stereosel-
    ection during reaction with bromofluoro-
    ketene silyl acetal, 43–44

effects of reaction temperature on stereoselection during reaction with difluoroketene silyl acetal, 42–43
extended open transition states, 45
^{19}F NMR of 1:1 mixture of acetal to catalyst, 45f
molecular orbital calculation, 40–41
preparation of difluoroketene and bromofluoroketene trimethylsilyl ethyl acetals, 39
reaction mechanism, 44–46
uncatalyzed aldol reaction, 40
Ketomethylene peptide isosteres
general strategy for synthesis, 53
See also Monofluoroketomethylene peptide isosteres
Kiyooka's catalyst
boron complex, 41
See also Ketene silyl acetals, fluorine-substituted

L

Lactate derivatives of aryl-pyrazole herbicides. *See* Aryl-pyrazole herbicides
Lanthanoid-lithium-BINOL complexes (LLB)
assessment of rare-earth elements for Henry reaction of 2,2-difluoroaldehydes, 46–47
catalysts for nitroaldol (Henry) reaction, 46
See also Nitroaldol reaction of 2,2-difluoroaldehydes
Lewis acids
catalyzing asymmetric aldol reaction of bromofluoroketene silyl acetal, 42
catalyzing asymmetric aldol reaction of difluoroketene silyl acetal, 41–42
Lipase-catalyzed hydrolysis, cyanohydrin acetates, 199
Lipids, ω-fluorinated
natural product, 212
organic fluorine in waxy seed kernel, 213f
Liquid crystals (LCs), fluorine-containing
absolute configuration by Mosher method, 229
absolute configuration by Trost method, 230
alkylation of CF$_3$CH(OCH$_2$Ph)OTs with aluminum reagents, 233t
alkylation of hemiacetal tosylates with aluminates derived from 1-alkenes, 234
alkylation with aluminates derived from 1-alkenes, 235t
asymmetric hemiacetal synthesis, 228, 229t
asymmetric synthesis of 1-alkoxy-2,2,2-trifluoroethanol derivatives, 228–230
comparison of organometallic reagents for ethyl substitution, 233t

determination of absolute configurations of CF$_3$CH(OCH$_2$Ph)OCOPh, 229–230
dichiral LC compound and phase transition temperatures, 235f
effect of alkoxy ligand, 230
ethylation of CF$_3$CH(OCH$_2$Ph)OSO$_2$R with aluminum reagents, 232t
nucleophilic substitution of hemiacetal sulfonates, 231–234
nucleophilic substitution of optically active 1-alkoxy(polyfluoro)alkyl sulfonates, 231–234
optical resolution of CF$_3$CH(OCH$_2$Ph)-OSO$_2$R, 231
optimization of asymmetric hemiacetal synthesis, 228
research plan, 227
resolution of trifluoroacetaldehyde hemiacetal sulfonates, 231
structural correlation, 230
structural elucidation of optically isotropic (IsoX) phase, 237
synthesis of carboxylic acid part of dichiral LCs, 236
synthesis of dichiral LCs containing two fluorines at two chiral centers, 234, 236
synthesis of phenolic part of dichiral LCs, 236
typical LC molecules, 227f
unusual endothermic transition, 237
use of aluminate reagents derived from 1-alkenes, 234
See also Diastereomeric α- or β-CF$_3$ liquid crystalline molecules
Lithium alkyl aryl sulfoxides, synthesis of β-fluoroalkyl β-amino alcohols, 135–136

M

Masamune's catalyst
boron complex, 41
See also Ketene silyl acetals, fluorine-substituted
Medicinal chemistry, emphasis on chiral compounds and asymmetric syntheses, 1
Methyl groups, fluorinated. *See* Sugars containing fluorinated methyl groups
Methyl vinyl ketone, DABCO-catalyzed dimerization, 120
Metoprolol, preparation of fluorinated analog, 48–49
Mitsunobu-like azidation, synthesis of monofluorinated amino compounds, 99, 111
Monofluoro-amino compounds
approach to enantiopure quaternary fluoro amino compounds via sulfinyl epoxide chemistry, 111–112
1,3-dipolar cycloadditions involving fluorinated sulfinyl dipolarophiles, 112–113

stereoselective synthesis of 3-fluoro-D-ala-
nine, 111
syntheses, 111–113
synthesis and elaboration of 6-fluoromethyl
furo[3,4-c]isoxazolidines, 113
synthesis of monofluorinated nucleosides,
112
Monofluoroketomethylene peptide isosteres
allylic stereocontrol by N-tritylamine
group, 58f
comparison of nucleophilic carbonyl addition with electrophilic enol addition, 55f
conversion of N-trityl aminoketones to E-
TMS enol ethers with lithium tert-butyl
trimethylsilyl amide (LTBTMS/TMSCl),
55
electrophilic fluorination of 2S epimers, 57
electrophilic fluorination of E-TMS enol
ethers of N-trityl amino ketones, 55
electrophilic fluorination of Z-TMS enol
ethers of N-trityl amino ketones, 55
enolate formation in N-tritylated aminoketones, 54
general synthetic strategy, 53
methods to introduce fluorine into densely
functionalized molecule, 52–53
possible diastereomers, 56, 57f
stereocontrol in syn and anti Z-TMS enol
ethers, 58, 59f
structural rationale for stereocontrol by N-
tritylamino group in fluorinations of
TMS-enol ethers, 58
structure, 56f
synthesis of β-fluoro-α-ketoesters, 54
synthesis of hydroxyethylene peptide isosteres, 54
synthesis of series of, 56
Morita–Baylis–Hillman (MBH) reaction
α-acetylenic α'-trifluoromethyl ketones, 121
addition of heteroaryl trifluoromethyl ketones, 121
aromatic trifluoromethyl ketones, 120–121
aryl trifluoromethyl ketones, 120
asymmetric, 122–124
camphor sultam based asymmetric, 123
chiral auxiliaries tested in literature for
asymmetric, 123f
chiral catalysts tested in literature for asymmetric, 124f
2-chloro-2,2-difluoroacetophenone, 121
diastereomeric benzaldehydes, 123
1,4-diazabicyclo[2.2.2]octane (DABCO)-cat-
alyzed dimerization of methyl vinyl ketone, 120
exploratory study of terpenyl alcohols for
reaction of fluoro-benzaldehydes, 124
fluorinated benzaldehydes, 118
fluorocarbonyls, 118–121
general scheme, 117

hexafluoroacetone, 120
mechanism, 117–118
pentafluorobenzaldehyde, 119
perfluoro-aldehydes, 119
perfluoro-aldehydes with ethyl acrylate, 119
perfluoro-aliphatic aldehydes, 118–119
polymerization of fluoral, 119
trimerization of 1,1,1-trifluoroacetone, 120
Mosher method, absolute configuration of
CF₃CH(OCH₂Ph)OCOPh, 229
Mukaiyama-aldol reaction
possible mechanism, 66–67
See also Ketene silyl acetals, fluorine-substituted

N

Natural products, fluorinated
asymmetric synthesis of 4-fluorothreonine,
214f
¹⁹F-MRI image of seed nut of Dichapetalum toxicarium, 213f
ω-fluorinated lipids, 212
fluoroacetate, 211–212
fluoroacetone, 212–213
fluorocitrate, 212
4-fluorothreonine, 213–214
nucleocidin, 213
structures, 211
See also Fluorometabolite biosynthesis by
Streptomyces cattleya
Nitroaldol reaction of 2,2-difluoroaldehydes
assessment of rare-earth elements for
Henry reaction, 46–47
asymmetric reaction of 5-phenyl-2,2-diflu-
oropentanal, 47t
enantioface selection, 47f
lanthanoid-lithium-BINOL complexes
(LLB catalysts), 46
LLB catalysts for asymmetric nitroaldol
(Henry) reaction, 46
preparation of fluorinated analog of β-adrenergic blocking agent, metoprolol,
48–49
various 2,2-difluoroaldehydes, 47–48
Norepinephrine (NE)
asymmetric carbonyl reduction, 199, 202
asymmetric cyanohydrin synthesis, 202, 203
binding affinities, 207
effects of fluorine substitution, 196
fluorostyrene by Wittig olefination of aldehyde, 199, 200
functioning through interactions with membrane-bound adrenergic receptors, 194,
196
limitation of fine-tuning to enhance selectivity, 198
lipase-catalyzed hydrolysis of cyanohydrin
acetates, 199

naturally occurring catecholamine, 194
oxynitrilase-catalyzed addition of HCN to
  fluorinated benzaldehydes, 199
production of chloroketones from aldehydes, 201
ring fluorinated derivatives, 196, 197
Sharpless asymmetric aminohydroxylation,
  199
structure, 195
synthetic approaches to enantiomers of fluorinated, 198–202
Nucleocidin, fluorinated natural product, 213
Nucleophilic substitution
  hemiacetal sulfonates, 231–234
  optically active 1-alkoxy(polyfluoro)alkyl
    sulfonates, 231–234
  use of aluminate reagents from 1-alkenes,
    234

O

Organoboranes. *See* Fluoroorganic compounds via chiral organoboranes
Organofluorine compounds
  enzymatically controlled reactions, 18
  range of research studies, 1–3
  unique characteristics, 255
Oxynitrilase-catalyzed addition, HCN to fluorinated benzaldehydes, 199

P

P450 enzymes
  metabolism of fluorine-containing taxoids
    by, 163, 165*f*
  primary sites of hydroxylation on paclitaxel
    and docetaxel, 163, 164*f*
  See also Fluorine-containing taxoids
Paclitaxel
  acting as spindle poisons, 159
  anticancer agents, 11
  bioassay for new fluorine-containing tax-

  oids, 161
  conformational analysis, 171, 172*f*
  cytotoxicities, 163*t*, 169*t*, 171*t*
  fluorine-containing analog, 12*f*
  primary sites of hydroxylation by P450 fam-
    ily of enzymes, 164*f*
  structure, 159
  treatment of ovarian and metastatic breast
    cancer, 158
  See also Fluorine-containing taxoids
Pentafluorobenzaldehyde, Morita–Baylis–
  Hillman reaction, 119
Peptidomimetic units. *See* Fluoroalkyl peptidomimetic units
Perfluoro-aldehydes, Morita–Baylis–Hillman
  reaction, 119

Perfluoro-aldehydes with ethyl acrylate,
  Morita–Baylis–Hillman reaction, 119
Perfluoro-aliphatic aldehydes, Morita–Baylis–
  Hillman reaction, 118–119
Perfluorocarbon (PFC) compounds
  association equilibrium between quinuclid-
    ine and 1-iodoheptafluoropropane, 249
  crystal packing of system from (-)-sparteine
    hydrobromide and *(S)*-1,2-dibromohex-
    afluoropropane, 252*f*
  differential scanning calorimetry (DSC),
    245–247
  DSC of α,ω-diiodoperfluoroalkanes and or-
    ganic base Kryptofix.2.2.2, 245*f*
  DSC of adducts from Kryptofix.2.2.2. and
    α,ω-diiodoperfluoroalkanes, 246*f*
  electron donor-acceptor complexes from ali-
    phatic halides and organic bases, 240*t*
  electronic and steric effects of $\Delta\delta_{CF2I}$ values
    by 1,2-diiodotetrafluoroethane, 251*t*
  ^{19}F-NMR chemical shift differences ($\Delta\delta_{CF2}$)
    of 1,2-diiodo- and 1,2-dibromotetrafluor-
    oethane in different solvents, 250*t*
  general aspects of Rf–X···El interactions
    (X=Cl,Br,I and El=N,O,S), 240–241
  infinite 1D network from 1,2-diiodote-
    trafluoroethane and *N,N,N',N'*-tetrameth-
    ylethyldiamine (TMEDA), 241
  infinite 1D networks from α,ω-diiodoper-
    fluoroalkanes and various organic bases,
    242
  infrared (IR) and Raman spectroscopies,
    247–249
  IR spectra of TMEDA, 1,4-diiodoperfluor-
    obutane, and their adduct, 247*f*
  Raman spectrum of 1,4-diiodoperfluorobu-
    tane, 248*f*
  Raman spectrum of adduct between 1,4-dii-
    odoperfluorobutane and TMEDA, 248*f*
  requiring specifically tailored approach,
    239–240
  resolution of 1,2-dibromohexafluoropro-
    pane, 251–253
  Rf–X···El interaction in solid phase,
    241–249
  study of Rf–X···El interaction in liquid
    phase, 249–251
  system from (-)-sparteine hydrobromide
    and *(S)*-1,2-dibromohexafluoropropane,
    252
  X-ray packing of co-crystal from 1,2-diiodo-
    tetrafluoroethane and base Kryp-
    tofix.2.2.2, 244*f*
  X-ray packing of co-crystal from 1,2-diiodo-
    tetrafluoroethane and TMEDA, 243*f*
  X-rays, 243–245
Phase transition temperatures, dichiral liquid
  crystalline compound, 234, 235*f*
α-Phenethanols, preparation of ring-fluorinated, 30

5-Phenyl-2,2-difluoropentenal, asymmetric nitroaldol reaction, 46–47
Pictet–Spengler cyclization, stereoselective under TFA catalysis, 104
α-Pinene, application of diisopinocampheylborane reagent, 30
Pivalophenone, asymmetric reduction, 23
Polarization, spontaneous (Ps), diastereomeric liquid crystal molecules with CF_3-group, 260
Protease inhibitors
  HIV-1 inhibition, 91
  regioisomers in synthesis, 84
Prozac (fluoxetine)
  chemoenzymatic route to (R)-fluoxetine, 15f
  depression treatment, 11, 14
Pummerer reaction
  β-allyloxy-sulfoxides, 113
  mechanism of non-oxidative, 130–131
  non-oxidative, 101
  step in sequence leading to pyruvaldehyde N,S-ketals, 105–106
  stereoselective application of non-oxidative, 132–133
  sulfur-free oxiranes, 112
  synthesis of 1-trifluoromethyl tetrahydroisoquinoline alkaloids, 104
  transforming sulfinyl oxiranes, 111–112
Pyrazoles. See Aryl-pyrazole herbicides
Pyruvate
  incorporation in fluorometabolites, 215, 218, 219f
  See also Fluorometabolite biosynthesis by Streptomyces cattleya

R

Racemic perfluorocarbons. See Perfluorocarbon (PFC) compounds
Reductions, asymmetric
  α-acetylenic α'-fluoroalkyl ketones with DIP-Chloride, 27f
  α-acetylenic α'-perfluoroalkyl ketones with DIP-Chloride, 27f
  alkyl fluoroalkyl ketones, 25–26
  alkyl trifluoromethyl ketones with B-chlorodiisopinocampheylborane (DIP-Chloride), 26f
  α-alkynyl α'-fluoroalkyl ketones, 26–27
  aryl fluoroalkyl ketones, 24–25
  bisperfluoroalkyl aryl diketones, 27–28
  bulk comparison of F and CH_3 groups, 25f
  comparing physical nature of CH_3, CF_3, and $C(CH_3)_3$ groups, 24
  comparison of acetophenone, pivalophenone, and 2,2,2-trifluoroacetophenone with DIP-Chloride, 23f
  DIP-Chloride reagent, 23

B-isopinocampheyl-9-borabicyyclo[3.3.1]nonane (Alpine-Borane) reagent, 24
  mechanism of 2,2,2-trifluoroacetophenone with (-)-DIP-Chloride, 24f
  preparation of ring-fluorinated a-phenethanols via reduction with DIP-Chloride, 30f
  prochiral fluoroketones, 23–24
  ring-substituted acetophenones, 29–30
  step in synthesis of fluorinated norepinephrines, 199, 202
  synthesis and ring-cleavage reactions of trifluoromethyloxirane, 28–29
  synthesis of enantiomerically pure fluoro-organic molecules, 22–30
  trifluoromethyloxirane preparation, 28
  See also Fluoroorganic compounds via chiral organoboranes
Reductive amination of fluoro-carbonyl compounds
  approaches to fluorinated amino acids, 74–76
  asymmetric biomimetic transamination, 77–81
  biological transamination and previous synthetic models, 76–77
  general synthetic application of [1,3]-proton shift reaction ([1,3]-PSR), 75–76
  isomerization giving rise to targeted Schiff base, 78
  isomerization of enaminolizable trifluoromethyl benzyl imine, 79
  isomerization of methyl trifluoromethyl imine, 78–79
  possible intermediates in isomerizations under study, 80–81
  reaction between per(poly)fluoroalkyl carbonyl compounds and N-(benzyl)triphenylphosphazene, 74–75
  synthesis of starting chiral fluorinated Schiff bases, 77
  synthetically useful reagents for biomimetic transformation of amines to carbonyl compounds, 76–77
  transamination of perfluoroalkyl β-keto carboxylic esters, 79–80
Redux [(S)-(+) Dexfenfluramine], CNS agent, 11

S

Self-assembly
  spontaneous optical resolution of liquid crystalline racemates, 263–265
  See also Perfluorocarbon (PFC) compounds
Serine
  incorporation in fluorometabolites, 215, 218, 219f
  See also Fluorometabolite biosynthesis by Streptomyces cattleya

Sharpless asymmetric aminohydroxylation, step in synthesis of fluorinated norepinephrines, 199

Sharpless oxidation, enantiopure β-fluoro alanines, 102–103

Sharpless protocol, trifluoroalanine by oxidation of corresponding alaninol, 101

Silyl enol ethers
catalytic reactions with fluoral, 68*t*
Friedel–Crafts reactions, 66–69
^{1}H NMR spectra, 69*f*
possible mechanism of Mukaiyama-aldol reactions, 66–67
sequential diastereoselective reactions of resultant, 69
transition state for Friedel–Crafts product, 69

Staudinger reaction
preparation of *N*-aryl, *N*-alkoxycarbonyl, and *N*-unsubstituted polyfluoroalkyl enamines and imines, 100
producing sulfinimines, 109–110

Streptomyces cattleya. See Fluorometabolite biosynthesis by *Streptomyces cattleya*

Succinate
incorporation in fluorometabolites, 218, 220*f*

See also Fluorometabolite biosynthesis by *Streptomyces cattleya*

Sugars
non-carbohydrate synthesis, 9*f*
See also Fluoro sugars

Sugars containing fluorinated methyl groups
ab initio calculations, 148–149
aldol structures as targets for fluorine modification, 142
conformation analysis using MOPAC AM1, 146–147
hydrogenation mechanism, 152–153
hydrogenation of 2- and 3-CF$_3$ olefins, 149*f*
hydrogenation of 4-CF$_3$ olefin, 150*f*
hydrogenation of difluoromethylenated materials, 145–146
hydrogenation of *exo*-monofluoromethylenated materials, 148*f*
most stable conformer by AM1 calculations, 152*f*
most stable conformer from AM1 calculations, 147*f*
NBO charges and frontier orbital energy levels of ethylenes by *ab initio* calculations, 148*f*
possible mechanism on hydrogenation, 153*f*
preparation of CF$_3$-containing substances, 149–150
preparation of difluoromethyl compounds, 144–147
preparation of mono- and trifluoromethylated materials, 147–152

preparation of sugars containing difluoromethylene group at 2, 3, and 4 positions, 145*f*
retrosynthetic scheme to target aldol structures, 143*f*
ring opening of CHF$_2$ compound, 154*f*
ring opening of cyclic compounds, 153–154
substituent effect of hydrogenation selectivity, 151*t*
synthetic plans for fluorine-possessing aldol structures, 144

Sulfonates. *See* 1-Alkoxy(polyfluoro)alkyl sulfonates

Sulfoxides, chiral. *See* Fluoroalkyl amino compounds

T

Taxoids. *See* Fluorine-containing taxoids
Taxol. *See* Paclitaxel
Taxotere. *See* Docetaxel
Terpenyl alcohols
exploratory study for Morita–Baylis–Hillman (MBH) reaction of fluorobenzaldehydes, 124
exploratory study for vinylalumination reactions, 124–125

Titanium catalysts, binaphthol-derived. *See* Friedel–Crafts reactions

Transamination
asymmetric synthesis of β-fluoroalkyl-β-amino acids, 6, 8*f*
biological, 76–77
See also Reductive amination of fluoro-carbonyl compounds

Tribactams, fluorinated
antimicrobial activity, 184–185
characterization using 2D NMR, 189–190
chromosomal β-lactamases, 184
confirmation of structure using 2D totally correlated spectroscopy (TOCSY), 190
cycloaddition reaction for β-lactam nucleus, 185
imipenem, 183, 184
β-lactamase sensitivity, 184–185
NMR spectroscopic data for series of fluorine-containing trinems as mixture of diastereomers, 190*t*
nonselectivity of Hunsdiecker decarboxylation reaction, 188
stereochemistry of addition, 188–189
stereoselectivity of cyclohexanone addition, 189
synthesis, 185–189
synthetic step using Hunsdiecker oxidative decarboxylation, 187
trinem family of synthetic β-lactams, 182–183
1,1,1-Trifluoroacetone, trimerization, 120

2,2,2-Trifluoroacetophenone
  asymmetric reduction, 23
  mechanism of reduction with DIP-Chloride, 23, 24*f*
α-Trifluoromethyl-alanine, synthesis of both enantiomers, 134–135
α-Trifluoromethyl-allo-threoninate, synthesis of enantiopure, 134–135
β-Trifluoromethyl-β-amino acids, chemo-enzymatic approach to chiral, 6, 8*f*
Trifluoromethylation, synthesis of enantiomerically pure (2R)-N-Boc-2-amino-4,4,4-trifluorobutanoic acid, 3, 5*f*
α-Trifluoromethyl-butyrine, synthesis of enantiopure, 134–135
γ-Trifluoromethyl-GABOB (γ-Tfm-GABOB), stereoselective synthesis, 133, 134
α-Trifluoromethyl-β-hydroxy aspartic units, synthesis of enantiopure, 136, 138
Trifluoromethyloxirane
  preparation, 28
  ring cleavage reactions, 28–29
  synthesis of ferroelectric liquid crystals, 29
α-Trifluoromethyl-phenylalanine, synthesis of both enantiomers, 134–135
α-Trifluoromethyl-serine, synthesis of both enantiomers, 134–135
2-Trifluoromethyl-sphingolipids, retrosynthetic approach to, 136, 137
α-Trifluoromethyl-threoninate, synthesis of enantiopure, 134–135
Trifluoronorephedrine, stereoselective synthesis, 133
Trifluoropyruvate, Evans aldol reaction with imines of, 136–138
Trinems
  family of synthetic β-lactams, 182–183
  See also Tribactams, fluorinated
Trost method, absolute configuration of CF₃CH(OCH₂Ph)OCOPh, 230

V

Vinylalumination
  exploratory study of terpenyl alcohols, 124–125
  fluoro-aldehydes and ketones, 121–122
Vinylcupration, fluoro-aldehydes and ketones, 122
Vinyl ethers
  catalytic reactions with fluoral, 64*t*
  diastereoselective oxidation reaction, 66
  Friedel–Crafts reactions, 63–66
  ¹H NMR spectra of, and β-methylstyrene, 65*f*
  sequential diastereoselective reactions of resultant, 65–66
  transition state of diastereoselective oxidation reaction, 66*f*
Vinylmetalation
  exploratory study of terpenyl alcohols, 124–125
  fluoro-carbonyls, 121–122
  vinylalumination of fluoro-aldehydes and ketones, 121–122
  vinylalumination of fluoro-carbonyl compounds, 122
  vinylcupration of fluoro-aldehydes and ketones, 122

W

Wittig olefination, fluorostyrene preparation, 199, 200

Z

Zimmerman-Traxler, chair-like transition state, 132